Cohomology Theories for Compact Abelian Groups

Karl H. Hofmann · Paul S. Mostert

With an Appendix by Eric C. Nummela

Springer-Verlag New York Heidelberg Berlin 1973

Karl H. Hofmann · Paul S. Mostert
Eric C. Nummela
Tulane University, New Orleans/USA
Dept. of Mathematics

Originalausgabe erschien im VEB Deutscher Verlag der Wissenschaften, Berlin
Vertrieb ausschließlich für die DDR und die sozialistischen Länder

Lizenzausgabe
im Springer-Verlag Berlin/Heidelberg/New York
Vertrieb für alle übrigen Länder einschließlich BRD

AMS Subject Classifications (1970)
Primary 35J20, 35J25, 35J30, 35J35, 35J40, 35K20, 35K35, 35L20
Secondary 46E35, 49B25

ISBN 0-387-05730-7 Springer-Verlag New York-Heidelberg-Berlin
ISBN 3-540-05730-7 Springer-Verlag Berlin-Heidelberg-New York

Printed in the German Democratic Republic
Lizenz-Nr. 206 · 435/176/73
VEB Druckerei »Gottfried Wilhelm Leibniz«, Gräfenhainichen

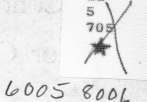

Acknowledgements

The first version of this tract was written while the authors were members of the Institute for Advanced Study in Princeton on leave from Tulane University in 1967–1969. During this period the first author was a fellow of the Alfred P. Sloan Foundation and the second held a National Science Foundation Senior Postdoctoral Fellowship. While at the Institute, the authors had the benefit of inspiring conversations with A. Borel, D. Montgomery, and J. P. Serre. Announcements of some of the results were made in the Bulletin of the American Mathematical Society [24] and [25].

The present version was completed during the summer of 1970 after a one year special program in category theory at Tulane University with visits by S. MacLane and many others. Even before that occasion S. MacLane had taken an interest in our work and had contributed valuable advice. The Chapter on Kan extensions, which reflects considerations not available two years ago, is typical of the new material added.

Correspondence with L. Evens and S. Priddy contributed to our understanding of the cohomology ring of a finite group and certain resolutions. Concerning the computation of the cohomology of a classifying space, we had important conversations with A. E. Dold and D. Puppe, who also contributed remarks about the singular cohomology of $K(\pi, n)$ spaces. During the summer of 1970 the authors were supported by the National Science Foundation.

It is a pleasure to acknowledge the cooperation of Eric Nummela who accepted our invitation to write Chapter VI and to contribute to our discussion the insights he had obtained in his dissertation about classifying spaces for compact monoids. He also contributed to Section 2 of Chapter III.

Karl H. Hofmann
Paul S. Mostert

Table of contents

Introduction

Of all topological algebraic structures compact topological groups have perhaps the richest theory since so many different fields contribute to their study: Analysis enters through the representation theory and harmonic analysis; differential geometry, the theory of real analytic functions and the theory of differential equations come into the play via Lie group theory; point set topology is used in describing the local geometric structure of compact groups via limit spaces; global topology and the theory of manifolds again play a role through Lie group theory; and, of course, algebra enters through the cohomology and homology theory.

A particularly well understood subclass of compact groups is the class of compact abelian groups. An added element of elegance is the duality theory, which states that the category of compact abelian groups is completely equivalent to the category of (discrete) abelian groups with all arrows reversed. This allows for a virtually complete algebraisation of any question concerning compact abelian groups.

The subclass of compact abelian groups is not so special within the category of compact groups as it may seem at first glance. As is very well known, the local geometric structure of a compact group may be extremely complicated, but all local complication happens to be "abelian". Indeed, via the duality theory, the complication in compact connected groups is faithfully reflected in the theory of torsion free discrete abelian groups whose notorious complexity has resisted all efforts of complete classification in ranks greater than two.

All of these features qualify the category of compact abelian groups as a very suitable test category for the applicability of various tools; a tool which does not work rather explicitly in the category of compact abelian groups is not likely to be very useful in the more general circumstances of arbitrary compact or even more general topological groups.

The present monograph is an essay to test the tools of cohomology on the category of compact abelian groups. The usefulness of cohomology does not need, of course, any corroboration. No comprehensive study of the explicit nature of the cohomology functor on this category has, however, been undertaken until now. The current treatise attempts to fill this gap — although its authors lay no claim to having completely succeeded.

Since the structure of a compact abelian group has both topological and algebraic underlying structures, it should come as no surprise that we have at least *two* cohomology theories to investigate:

(A) The Čech cohomology theory of the underlying compact space of the group;

(B) The algebraic cohomology of the underlying group.

Neither could be expected to give complete information about the topological group, and indeed the two together would be expected also to fail, since the interplay of algebra and topology through the continuity of multiplication is lost. Thus, a third theory, the *algebraic* cohomology of a *compact* group needs to be considered. It requires the construction of a functor which is defined on the category of all compact groups and which yields, when restricted to the subcategory of finite groups, exactly the algebraic cohomology.

The various possible constructions which lead to such a functor are by no means unknown to the topologist. At least they are quite familiar in the case of Lie groups. Indeed suppose for the moment, that G is a compact Lie group. It is in fact one of the basic problems in the theory of fiber bundles to associate with G (in a functorial fashion, if possible) a fiber bundle with fiber G and fiber projection $E(G) \to B(G)$ in such a fashion that every continuous map $\varphi\colon B \to B(G)$ of the base space B of an arbitrary bundle $E \to B$ with fiber G lifts to a bundle map $E \to E(G)$ and that the isomorphy types of bundles with fiber G and base space B are in this fashion completely classified by the homotopy classes of the maps φ. The bundle $E(G) \to B(G)$ is called a *universal bundle* and the base space is called a *classifying space* for G. Functorial constructions of such bundles have been given by several authors, the first one being Milnor [28]; others following suit with constructions having various advantages and drawbacks were Dold and Lashof [12], Milgram [43], and McCord [34], to name a few. Before these construction were known, one used bundles which had the properties described above only up to certain dimensional limits which in each case were taken arbitrarily large. This was the method used by Borel [3] in his key work concerning the cohomology of Lie groups. It turns out that the fiber space $E(G)$ is aspherical, hence in particular acyclic, and that in all the functorial constructions it has a natural filtration which occasionally turns out to be useful. What is essential for our consideration is that the Čech cohomology of a classifying space $B(G)$ is a contravariant functor of G which yields the algebraic cohomology if it is restricted to finite groups. Thus we may elaborate the description of the third type of cohomology theory which will concern us; indeed it is

(C) The Čech cohomology of a classifying space $B(G)$ of the given group G.

It is, however, not our principal goal to elaborate extensively on the construction of classifying spaces and the computation of their cohomology. Our aim is to take the functor $G \mapsto hG = H(B(G))$, where H is the Čech cohomology functor (with integral coefficients) of topological spaces, and compute it as explicitly as we can for compact abelian groups G. Nevertheless, we indicate how the Milnor construction of $B(G)$ for a compact group G is built up inductively through its filtration, and how the cohomology of $B(G)$ is computed. This involves a limiting process for which we use the "telescope device" for filtered spaces with cofibrations as the inclusion maps of the filtration, which also goes back to Milnor. A sidelight of interest for the recently growing field of compact transformation monoids is that the Milnor construction is unsuitable for compact monoids (semigroups with identity) in place of groups. More recent constructions of the classifying space do not have this drawback, although the Milgram construction, which by many is considered the optimal approach, would not even work for all compact groups, since it requires that the inclusion map of the identity into the group is a cofibration. The Dold-Lashof process has neither of these disadvantages. An appropriate version of this construction thus permits the extension of the functor h to all compact monoids. An account of this construction has been written by E. Nummela and included as an appendix (Chapter VI). It is independent of any other considerations.

Our description of the Milnor construction comes in Chapter III. We do not talk about topology before this point and hence this discussion of classifying spaces is independent of the preceding portion.

As is usually the case in any investigation about compact groups we have, or obtain in the process, a very good understanding of the functors $G \mapsto HG$ and $G \mapsto hG$ for Lie groups G. The question then becomes one of a general functorial nature: If we have a category, such as, in our example, the category *Comp* of all compact abelian groups, and a functor F defined on it with values in some suitable category $\mathfrak{A}$, which, in the current case, is the category of graded commutative algebras with arrows reversed, and if we know the action of the functor on a subcategory, such as the category *Lie* of all Lie groups in *Comp*, can we then infer the complete nature of the functor F? If one expects a positive answer to this question, then such optimism can only be based on the fact, that the subcategory *Lie* is large enough in *Comp* to determine the behavior of the functor. Indeed, since every compact group is a projective limit of Lie groups, we can say that the category Lie is *dense* in *Comp*. The full story is more complicated, but the point here is that dense subcategories do determine functors which are *continuous* in the sense that they preserve certain limits. The crucial application in our case is that the functors H and h preserve limits over directed index sets (but by no means all limits), and that the functorial theory we develop is exactly applicable to such a situation.

In fact our theory will yield more than what would be required for the application we have in mind. If we have a subcategory (like *Lie* in the category *Comp*) we may, sometimes, be given a functor on the subcategory, and we may then raise the question whether or not there exists an extension of the functor to a functor defined on the bigger category. And, if such extensions exist, we may ask if there is a *best* such extension in that it has the *universal* property that there is a unique natural transformation from it to any other which when restricted to the subcategory is the identity. This question has become known in category theory in recent years under the heading of "Kan extension of functors". The theory we develop indeed results in an existence (and, automatically, a uniqueness) theorem for Kan extensions of certain functors. In particular it follows, that the Čech cohomology of compact groups is the Kan extension of the manifold cohomology on Lie groups (on which singular, Čech, or indeed anybody's cohomology theories agree). This assertion, for example fails to be correct for singular cohomology, which is, of course an extension of its own restriction to Lie groups, but it is by no means the universal one. This is a more or less formal proof of the generally accepted fact that for global topological investigations on compact groups or monoids the Čech cohomology is superior to singular cohomology. Our theory of Kan extensions is given in Chapter IV. It is again independent of anything that precedes it and may be read independently of the rest of the treatise.

The greater value of the functor h over the functor H remains invisible as long as we restrict our considerations to compact connected abelian groups since both determine the group completely in that case. Indeed, the results about H and h concerning the connected case, described in Section 1 of Chapter V, could be read without Chapters I and II. They are as follows:

Theorem. *If G is a compact connected abelian group, then both space cohomology HG and algebraic cohomology hG are commutative graded Hopf algebras in a natural fashion. Moreover HG is naturally isomorphic to $\bigwedge \hat{G}$, the exterior algebra generated* *by the character group $\hat{G}$ of G in degree 1, and hG is naturally isomorphic to $P\hat{G}$, the symmetric algebra generated by $\hat{G}$ in degree 2.*

However, the situation becomes strikingly more difficult with h when one leaves the realm of connected groups.

In the non-connected case, the space cohomology may still be described completely as follows:

Theorem. *Let R be any commutative ring with identity and G be a compact abelian group. Then there are natural isomorphisms of R-Hopf algebras*

$$H(G, R) \cong C(G, R) \otimes \wedge \hat{G}_0 \cong R \otimes C(G, \mathbf{Z}) \otimes \wedge \hat{G}_0,$$

where $C(G, R)$ denotes the Hopf algebra of constant functions $G \to R$ (considered as graded having only non-zero contribution in degree 0).

Greater complexity in the non-connected case is to be expected since any theory covering the general case must in particular cover the case of finite abelian groups. The space cohomology, in this case (as is evident from the theorem above) reduces to a virtually trivial proposition, and, as far as the algebraic cohomology is concerned, it is here that the true difficulties start — if one insists on explicit functorial statements about the algebraic cohomology ring. There exist expositions of the algebraic cohomology of finite abelian groups in the more general context of computing the cohomology and homology of Eilenberg-MacLane spaces, most notably the Cartan Seminar Report covering this topic [10]. But the emphasis there is different, and the information available is not very explicit for our purposes. For this reason, our first two chapters are devoted to the explicit structure of the cohomology ring of a finite abelian group, and may be read independently of the rest of the treatise.

The gist of our study in these two chapters is as follows. For a finite abelian group G and its integral group ring S there is a resolution $0 \leftarrow \mathbf{Z} \leftarrow X^*$ of $\mathbf{Z}$ by finitely generated free S-modules which is much smaller than the standard bar resolution. It is essential to obtain a resolution that is small enough to work with since the desired algebraic cohomology with coefficients in a G-module R is given by the homology of the cochain complex $\mathrm{Hom}_S(X^*, R)$. The crucial fact about this resolution is, however, that in the case of trivial action on R, which is the one that interests us here, the cochain complex $\mathrm{Hom}_S(X^*, R)$ is in fact a bigraded differential Hopf algebra of the form $P_R A \otimes_R \wedge_R A$ (taking the relevant case in which R is a ring), where $P_R A$ is the symmetric R-algebra generated by the R-module A in degree 2, and where $\wedge_R A$ is the exterior R-algebra generated by A in degree 1, and where A is a finitely generated free R-module with basis $a_1, \ldots, a_n$; the differential and derivation d is given by $d(P_R A \otimes 1) = 0$

and $d(1 \otimes a_k) = z_k(a_k \otimes 1)$, where $z_k \in R$ with $z_1 | z_2, \ldots, z_{n-1} | z_n$. The z_k are the elementary divisors of G. This is a special case of a so-called *Koszul-complex*. The schematic display of the bigraded differential algebra is given in the diagram

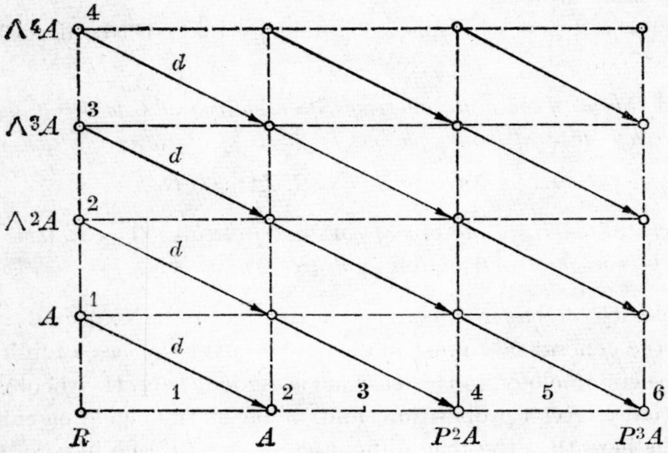

This complex is the E_2-term of a spectral algebra functorially associated with the morphism $\varphi\colon A \to A$ defined by $\varphi(a_k) = z_k a_k$, and we denote it, therefore, by $E_2(\varphi)$. The homology algebra $E_3(\varphi)$ in the present case is again a bigraded commutative algebra, and, relative to its total degree, is isomorphic to the desired cohomology algebra $h(G, R)$. This fact, together with the detailed description of what has been outlined here, is given in Chapter II, Section 2.

So here is, in any event, a resolution, from which one may hope to gain explicit insight about $h(G, R)$. This motivates the entire first Chapter. The core of the first Chapter is Section 2, which is an analysis of the spectral algebras $E_2(\varphi)$, $E_3(\varphi)$. (Section 1 and Section 3 are, in a sense, accessories.) This section is extremely technical and, on occasions, exasperatingly computational, but virtually all of the more explicit theorems about the cohomology ring $h(G, R)$ can be traced back to results about the spectral algebra $E_3(\varphi)$.

The material in Chapter I is then exploited for the cohomology theory of finite abelian groups in Chapter II, Section 3. An idea of some of the results that we obtain, most of which are extended to arbitrary compact abelian groups in the later parts of the treatise by methods outlined earlier, is as follows:

Theorem. a) *Let G be a compact abelian group and $\hat{G}$ its character group. Let R be a principal domain (such as* $\mathbb{Z}$ *or a field). Let $P\hat{G}$ be the symmetric graded ring generated by $\hat{G}$ in degree 2. Then there is a natural coretraction of graded commutative* *algebras*

$$\tau_{G,R}: R \otimes P\hat{G} \to h(G, R)$$

(*i.e. an inclusion map which splits with an algebra morphism*) *which makes* $h(G, R)$ *into an augmented* $R \otimes P\hat{G}$*-algebra in a natural way. The morphism* τ *is an isomorphism if* G *is connected, and, in case that* $R = \mathbf{Z}$, *only if* G *is connected.*

b) *Let* $\wedge G$ *denote the exterior algebra over the compact group* G *in degree* 1 (*which exists and has the familiar properties of the discrete situation*). *Then there is a natural coretraction of commutative algebras*

$$\varrho_{G,R}: \mathrm{Hom}\,(\wedge G, R) \to h(G, R)$$

where Hom $(\wedge G, R)$ *denotes the commutative graded ring of all locally constant morphisms into the discrete ring* R.

c) *The resulting morphism of graded commutative* R*-algebras*

$$\omega_{G,R}: P\hat{G} \otimes \mathrm{Hom}\,(\wedge G, R) \to h(G, R)$$

is bijective in degrees 0, 1, 2 (*but is not bijective in general*).

This theorem implies more than appears on the surface, since its proof contains deeper insights into the nature of the natural coretractions.

The map ϱ in b) is always trivial if R has zero characteristic or if G is connected. Further, the theorem maintains with an arbitrary abelian group R in place of a ring, in which case the phrase "graded commutative ring" in the theorem must be replaced consistently by "graded commutative group". The power of the theorem is best seen when R is a field. Then we have

Theorem. *Let* G *be a compact abelian group and let* R *be a field. Then the morphism*

$$\omega_{G,R}: P\hat{G} \otimes \mathrm{Hom}\,(\wedge G, R) \to h(G, R)$$

is an isomorphism of algebras. There are also natural isomorphisms of R*-Hopf algebras*

$$P_R(R \otimes \hat{G}) \otimes_R \wedge_R \mathrm{Hom}\,(G, R) \cong R \otimes P\hat{G} \otimes \wedge\, \mathrm{Tor}\,(\hat{G}, K) \cong h(G,R)$$

where K *is the prime field of* R.

This theorem then gives a complete structure theorem for the functor h for field coefficients. For other coefficient rings the situation is, in general, more complicated.

A typical situation which illustrates other circumstances, under which the useful morphism ω is an isomorphism is the following:

Proposition. *Let* $R = \mathsf{Z}/m\mathsf{Z}$ *with a natural number* m *and suppose that* $G = \mathsf{Z}(z_1) \otimes \cdots \otimes \mathsf{Z}(z_n)$ *with* $m | z_1 | z_2 \cdots z_{n-1} | z_n$. *Then*

$$\omega_{G,R}: P\hat{G} \otimes \mathrm{Hom}(\wedge G, R) \to h(G, R)$$

is an isomorphism of graded algebras.

This applies in particular to the case $z_1 = \cdots = z_n$, i.e., when $G = \mathsf{Z}(z)^n$. There are two basic ideas which emanate from this proposition:

(A) The exact coefficient sequence

$$0 \to \mathsf{Z}/m\,\mathsf{Z} \to \mathsf{Z}/m^2\mathsf{Z} \to \mathsf{Z}/m\,\mathsf{Z} \to 0$$

yields, via the usual long exact cohomology sequence arising from it, a degree one map $d: h(G, \mathsf{Z}/m\,\mathsf{Z}) \to h(G, \mathsf{Z}/m\mathsf{Z})$ which turns out to be a derivation with respect to the algebra structure of $h(G, R)$. This is the so-called *Bockstein* differential. Its significance und er current circumstances is that it is exact, and that the image of d is isomorphic to the ideal $h^+(G, \mathsf{Z}) = h^1(G, \mathsf{Z}) + h^2(G, \mathsf{Z}) + \cdots$. This essentially gives us the ring $h(G, \mathsf{Z})$ as a well described subring of a well-understood ring. For this reason, we give the formalism of the Bockstein differential a good deal of attention. The general background for it is provided in Chapter I, Section 4, which is independent of the rest of the work and may serve as a self-contained introduction to this circle of ideas.

(B) The integral cohomology of a sum of cyclic groups of the same order is amenable to computation as explained above. Any finite abelian group is built up from groups of this special type, and it is possible to compute the cohomology inductively from the known cohomology of these portions. The actual execution, which is carried out in Proposition 9 in Section 3 of Chapter II, is technically complicated. (See also Theorem 11 of the same Section.) Some consequences which maintain for the case of compact groups are collected in the following:

Theorem. *Let* G *be a compact abelian group and let*

$$b^i: h^i(G, \mathsf{R}/\mathsf{Z}) \to h^{i+1}(G, \mathsf{Z})$$

be the connecting morphism stemming from the coefficient sequence

$$0 \to \mathsf{Z} \to \mathsf{R} \to \mathsf{R}/\mathsf{Z} \to 0.$$

If G *is totally disconnected, then* b^i *is an isomorphism for all positive* i. *Let*

$$\varrho_{G,\mathsf{R}/\mathsf{Z}}: \mathrm{Hom}(\wedge G, \mathsf{R}/\mathsf{Z}) = (\wedge \hat{G}) \to h(G, \mathsf{R}/\mathsf{Z})$$

be the coretraction given in a previous theorem. The composite map

$$(\wedge^+ \hat{G}) \xrightarrow{\varrho_{G,\mathsf{R}/\mathsf{Z}}} h(G, \mathsf{R}/\mathsf{Z}) \xrightarrow{b} h(G, \mathsf{Z})$$

is an injection of degree $+1$ *of graded abelian groups. We denote the group* $h^0(G, \mathbf{Z})$ $+$ *image of this map by* $M(G)$. (*Then* $M^0(G) = \mathbf{Z}$, $M^1(G) = 0$, $M^2(G) \cong \hat{G}$, $M^3(G) \cong (\bigwedge^2 G)^{\wedge}, \ldots, M^p(G) \cong (\bigwedge^{p-1} G)^{\wedge}$.) *We consider* $h(G, \mathbf{Z})$ *as a* $P\hat{G}$*-module via the coretraction*

$$\tau_{G,\mathbf{Z}}: P\hat{G} \to h(G, \mathbf{Z}).$$

Then:

(a) $h(G, \mathbf{Z})$ *is a torsion free, but not in general a free* $P\hat{G}$*-module.*

(b) $h(G, \mathbf{Z}) = P\hat{G} \cdot M(G)$, *and* $M(G)$ *is a minimal generating subgroup of the* $P\hat{G}$*-module* $h(G, \mathbf{Z})$.

(c) *The subgroup* $M(G)$ *generates* $h(G, \mathbf{Z})$ *as a ring.*

It is instructive to visualize the situation via a certain bigradation which exists on $h(G, \mathbf{Z})$ (although not in a completely functorial fashion in G):

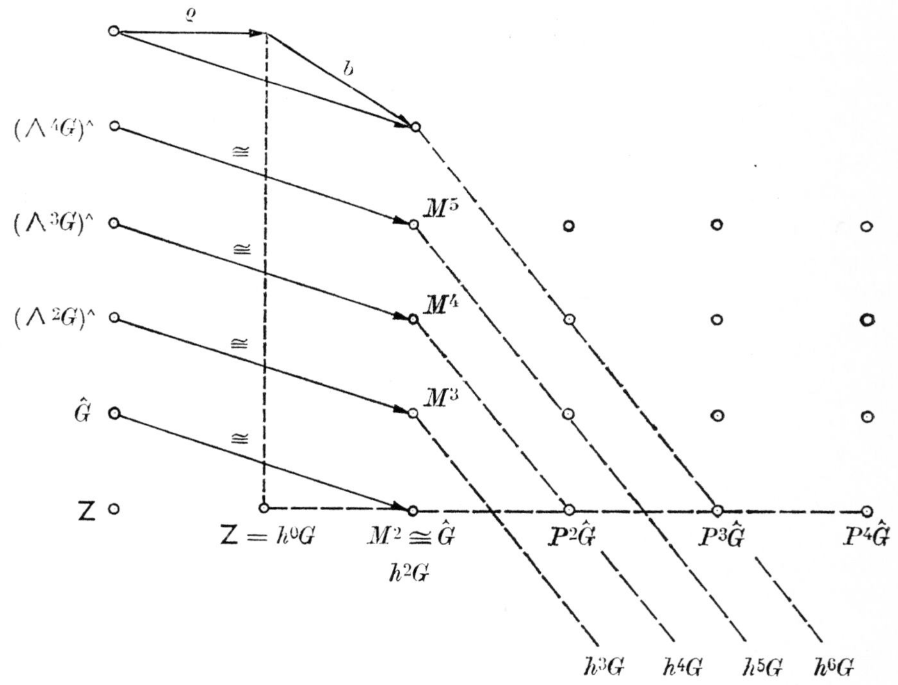

Interdependence of Chapters

It might help the reader to have a chart of the logical interdependence of the segments of the treatise and in particular to observe that some of the later sections may be read without going through earlier portions. It should be used together with the table of contents.

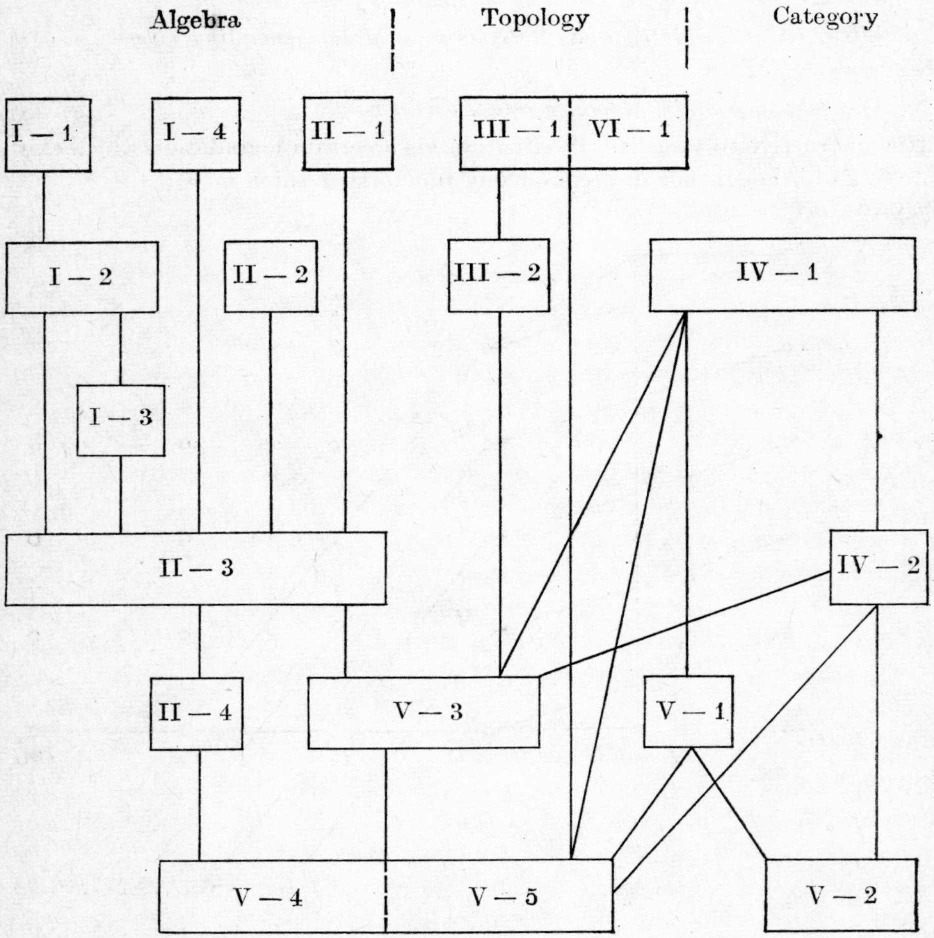

Chapter I

Algebraic background

Section 1

On exponential functors

The pair of functors $\bigwedge$ and P, which assign to an R-module its exterior algebra and its polynomial algebra, respectively, plays an important role in all that follows. For a unified treatment, we consider PM, for any module M, as a commutative graded R-algebra in which all elements of M have degree 2, so that in fact all homogeneous components of PM of odd degree are zero. The functors $\bigwedge$ and P are special instances of functors which we will call *exponential functors* (see also Hofmann [20]). Our functorial considerations will, in particular, give a conceptual proof of the following fact: Let $\mathrm{Hom}(\bigwedge G, R)$, for an abelian group G and a commutative ring R with identity, be the naturally graded R-module of all group morphisms $\bigwedge G \to R$. Then $\mathrm{Hom}(\bigwedge A, R)$ is in a natural fashion a graded commutative R-algebra. The natural inclusion map $\mathrm{Hom}(G, R) \to \mathrm{Hom}(\bigwedge G, R)$ then extends uniquely to a morphism of graded R-algebras $\bigwedge \mathrm{Hom}(G, R) \to \mathrm{Hom}(\bigwedge G, R)$. In order to familiarize ourselves with the situation, we will calculate explicitly this morphism. Similar considerations pertain to the case of P in place of $\bigwedge$.

Definition 1.1. A category $\mathfrak{M}$ is called *multiplicative*, if it is equipped with a functor $-\otimes-: \mathfrak{M} \times \mathfrak{M} \to \mathfrak{M}$ of two arguments satisfying certain properties which we will describe somewhat informally; this description will be adequate since many of the formal features of a multiplicative category are self-understood in the concrete instances to which we shall apply the concept. For a more detailed exposition see e.g. MacLane [33].

(a) $\otimes$ is associative; i. e. there is a natural isomorphism

$$\alpha_{A,B,C}: A \otimes (B \otimes C) \to (A \otimes B) \otimes C.$$

(b) $\otimes$ is commutative; i. e. there is a natural involution

$$\gamma_{A,B}\colon A \otimes B \to B \otimes A .$$

(c) $\otimes$ has an identity; i.e. $\mathfrak{M}$ has a ground object E such that there are natural isomorphisms

$$\varepsilon_A\colon E \otimes A \to A \quad \text{and} \quad \varepsilon'_A\colon A \otimes E \to A .$$

(d) The multiplicative structure is coherent, i.e. the following diagrams are commutative:

$$\begin{array}{ccccc} A \otimes (B \otimes (C \otimes D)) & \longrightarrow & (A \otimes B) \otimes (C \otimes D) & \longrightarrow & ((A \otimes B) \otimes C) \otimes D \\ \big\downarrow & & & & \big\uparrow \\ A \otimes ((B \otimes C) \otimes D) & & \longrightarrow & & (A \otimes (B \otimes C)) \otimes D \end{array}$$

(all morphisms being obvious α-morphisms)

$$\begin{array}{ccccc} A \otimes (B \otimes C) & \xrightarrow{\alpha} & (A \otimes B) \otimes C & \xrightarrow{\gamma} & C \otimes (A \otimes B) \\ \big\downarrow{\scriptstyle A \otimes \gamma} & & & & \big\downarrow{\scriptstyle \alpha} \\ A \otimes (C \otimes B) & \xrightarrow{\alpha} & (A \otimes C) \otimes B & \xrightarrow{\gamma \otimes B} & (C \otimes A) \otimes B \end{array}$$

$$\begin{array}{ccc} E \otimes (B \otimes C) & \xrightarrow{\alpha} & (E \otimes B) \otimes C \\ & {\scriptstyle \varepsilon} \searrow \quad \swarrow {\scriptstyle \varepsilon \otimes C} & \\ & B \otimes C & \end{array}$$

$$\begin{array}{ccccc} A \otimes (E \otimes C) & \xrightarrow{\alpha} & (A \otimes E) \otimes C & \xrightarrow{\gamma \otimes C} & (E \otimes A) \otimes C \\ & {\scriptstyle A \otimes \varepsilon} \searrow & & \swarrow {\scriptstyle \varepsilon \otimes C} & \\ & & A \otimes C & & \end{array}$$

Some authors call a multiplicative category *monoidal*. MacLane has shown that coherence implies the commutativity of all diagrams which are built up from the associativity, the commutativity, and the neutrality morphism in a feasible way. A functor $M\colon \mathfrak{M} \to \mathfrak{M}'$ between multiplicative categories is called *multiplicative* (*monoidal*) if there is a natural isomorphism

$$\beta_{X,Y}\colon MX \otimes' MY \to M(X \otimes Y),$$

which we will always tacitly assume to be compatible with the coherence morphisms in whichever way necessary.

If $\mathfrak{A}$ is an additive category with biproducts (i. e. products which are simultaneously coproducts under the natural morphism $A \oplus B \to A \times B$ which exists in every pointed category) then $\mathfrak{A}$ is a multiplicative category relative to $\oplus$ and a monoidal functor $(\mathfrak{A}, \oplus) \to (\mathfrak{M}, \otimes)$ into a multiplicative category is called an *exponential functor*. Thus, for an exponential functor E, we have a natural isomorphism $\varepsilon_{A,B}\colon E(A \oplus B) \to EA \otimes EB$ (which is compatible with coherence).

A typical example of an additive category is the category Ab of abelian groups, an example of a multiplicative category is the category $\mathfrak{R}$ of graded commutative rings; examples of exponential functors are the functors

$$\wedge\colon Ab \to \mathfrak{R} \quad \text{and} \quad P\colon Ab \to \mathfrak{R},$$

where $\wedge A$ is the exterior algebra generated by the abelian group A in degree 1 and where PA is the symmetric algebra generated by the group A in degree 2. Clearly, if E_i, $i = 1, \ldots, n$ are exponential functors, then so is $E_1 \otimes \cdots \otimes E_n$. If E is exponential and F is additive, then EF is exponential.

There is no real danger of confusion if we omit, in diagrams involving products in a monoidal category, the parentheses and the coherence morphisms. In case of doubt it is our hypothesis that all natural morphisms occurring are compatible with the coherence morphisms.

If $\mathfrak{A}$ has biproducts (i. e., products which are simultaneously coproducts), E is exponential, and if, for some object A in $\mathfrak{A}$, the morphisms

$$\Delta_A\colon A \to A \times A = A \oplus A \quad \text{and} \quad \Delta^A\colon A \oplus A \to A$$

are the diagonal and codiagonal morphisms, then the following diagram defines on EA the structure of a Hopf algebra (EA, μ_A, γ^A):

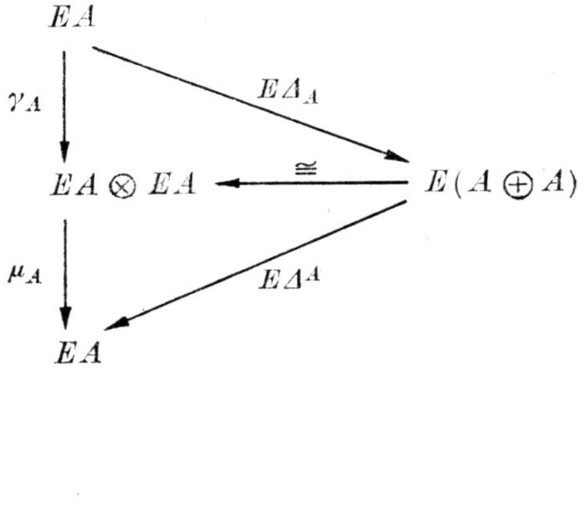

Definition 1.2. Let $\mathfrak{A}_i$, $i = 1, 2$, be additive and $\mathfrak{M}_i$, $i = 1, 2$, be multiplicative categories. A functor $S\colon \mathfrak{A}_1 \to \mathfrak{A}_2^*$ (resp. a functor $T\colon \mathfrak{M}_1 \to \mathfrak{M}_2^*$) is called *subadditive* (resp. *submultiplicative*) if there is a natural transformation of bifunctors $\alpha_{A,B}\colon SA \oplus SB \to S(A \oplus B)$ (resp.

$$\beta_{X,Y}\colon TX \otimes TY \to T(X \otimes Y)).$$

If $E_i\colon \mathfrak{A}_i \to \mathfrak{M}_i$, $i = 1, 2$, are exponential functors, then we call a natural transformation $\varphi\colon E_2 S \to TE_1$ *compatible*, if the following diagram commutes:

$$\begin{array}{ccc} E_2SA \otimes E_2SB & \xrightarrow{\varphi_A \otimes \varphi_B} & TE_1A \otimes TE_1B \\ \downarrow{\scriptstyle (\varepsilon_2 S_{A,B})^{-1}} & & \downarrow{\scriptstyle (\beta E_1)_{A,B}} \\ E_2(SA \oplus SB) & & T(E_1A \otimes E_1B) \\ \downarrow{\scriptstyle E_2 \alpha_{A,B}} & & \downarrow{\scriptstyle (T\varepsilon_1)_{A,B}} \\ E_2S(A \oplus B) & \xrightarrow{\varphi_{A \oplus B}} & TE_1(A \oplus B) \end{array}$$

Note that by the naturality of φ the following diagram commutes:

$$\begin{array}{ccc} E_2S(A \oplus A) & \xrightarrow{\varphi_{A \oplus A}} & TE_1(A \oplus A) \\ \downarrow{\scriptstyle E_2 S\Delta_A} & & \downarrow{\scriptstyle TE_1 \Delta_A} \\ E_2SA & \xrightarrow{\varphi_A} & TE_1A \end{array}$$

Lemma 1.3. *If $\mathfrak{A}_i$, is an additive, $\mathfrak{M}_i$ is a multiplicative category, $i = 1, 2$, $S\colon \mathfrak{A}_1 \to \mathfrak{A}_2^*$ a subadditive, $T\colon \mathfrak{M}_1 \to \mathfrak{M}_2^*$ a submultiplicative, and $E_i\colon \mathfrak{A}_i \to \mathfrak{M}_i$, $i = 1, 2$, exponential functors, then E_2SA is an algebra relative to the multiplication*

$$E_2SA \otimes E_2SA \xrightarrow{(\varepsilon_2 S_{A,A})^{-1}} E_2(SA \oplus SA) \xrightarrow{E_2 \alpha_{A,A}}$$
$$E_2S(A \oplus A) \xrightarrow{E_2 S\Delta_A} E_2SA$$

and TE_1A is an algebra relative to the multiplication

$$TE_1A \otimes TEA \xrightarrow{(\beta E_1)_{A,A}} T(E_1A \otimes E_1A) \xrightarrow{T\varepsilon_{1A,A}^{-1}}$$
$$TE_1(A \oplus A) \xrightarrow{TE_1 \Delta_A} TE_1A.$$

If the multiplications $\otimes$ are commutative, resp., associative, then these algebras are commutative, resp., associative provided that the following two diagrams commute:

$$\begin{array}{ccc} SA \oplus SB \oplus SC & \xrightarrow{1_{SA} \oplus \alpha_{B,C}} & SA \oplus S(B \oplus C) \\ \downarrow{\scriptstyle \alpha_{A,B} \oplus 1_{SC}} & & \downarrow{\scriptstyle \alpha_{A,B\oplus C}} \\ S(A \oplus B) \oplus SC & \xrightarrow{\alpha_{A\oplus B,C}} & S(A \oplus B \oplus C) \end{array}$$

$$\begin{array}{ccc} TA \otimes TB \otimes TC & \xrightarrow{1_{TA} \otimes \beta_{B,C}} & TA \otimes (TB \otimes TC) \\ \downarrow{\scriptstyle \beta_{A,B} \otimes 1_C} & & \downarrow{\scriptstyle \beta_{A,B\otimes C}} \\ T(A \otimes B) \otimes TC & \xrightarrow{\beta_{A\otimes B,C}} & T(A \otimes B \otimes C) \end{array}$$

Proof. We prove, e.g., the associativity of the multiplication on E_2S. The other proofs are similar. Let μ be the multiplication on E_2SA. We prove the commutativity of the following diagram by diagram chasing, where we write E for E_2:

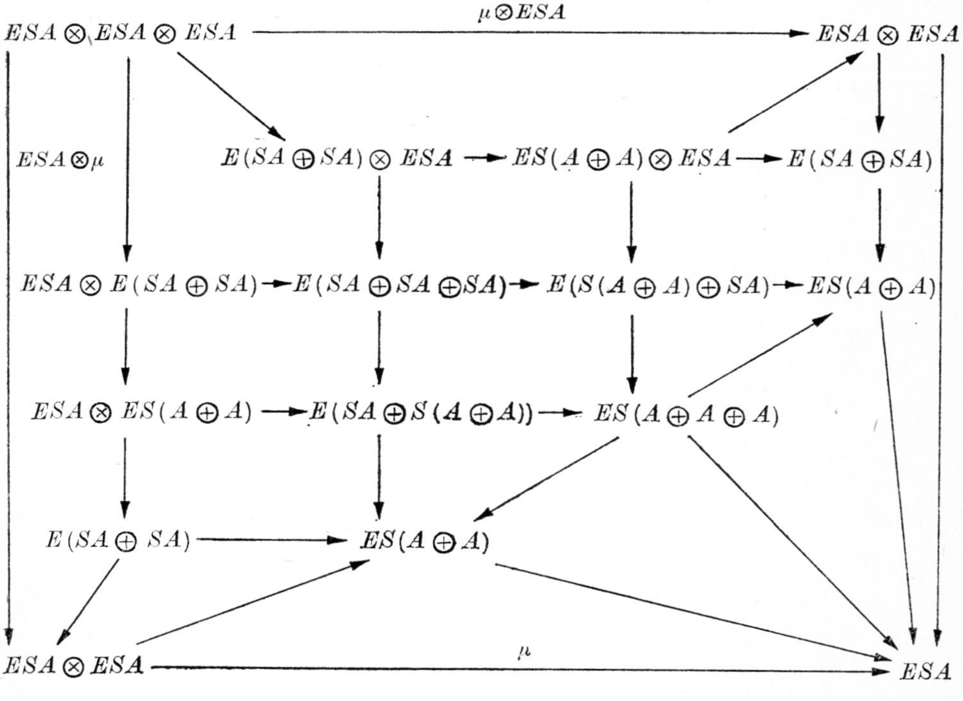

Lemma 1.4. *If, under the conditions of Lemma 1.3, $\varphi\colon E_2S \to TE_1$ is compatible, then φ_A is a morphism of algebras.*

Proof. The assertion is equivalent to the commutativity of the outer rectangle in the following diagram, in which the top rectangle commutes by hypothesis and the bottom rectangle by the naturality of φ:

$$\begin{array}{ccc}
E_2SA \otimes E_2SA & \xrightarrow{\varphi_A \otimes \varphi_A} & TE_1A \otimes TE_1A \\
\big\downarrow {\scriptstyle (\varepsilon_2 S_{A,A})^{-1}} & & \big\downarrow {\scriptstyle \beta E_{1A,A}} \\
E_2(SA \oplus SA) & & T(E_1A \otimes E_1A) \\
\big\downarrow {\scriptstyle E_2\alpha_{A,A}} & & \big\downarrow {\scriptstyle (T\varepsilon_1)_{A,A}} \\
E_2S(A \oplus A) & \xrightarrow{\varphi_{A\oplus A}} & TE_1(A \oplus A) \\
\big\downarrow {\scriptstyle E_2S\Delta_A} & & \big\downarrow {\scriptstyle TE_1\Delta_A} \\
E_2SA & \xrightarrow{\varphi_A} & TE_1A \quad .
\end{array}$$

Lemma 1.5. *Let $\mathfrak{C}$, $\mathfrak{M}$ be categories with multiplication. Suppose that in $\mathfrak{M}$ there is a natural transformation*

$$j_{A,B}\colon (A \otimes A) \otimes (B \otimes B) \to (A \otimes B) \otimes (A \otimes B)$$

and in $\mathfrak{C}$ a natural transformation

$$i_{A,B}\colon (A \otimes B) \otimes (A \otimes B) \to (A \otimes A) \otimes (B \otimes B).$$

(These conditions are certainly satisfied if the multiplications are associative and commutative.)

(a) *If $\gamma_A\colon A \to A \otimes A$ and $\gamma_B\colon B \to B \otimes B$ are coalgebras in $\mathfrak{C}$, then $A \otimes B$ is a coalgebra via the diagram*

$$\begin{array}{ccc}
A \otimes B & & \\
\big\downarrow {\scriptstyle \gamma_A \otimes \gamma_B} & \searrow {\scriptstyle \gamma_{A \otimes B}} & \\
(A \otimes A) \otimes (B \otimes B) & \xrightarrow[i_{A,B}]{} & (A \otimes B) \otimes (A \otimes B)
\end{array}$$

(b) *Suppose that $T\colon \mathfrak{C} \to \mathfrak{M}$ is a submultiplicative cofunctor, i. e. there is a natural transformation $\beta_{A,B}\colon\ TA \otimes TB \to T(A \otimes B)$. Under the conditions of* (a), *$T(A \otimes B)$ and $TA \otimes TB$ are coalgebras, via the diagrams*

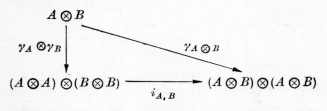

$$
\begin{array}{ccc}
T(A\otimes B)\otimes T(A\otimes B) & & \\
\downarrow{\scriptstyle \beta_{A\otimes B, A\otimes B}} & \searrow{\scriptstyle \nu} & \\
T((A\otimes B)\otimes(A\otimes B)) & \xrightarrow[\gamma_{A\otimes B}]{} & T(A\otimes B)
\end{array}
$$

$$
\begin{array}{ccccc}
(TA\otimes TB)\otimes(TA\otimes TB) & & & \overset{\mu}{\searrow} & \\
\downarrow{\scriptstyle j_{TA,TB}} & & & & \\
(TA\otimes TA)\otimes(TB\otimes TB) & \xrightarrow[\beta_{A,A}\otimes\beta_{B,B}]{} & T(A\otimes A)\otimes T(B\otimes B) & \xrightarrow[T\gamma_A\otimes T\gamma_B]{} & TA\otimes TB
\end{array}
$$

(c) *Suppose that under the conditions of* (a) *and* (b) *one has*

$$\beta_{A\otimes A, B\otimes B}(\beta_{A,A}\otimes\beta_{B,B})\,j_{TA,TB} = Ti_{A,B}\,\beta_{A\otimes B, A\otimes B}(\beta_{A,B}\otimes\beta_{A,B}).$$

Then $\beta_{A,B}\colon TA\otimes TB \to T(A\otimes B)$ *is a morphism of algebras.*

Remark. If $\mathfrak{C}$ is the category of Z-modules and $\mathfrak{M}$ the category of R-modules, where R, Z are commutative rings and R is an Z-algebra, and if $T = \mathrm{Hom}_Z(-, M)$ for a fixed R-algebra M and β is defined by

$$\beta(f\otimes g)(a\otimes b) = f(a)\,g(b)$$

then all hypotheses in Lemma 1.5 are satisfied.

Proof of Lemma 1.5. The statements in (a) and (b) are clear. For a proof of (c), we consider the following diagram (p. 30).

The rectangle commutes by hypothesis, the parallelogram by the naturality of β and the triangle because it arises from the commutative diagram in (a) by application of T. Hence the whole diagram commutes. It follows that the diagram

$$
\begin{array}{ccc}
(TA\otimes TB)\otimes(TA\otimes TB) & \xrightarrow{\beta_{AB}\otimes\beta_{AB}} & T(A\otimes B)\otimes T(A\otimes B) \\
\downarrow{\scriptstyle \mu} & & \downarrow{\scriptstyle \nu} \\
TA\otimes TB & \xrightarrow{\beta_{AB}} & T(A\otimes B)
\end{array}
$$

commutes, and this is the assertion.

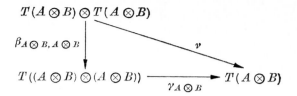

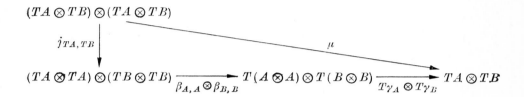

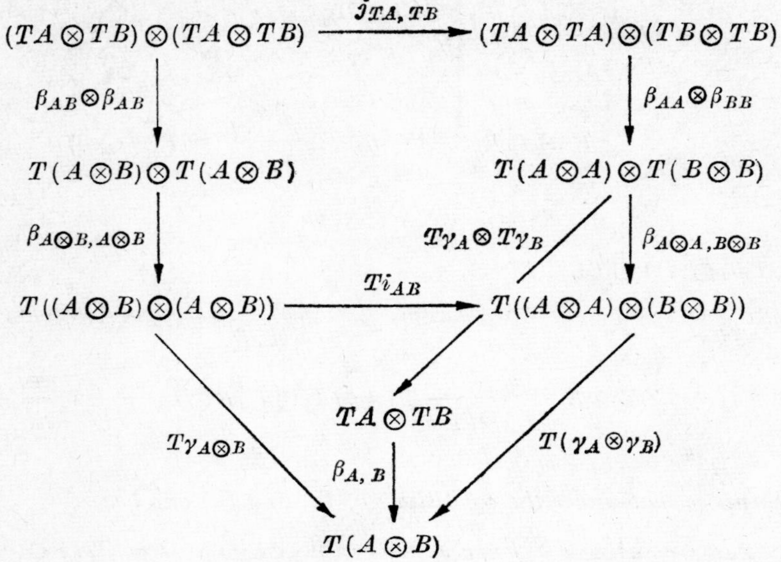

In the second part of this section we specialize the preceding purely functorial considerations to the case of categories of modules.

If R and Z are commutative rings with identity and R is a Z-algebra, and if M is an R-algebra, then M is automatically a Z-algebra under $z \cdot m = \zeta(z)\, m$ where $\zeta: Z \to R$ is defined by $\zeta(z) = z \cdot 1$.

If B is any Z-module, then $\mathrm{Hom}_Z\,(B, M)$ is an R-module under the operation defined by $(r \cdot f)\,(b) = r \cdot f(b)$. If B is a graded Z-module then $\mathrm{Hom}_Z(B, M)$ is a graded R-module whose homogeneous component of degree n is

$$\mathrm{Hom}_Z(B^n, M).$$

If M is in fact an R-algebra and A and B are Z-modules, then there is a natural transformation

$$\beta_{A,B}\colon \mathrm{Hom}_Z(A, M) \otimes_R \mathrm{Hom}_Z(B, M) \to \mathrm{Hom}_Z(A \otimes_Z B, M)$$

defined by $\beta_{A,B}(f \otimes g)\,(a \otimes b) = f(a)\, g(b)$. If A and B are finitely generated free Z-modules and $M = R$, then $\beta_{A,B}$ is an isomorphism by the additivity of $\mathrm{Hom}_Z\,(-, M)$ since it is an isomorphism for $A = B = Z \cong A \otimes_Z B$. This means that $\mathrm{Hom}_Z\,(-, M)$ is submultiplicative and indeed multiplicative on the subcategory of finitely generated free Z-modules.

Proposition 1.6. *Let R and Z be commutative rings with identity and suppose that R is a Z-algebra. (Frequently $Z = \mathbf{Z}$.)*

Let

$$\mathfrak{A}_Z = \textit{category of } Z\textit{-modules},$$
$$\mathfrak{A}_R = \textit{category of } R\textit{-modules},$$
$$\mathfrak{M}_Z = \textit{category of graded commutative algebras over } Z,$$
$$\mathfrak{M}_R = \textit{category of graded commutative algebras over } R.$$

Let $E_i\colon \mathfrak{A}_i \to \mathfrak{M}_i$, $i = R, Z$, *be either* $\wedge_i$ *or* P_i. *Then we have the following statements:*

(a) *There is a compatible natural transformation of graded commutative R-algebras* $\varphi\colon E_R \operatorname{Hom}_Z(-, M) \to \operatorname{Hom}_R(E_Z -, M)$ *for any fixed R-algebra M.*

(b) *The morphism φ is also natural in M and is uniquely determined by the requirement that*

$$\varphi_A^1\colon \operatorname{Hom}_Z(A, M) \to \operatorname{Hom}_Z(A, M) \quad \textit{for} \quad E = \wedge$$

and

$$\varphi_A^2\colon \operatorname{Hom}_Z(A, M) \to \operatorname{Hom}_Z(A, M) \quad \textit{for} \quad E = P$$

is the identity isomorphism.

Proof. Since there is no danger of confusion, we omit the subscripts at Hom_Z, E_Z, E_R. By Lemma 1.3, $E \operatorname{Hom}(A, M)$ and $\operatorname{Hom}(EA, M)$ are graded commutative associative R-algebras when M is an R-algebra. There is an injection of R-modules $\operatorname{Hom}(A, M) \to \operatorname{Hom}(E^i A, M) \to \operatorname{Hom}(EA, M)$, where $i = 1$ if $E = \wedge$ and $i = 2$ if $E = P$. By the universal property of the functors $\wedge$ and P, there is a unique morphism $\varphi_A\colon E \operatorname{Hom}(A, M) \to \operatorname{Hom}(EA, M)$ of graded R-algebras extending this injection, and this morphism is clearly natural in A. It remains to show that φ_A is compatible. The diagram

$$\begin{array}{ccc}
E \operatorname{Hom}(A_j, M) & \xrightarrow{\varphi_{A_j}} & \operatorname{Hom}(EA_j, M) \\
\downarrow & & \downarrow \\
E(\operatorname{Hom}(A_1, M) \oplus \operatorname{Hom}(A_2, M)) & & \operatorname{Hom}(EA_1 \otimes EA_2, M) \\
\downarrow & & \downarrow \\
E(\operatorname{Hom}(A_1 \oplus A_2, M)) & \xrightarrow{\varphi_{A_1 \oplus A_2}} & \operatorname{Hom}(E(A_1 \oplus A_2), M)
\end{array}$$

commutes for $j = 1, 2$. It follows, then, that the first diagram defining the compatibility of φ commutes. Thus φ is compatible.

Remark. Note that $\varphi_A^0\colon E_R^0 \operatorname{Hom}(A, M) \to \operatorname{Hom}(E_Z^0 A, M)$ is the morphism $R \to M$ mapping r onto $r \cdot 1$.

It is possible to determine the natural transformation φ explicitly:

Corollary 1.7. *Let φ_A be as in Proposition* 1.6. *Let f_i, $i = 1, \ldots, p$ be elements of* $\operatorname{Hom}_Z(A, M)$ *and x_i, $i = 1, \ldots, p$, be elements in A. Then, if $E = \wedge$ we have*

(a) $\varphi_A(f_1 \wedge \cdots \wedge f_p)(x_1 \wedge \cdots \wedge x_p) = \det((f_i(x_j))_{i,j=1,\ldots,p})$.

If $E = P$, we have

(b) $\varphi_A(f_1 \cdots f_p)(x_1 \cdots x_p) = \operatorname{perm}((f_i(x_j))_{i,j=1,\ldots,p})$, *where the permanent of a square matrix a_{ij}, $i = 1, \ldots, p$ is the sum of all $a_{1\sigma(1)} \cdots a_{p\sigma(p)}$ taken over all permutations σ of $\{1, \ldots, p\}$.*

Proof. We establish the claim by showing that the φ defined by (a), resp. (b), is compatible. It is easily seen to be natural. Then φ_A is a morphism of graded algebras and obviously satisfies the initial condition by which φ was defined in Proposition 1.6. The uniqueness of φ will then imply the assertion.

We treat the case $E = \wedge$; the remaining case $E = P$ is completely analogous. Let A, B be two R-modules, let $f^p = f_1 \wedge \cdots \wedge f_p$, $g^q = g_1 \wedge \cdots \wedge g_q$ with $f_i \in \operatorname{Hom}(A, M)$, $g_i \in \operatorname{Hom}(B, M)$, and define $x^p = x_1 \wedge \cdots \wedge x_p$,

$$y^q = y_1 \wedge \cdots \wedge y_q.$$

We identify $(f, g) \in \operatorname{Hom}(A, M) \oplus \operatorname{Hom}(B, M)$ with its image in

$$\operatorname{Hom}(A \oplus B, M).$$

If we let

$$(a, b)^{p+q} = (a_1, b_1) \wedge \cdots \wedge (a_{p+q}, b_{p+q}) \in \overset{p+q}{\wedge} (A \oplus B),$$

then (with $S = \operatorname{Hom}$ on $\mathfrak{A}$ and $T = \operatorname{Hom}$ on $\mathfrak{M}$) we have

$$\begin{aligned}&\{[\varphi_{A\oplus B}(E\,\alpha_{A,B})(\varepsilon\, S_{A,B})^{-1}](f^p \otimes g^q)\}(a, b)^{p+q} \\ &= \det((f_i(a_j))_{i,j=1,\ldots,p}) \det((g_i(b_j))_{i,j=1,\ldots,q}).\end{aligned}$$

It follows now that

$$\begin{aligned}&\{[(T\,\varepsilon_{A,B})^{-1}\varphi_{A\oplus B} E\,\alpha_{A,B}(\varepsilon\, S_{A,B})^{-1}](f^p \otimes g^q)\}(x^p \otimes y^q) \\ &= \{[\varphi_{A\oplus B}(E\,\alpha_{A,B})(\varepsilon\, S_{A,B})^{-1}](f^p \otimes g^q)\}((x_1, 0) \wedge \cdots \wedge (x_p, 0) \\ &\quad \wedge (0, y_1) \wedge \cdots \wedge (0, y_q)) \\ &= \det((f_i(x_j))_{i,j=1,\ldots,p}) \det((g_i(y_j))_{i,j=1,\ldots,q}).\end{aligned}$$

On the other hand,

$$\begin{aligned}&\{[(\beta\, E_{A,B})(\varphi_A \otimes \varphi_B)](f^p \otimes g^q)\}(x^p \otimes y^q) = (\varphi_A f^p)(x^p)(\varphi_A f^q)(y^q) \\ &= \det((f_i(x_j))_{i,j=1,\ldots,p}) \det((g_i(y_j))_{i,j=1,\ldots,q}).\end{aligned}$$

This shows that φ_A is compatible.

If A and M are of a special nature we can obtain more detailed information about the morphisms

$$\varphi_A\colon \wedge_R \operatorname{Hom}(A, M) \to \operatorname{Hom}(\wedge_Z A, M)$$

and

$$\varphi_A\colon P_R \operatorname{Hom}(A, M) \to \operatorname{Hom}(P_Z A, M).$$

We will assume the following hypotheses about A and M:

(1) A is a finite direct sum of cyclic Z-modules, i. e. $A \cong Z/I_1 \oplus \cdots \oplus Z/I_n$ for suitable ideals I_i of Z.

(2) The Z-algebra morphism $Z \to M$ given by $z \to z \cdot 1$ vanishes on $I_1 + \cdots + I_n$.

Note that these hypotheses are satisfied if we have the following circumstances:

(3) A is a finitely generated free Z-module (i. e. $A \cong Z^n$) and $R = M = Z$.

Lemma 1.8. *Let A be a Z-module, R a commutative Z-algebra and M an R-algebra (hence in particular a Z-algebra), and suppose that the hypotheses* (1) *and* (2) *above are satisfied. Let a_i, $i = 1, \ldots, n$ be generators of the cyclic summands according to* (1). *Let $\sigma, s\colon \{1, \ldots, q\} \to \{1, \ldots, n\}$ be a non-decreasing, respectively a strictly increasing function, and write*

$$a_\sigma = a_{\sigma(1)} \cdots a_{\sigma(q)} \in P^q A \quad \text{and} \quad a_s = a_{s(1)} \wedge \cdots \wedge a_{s(q)} \in \wedge^q A.$$

Then

(i) *$P^q A$ is a direct sum of cyclic submodules generated by the a_σ and $\wedge^q A$ is a direct sum of cyclic submodules generated by the a_s.*

(ii) *There are unique elements $f_i \in \operatorname{Hom}(A, M)$, $i = 1, \ldots, n$ with $f_i(a_j) = 1$ if $i = j$ and $= 0$ otherwise, and, more generally, there are elements*

$$f \in \operatorname{Hom}(P^q A, M), \quad \textit{resp.} \quad f_s \in \operatorname{Hom}(\wedge^q A, M),$$

with $f_\sigma(a_\tau) = 1$ if $\sigma = \tau$ and 0 otherwise, respectively, $f_s(a_t) = 1$ if $s = t$ and 0 otherwise.

Proof. (i) follows from hypothesis (1), and (ii) from hypothesis (2).

Lemma 1.9. *Under the hypotheses of Lemma 1.8, with*

$$\varphi_A\colon \wedge_R \operatorname{Hom}(A, M) \to \operatorname{Hom}(\wedge_Z A, M)$$

and

$$\varphi_A\colon P_R \operatorname{Hom}(A, M) \to \operatorname{Hom}(P_Z A, M)$$

as in Proposition 1.6 *and Corollary* 1.7, *we have*

(a) $\varphi_A(f_{s(1)} \wedge \cdots \wedge f_{s(q)}) = f_s$ *for all* s,

(b) $\varphi_A(f_{\sigma(1)} \cdots f_{\sigma(q)}) = \prod_{i=1}^{n} e(i)!\, f_\sigma$ *for all* σ, *where* $e(i) = \operatorname{card} \sigma^{-1}(i)$ *for* $i = 1, \ldots, n$.

Proof. Case (a): We have

$$\varphi_A(f_{s(1)} \wedge \cdots \wedge f_{s(q)})(a_t) = \det\left(f_{s(i)}(a_{t(j)})_{i,j=1,\ldots,q}\right)$$

by Corollary 1.7. This expression is 0 for $t \neq s$ and 1 for $s = t$ by the definition of the f_i.

Case (b): The same argument holds with the permanent in place of the determinant and a non-decreasing function σ in place of a strictly increasing s. Let m_k be the *k-th* element in the u-element totally ordered set $\sigma(\{1, \ldots, q\})$, and let E_k be the $e(m_k) \times e(m_k)$ square matrix with all entries being 1. Then

$$\left(f_{\sigma(i)}(a_{\sigma(j)})_{i,j=1,\ldots,q}\right) = \begin{bmatrix} E_1 & \cdots & 0 \\ \vdots & & \vdots \\ 0 & \cdots & E_u \end{bmatrix}.$$

The permanent in question therefore is exactly

$$e(m_1)! \cdots e(m_u)! = e(1)! \cdots e(n)!.$$

Lemma 1.10. *Suppose that* $A = Z^n$ *is a free finitely generated Z-module, and that* $R = M$. *Then* $PA \cong PZ \otimes \cdots \otimes PZ$ *and*

$$\operatorname{Hom}(PA, R) \cong \operatorname{Hom}(PZ, R) \otimes_R \operatorname{Hom}(PZ, R) \otimes_R \cdots \otimes_R \operatorname{Hom}(PZ, R).$$

Further, the morphism

$$r \otimes f \to r \cdot f \colon R \otimes \operatorname{Hom}(PZ, Z) \to \operatorname{Hom}(PZ, R)$$

is an isomorphism of graded R-algebras. The structure of Hom (PA, R) *is therefore completely determined by the structure of* Hom (PZ, Z).

Proof straightforward.

Lemma 1.11. (a) *The graded ring PZ is isomorphic to the polynomial ring* $Z[X]$ *in one variable X of degree* 2.

(b) *The ring* $\hat{Z}[X] =$ Hom $(Z[X], Z)$ *has countably many generators* $1 = X^{(0)}$, $X^{(1)}, \ldots$ *forming a Z-basis, such that* $X^{(m)}(X^n) = 1$ *if* $m = n$ *and* $= 0$ *otherwise.*

The multiplication table of the $X^{(m)}$ *is given by*

$$X^{(m)} X^{(n)} = \binom{m+n}{m} X^{(m+n)}.$$

(c) *There is a commutative diagram*

$$\begin{array}{ccc} P_Z \operatorname{Hom}(Z, Z) & \xrightarrow{\varphi_Z} & \operatorname{Hom}(PZ, Z) \\ \downarrow \cong & & \| \\ Z[X] & \xrightarrow{\varphi_Z'} & \hat{Z}[X] \end{array}$$

with $\varphi_Z'(X^m) = m!\, X^{(m)}$, *and* φ_Z' *is an injection if* $Z = \mathsf{Z}$.

(d) *If* $Z = \mathsf{Z}$ *(the ring of integers) and if* Q *is the field of rationals, then there is a ring injection* $\operatorname{Hom}(Z[X], Z) \to Q[X]$ *sending* $X^{(m)}$ *onto* $\frac{1}{m!} X^m$.

(e) *If* $Z = \mathsf{Q}$ *then* $\varphi_Z \colon P_Z \operatorname{Hom}(Z, Z) \to \operatorname{Hom}(PZ, Z)$ *is an isomorphism.*

Proof. (a) is clear.

(b) We define the $X^{(m)}$ by the system of equations

$$X^{(m)}(X^n) = \begin{cases} 1, & \text{if } m = n, \\ 0, & \text{if } m \neq n. \end{cases}$$

The ring multiplication on $\operatorname{Hom}(Z[X], Z)$ is given by the following sequence of morphisms

$$\operatorname{Hom}(Z[X], Z) \otimes \operatorname{Hom}(Z[X], Z) \xrightarrow{\cong} \operatorname{Hom}(Z[X] \otimes Z[X], Z) \xrightarrow{\operatorname{Hom}(\psi, Z)} \operatorname{Hom}(Z[X], Z)$$

where $\psi \colon Z[X] \to Z[X] \otimes Z[X]$ is given by $\psi(X) = X \otimes 1 + 1 \otimes X$. It follows that

$$\psi(X^p) = (X \otimes 1 + 1 \otimes X)^p = \sum_{k=1}^{p} \binom{p}{k} X^k \otimes X^{p-k}.$$

If we identify $X^{(m)} \otimes X^{(n)}$ with its image in $\operatorname{Hom}(Z[X] \otimes Z[X], Z)$, then

$$[\operatorname{Hom}(\psi, Z)(X^{(m)} \otimes X^{(n)})](X^p) = (X^{(m)} \otimes X^{(n)})\, \psi(X^p) = \begin{cases} \binom{m+n}{m} & \text{if } p = m+n, \\ 0 & \text{otherwise.} \end{cases}$$

This finishes (b).

(c) is straightforward.

(d) follows from (c).

(e) The injectivity of φ_Z is clear and the surjectivity follows from (c) and the divisibility of $Z = Q$.

Notation. The functor $\operatorname{Hom}(P_Z -, Z)$ will be denoted $\hat{P}_Z$. For the symmetric algebra $P_Z Z = Z[X]$, we denote $\hat{P}_Z Z = \hat{Z}[X]$.

Lemma 1.12. *Let A satisfy hypothesis* (1) *and let*

$$\pi\colon Z^n \to A = Z/I_1 \oplus \cdots \oplus Z/I_n$$

be the obvious quotient map. Then we have injective morphisms of graded algebras

$$\begin{aligned}\operatorname{Hom}(P_Z A, M) &\xrightarrow{\operatorname{Hom}(P\pi, M)} \operatorname{Hom}(P_Z Z^n, M)\\ &\xrightarrow{\cong} \operatorname{Hom}(Z[X_1] \otimes \cdots \otimes Z[X_n], M)\\ &\longrightarrow M \otimes \operatorname{Hom}(Z[X_1] \otimes \cdots \otimes Z[X_n], Z)\\ &\xrightarrow{\cong} M \otimes \hat{Z}[X_1] \otimes \cdots \otimes \hat{Z}[X_n],\end{aligned}$$

$$\begin{aligned}\operatorname{Hom}(\wedge_Z A, M) &\xrightarrow{\operatorname{Hom}(\wedge\pi, M)} \operatorname{Hom}(\wedge_Z Z^n, M) \xrightarrow{\cong} M \otimes \operatorname{Hom}(\wedge Z^n, Z)\\ &\xrightarrow{M \otimes \varphi_{Z^n}^{-1}} M \otimes \wedge_Z \operatorname{Hom}(Z^n, Z) \longrightarrow M \otimes \wedge_Z Z^n.\end{aligned}$$

Proof. Since π is surjective, so are $P\pi$ and $\wedge\pi$, hence $\operatorname{Hom}(P\pi, M)$ and $\operatorname{Hom}(\wedge\pi, M)$ are injective. The isomorphisms of the first sequence are clear from the preceding. In the second sequence we need to know that

$$\varphi_{Z^n}\colon \wedge_Z \operatorname{Hom}(Z^n, Z) \to \operatorname{Hom}(\wedge_Z Z^n, Z)$$

is an isomorphism, which follows from lemmas 1.8 and 1.9.

Lemma 1.13. *Under the assumptions of the preceding lemma, there are natural injections*

$$\begin{aligned}P_R \operatorname{Hom}(A, R) &\xrightarrow{P_R \operatorname{Hom}(\pi, R)} P_R \operatorname{Hom}(Z^n, R)\\ &\xrightarrow{\cong} P_R R^n \longrightarrow R \otimes Z[X_1] \otimes \cdots \otimes Z[X_n]\end{aligned}$$

and

$$\begin{aligned}\wedge_R \operatorname{Hom}(A, R) &\xrightarrow{\wedge_R \operatorname{Hom}(\pi, R)} \wedge_R \operatorname{Hom}(Z^n, R) \xrightarrow{\cong} \wedge_R R^n\\ &\longrightarrow R \otimes \Lambda(n) \xrightarrow{\cong} R \otimes \Lambda(1) \otimes \cdots \otimes \Lambda(1),\end{aligned}$$

where $\Lambda(n) = \wedge_Z Z^n$ is the exterior algebra of Z in n generators.

The proof is similar.

In summing up the essential features of our discussion we arrive at the following description of the natural morphisms φ_A of 1.6 and 1.7 in the special case of conditions (1) and (2):

Proposition 1.14. *Let Z be a commutative ring with identity, R a commutative Z-algebra and M a commutative R-algebra. Suppose that Z contains ideals I_k, $k = 1, \ldots, n$ all of which are annihilated by the map $z \mapsto z \cdot 1\colon Z \to M$. Set $A = Z/I_1 \oplus \cdots \oplus Z/I_n$. Denote $\otimes_Z$ with $\otimes$. Then there are injections of graded algebras*

$$i_A\colon \mathrm{Hom}\,(P_Z A, M) \to M \otimes \hat{Z}[X_1] \otimes \cdots \otimes \hat{Z}[X_n]$$

and

$$i'_A\colon \mathrm{Hom}\,(\wedge_Z A, M) \to M \otimes \Lambda(n)$$

(with the free exterior algebra $\Lambda(n) = \wedge_Z Z^n$ in n generators over Z). If $M = R$, there are natural injections

$$j_A\colon P_R \,\mathrm{Hom}\,(A, M) \to R \otimes Z[X_1] \otimes \cdots \otimes Z[X_n],$$
$$j'_A\colon \wedge_R \mathrm{Hom}\,(A, M) \to R \otimes \Lambda(n),$$

and there are commuting diagrams

$$\begin{array}{ccc}
P_R\,\mathrm{Hom}\,(A, R) & \xrightarrow{\quad\varphi_A\quad} & \mathrm{Hom}\,(P_Z A, R) \\
\big\downarrow{\scriptstyle j_A} & & \big\downarrow{\scriptstyle i_A} \\
R \otimes Z[X_1] \otimes \cdots \otimes Z[X_n] & \xrightarrow{R \otimes \psi_1 \otimes \cdots \otimes \psi_n} & R \otimes \hat{Z}[X_1] \otimes \cdots \otimes \hat{Z}[X_n],
\end{array}$$

$$\begin{array}{ccc}
\wedge_R \mathrm{Hom}(A, R) & \xrightarrow{\quad\varphi'_A\quad} & \mathrm{Hom}(\wedge_Z A, R) \\
{\scriptstyle j'_A}\searrow & & \swarrow{\scriptstyle i'_A} \\
& R \otimes \Lambda(n) &
\end{array}$$

where $\psi_k\colon Z[X_k] \to \hat{Z}[X_k]$ is the morphism defined by $\psi_k(X_k^n) = n!\, X_k^{(n)}$. In particular, φ'_A is an isomorphism, and if R has characteristic zero, then φ_A is injective. If Z is a field of characteristic 0, then φ_A is an isomorphism.

Definition 1.15. The graded commutative rings $\hat{Z}[X]$ and their tensor powers are called *polynomial algebras with divided powers*. For any Z-module A

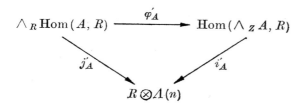

we will write $\hat{P}_Z A$ or just $\hat{P}A$ for the commutative algebra Hom (PA, Z), and we will call it the *symmetric divided algebra generated by* A. Note that $\hat{P}$ is a contravariant functor from the category of Z-modules into the category of graded commutative Z-algebras.

With this notation we have the following

Proposition 1.16. *Let Z be a commutative ring with identity, R a commutative Z-algebra and M a commutative R-algebra. (Frequently we will have $R = M$.) Then there is a diagram of natural transformations of graded commutative R-algebras*

$$\begin{array}{ccc} P_R \operatorname{Hom}_Z(A, M)\otimes_R \wedge_R \operatorname{Hom}_Z(A, M) & \xrightarrow{\varphi_{A,M}\otimes_R \varphi_{A,M}} & \operatorname{Hom}_Z(P_Z A, M)\otimes_R \operatorname{Hom}_Z(\wedge_Z A, M) \\ & \searrow^{\varkappa_{A,M}} \qquad \swarrow_{\beta_{A,M}} & \\ & \operatorname{Hom}_Z(P_Z A \otimes_Z \wedge_Z A, M) & \end{array}$$

The morphism $\beta_{A,M}$ is an isomorphism if $A = Z^n$ is finitely generated free. If $A \cong Z/I_1 \oplus \cdots \oplus Z/I_n$ for suitable ideals I_k of Z and if $z\cdot 1 = 0$ in M for $z \in I_k$ for all k, and if $R = M$ then $\varphi'_{A,M}$ is an isomorphism. If R has characteristic zero and A is finitely generated free, then $\varkappa_{A,M}$ is injective, and if Z is a field of characteristic zero, then $\varkappa_{A,M}$ is an isomorphism. For $M = R = Z$ and with $A =$ Hom (A, Z) *the diagram may be written*

$$\begin{array}{ccc} P\hat{A}\otimes \wedge \hat{A} & \longrightarrow & \hat{P}A \otimes (\wedge A)^\wedge \\ & \searrow \qquad \swarrow & \\ & (PA \otimes \wedge A)^\wedge & \end{array}$$

Proof. The proof has been accomplished in the previous discussion with the exception of the claim, that $\beta_{A,M}$ is a morphism of algebras. But this is a consequence of Lemma 1.5 and the accompanying remark.

We have seen how the symmetric divided algebras arise naturally as the duals of symmetric algebras. We now observe, at least for the case of finitely generated free modules, that symmetric algebras are dual to the symmetric divided algebras.

We first recall that for a fixed Z-module A there is a natural module morphism $\eta_A\colon A \to$ Hom (Hom (A, Z)), given by $\eta_A(a)(f) = f(a)$ for $a \in A$,

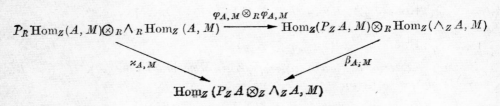

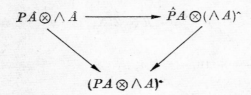

$f \in \operatorname{Hom}(A, Z)$. It is injective iff the relation $\eta_A(a)(f) = 0$ for all $f \in \operatorname{Hom}(A, F)$ entails $a = 0$; this is equivalent to saying that the morphisms $f: A \to Z$ separate the points of A.

Lemma 1.17. *Let A and B be Z-modules. The diagram*

$$\begin{array}{ccc} A \otimes A & \xrightarrow{\eta_A \otimes \eta_A} & \operatorname{Hom}(\operatorname{Hom}(A, Z), Z) \otimes \operatorname{Hom}(\operatorname{Hom}(A, Z), Z) \\ \downarrow{\scriptstyle \eta_{A \otimes A}} & & \downarrow{\scriptstyle \beta_{\operatorname{Hom}(A,Z)}} \\ \operatorname{Hom}(\operatorname{Hom}(A \otimes A, Z), Z) & \xrightarrow{\operatorname{Hom}(\beta_A, Z)} & \operatorname{Hom}(\operatorname{Hom}(A, Z) \otimes \operatorname{Hom}(A, Z), Z) \end{array}$$

with the natural morphism

$$\beta_B: \operatorname{Hom}(B, Z) \otimes \operatorname{Hom}(B, Z) \to \operatorname{Hom}(B \otimes B, Z)$$

commutes.

Proof. Let $a, b \in A$ and $f, g \in \operatorname{Hom}(A, Z)$,

$$[\beta_{\operatorname{Hom}(A,Z)}(\eta_A \otimes \eta_A)(a \otimes b)](f \otimes g) = \eta_A(a)(f)\, \eta_A(b)(g)$$

by the definition of β and this equals $f(a)\, g(b)$ by the definition of η. On the other hand

$$\begin{aligned}[\operatorname{Hom}(\beta_A, Z)\, \eta_{A \otimes A}(a \otimes b)](f \otimes g) &= \eta_{A \otimes A}(a \otimes b)(\beta_A(f \otimes g)) \\ &= \beta_A(f \otimes g)(a \otimes b)\end{aligned}$$

by the definition of η, but this again equals $f(a)\, g(b)$ by the definition of β.

Proposition 1.18. *Let G be a graded commutative algebra over Z such that $\operatorname{Hom}(G, Z)$ is a coalgebra via the commuting diagram*

$$\begin{array}{ccc} \operatorname{Hom}(G, Z) & \xrightarrow{\;c\;} & \operatorname{Hom}(G, Z) \otimes \operatorname{Hom}(G, Z) \\ & {\scriptstyle \operatorname{Hom}(m, Z)} \searrow \quad \swarrow {\scriptstyle \beta_G} & \\ & \operatorname{Hom}(G \otimes G, Z) & \end{array}$$

with the algebra multiplication $m: G \otimes G \to G$ and β_G as in the preceding Lemma 1.17. (Note that c always exists if β_G is an isomorphism, which is the case if all homogeneous components of G are finitely generated free Z modules.) Then $\eta_G: G \to \operatorname{Hom}(\operatorname{Hom}(G, Z), Z)$ is a morphism of algebras. If all homogeneous components of G are finitely generated free Z-modules, then η_G is an isomorphism of graded algebras.

Proof. We consider the following commutative diagram, in which we use $\hat{G} = \mathrm{Hom}\,(G, Z)$ as an abbreviation:

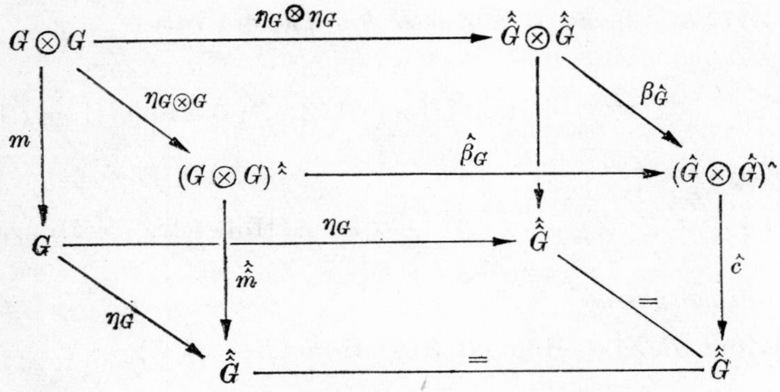

The back face must be shown to commute. The left face commutes by the naturality of η; the front face commutes since it arises from the triangle involving c and β_G upon applying $\mathrm{Hom}\,(-, Z)$; the top face commutes by the preceding Lemma 1.17, the right hand face commutes by the definition of the algebra multiplication $\hat{\hat{G}} \otimes \hat{\hat{G}} \to \hat{\hat{G}}$, the bottom commutes trivially. The assertion then follows.

The following is a crucial corollary:

Corollary 1.19. *Let Z be a commutative ring with identity and A a finitely generated free Z-module. Then PA and $\hat{P}A$ are duals of each other, i. e.*

$$\hat{P}A = \mathrm{Hom}\,(PA, Z) \quad \text{and} \quad PA \cong \mathrm{Hom}\,(\hat{P}A, Z).$$

In particular, $\mathsf{Z}[X]$ *and* $\hat{\mathsf{Z}}[X]$ *are duals of each other.*

Corollary 1.20. *Let E be a graded commutative algebra over Z satisfying the hypotheses of Proposition 1.18. Let R be a commutative Z-algebra. Then there are natural morphisms of algebras*

$$R \otimes E \xrightarrow{R \otimes \eta_E} R \otimes \mathrm{Hom}\,(\mathrm{Hom}\,(E, Z), Z) \longrightarrow \mathrm{Hom}\,(\mathrm{Hom}\,(E, Z), R)$$

where all $\otimes$ are over Z and where the last morphism is the usual natural morphism $R \otimes \mathrm{Hom}\,(B, Z) \to \mathrm{Hom}\,(B, R)$ given by $r \otimes f \to r \cdot f$ with $(r \cdot f)\,(b) = r \cdot f(b)$.

Lemma 1.21. *As functors from the category of Z-modules to the category of graded commutative R-algebras for a Z-algebra R, $R \otimes P_Z$ and $P_R(R \otimes -)$ are* *naturally isomorphic, and* $R \otimes \wedge_Z$ *and* $\wedge_R(R \otimes -)$ *are naturally isomorphic. Similarly,* $R \otimes P_Z \otimes \wedge_Z$ *and* $P_R \otimes_R \wedge_R$ *are naturally isomorphic.*

Proof. The functor P_Z from the category of Z-modules to the category of graded commutative algebras is left adjoint to the functor $(-)^2$ associating with a graded Z-algebra its homogeneous component of degree 2. The functor $R \otimes -$ from the category of commutative graded Z-algebras to the category of commutative graded R-algebras is left adjoint to the forgetful functor, similarly $R \otimes -$ as a functor from the category of Z-modules to the category of R-modules is a left adjoint of the forgetful functor. Thus both $R \otimes P_Z$ and $P_R(R \otimes -)$ are left adjoints to the functor which associates with a graded commutative R-algebra its homogeneous component of degree 2 as a Z-algebra. By the uniqueness of left adjoints, they have to be naturally isomorphic functors. The remaining assertions are similar.

We are now ready for the principal theorem which is used repeatedly in the latter parts of the discussion:

Theorem 1.22. *Let* Z *be a commutative ring with identity and* A *and* B *be* Z*-modules. Let* R *be a commutative* Z*-algebra. Let* $\hat{B} = \operatorname{Hom}(B, Z)$ *and let* $\hat{P}A$ *and* $\hat{\wedge} B$ *denote the graded commutative algebras* $\operatorname{Hom}(PA, Z)$ *and* $\operatorname{Hom}(\wedge B, Z)$, *respectively with* $PA = P_Z A$ *and* $\wedge B = \wedge_Z B$. *Then there are natural morphisms of graded (in fact bi-graded)* R*-modules*

$$
\begin{array}{ccc}
R \otimes PA \otimes \wedge B & \xrightarrow{R \otimes \eta_{PA} \otimes \eta_{\wedge B}} & R \otimes \operatorname{Hom}(PA, Z) \otimes \operatorname{Hom}(\wedge B, Z) \\
\cong \downarrow & & \downarrow \\
P_R A \otimes_R \wedge_R B & \longrightarrow & \operatorname{Hom}(\hat{P}A, R) \otimes_R \operatorname{Hom}(\hat{\wedge} B, R) \\
 & & \swarrow \beta \\
 & & \operatorname{Hom}(\hat{P}A \otimes \hat{\wedge} B, R) \\
 & & \downarrow \operatorname{Hom}(\hat{P}A \otimes \varphi'_B, R) \\
 & & \operatorname{Hom}(\hat{P}A \otimes \wedge \hat{B}, R)
\end{array}
$$

with $\varphi'_B \colon \wedge \hat{B} = \wedge \operatorname{Hom}(B, Z) \to \operatorname{Hom}(\wedge B, Z) = \hat{\wedge} B$ *as in Proposition* 1.6.

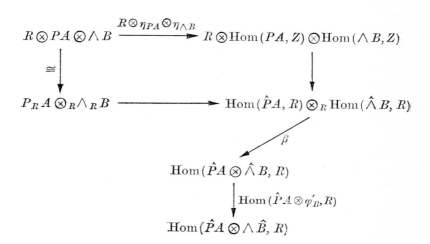

If A and B are finitely generated and free, then all morphisms in the diagram are isomorphisms of algebras.

$$\begin{array}{ccc}
\mathrm{Hom}_S(K\otimes L, A)\otimes_R \mathrm{Hom}(K\otimes L, A) & \xrightarrow{\xi_{K,L}\otimes\xi_{K,L}} & \mathrm{Hom}_S(L, \mathrm{Hom}(K, A))\otimes \mathrm{Hom}_S(L, \mathrm{Hom}(K, A)) \\
\downarrow & & \downarrow \\
\mathrm{Hom}_{S\otimes S}((K\otimes L)\otimes(K\otimes L), A\otimes_R A) & & \mathrm{Hom}_{S\otimes S}(L\otimes L, \mathrm{Hom}(K, A)\otimes_R \mathrm{Hom}(K, A)) \\
\downarrow & & \downarrow \\
\mathrm{Hom}_{S\otimes S}((K\otimes K)\otimes(L\otimes L), A\otimes_R A) & \xrightarrow{\xi_{K\otimes K, L\otimes L}} & \mathrm{Hom}_{S\otimes S}(L\otimes L, \mathrm{Hom}(K\otimes K, A\otimes_R A)) \\
\big\downarrow{\scriptstyle \mathrm{Hom}(\gamma_K\otimes\gamma_L, A\otimes_R A)} & & \big\downarrow{\scriptstyle \mathrm{Hom}(\gamma_L, \mathrm{Hom}(\gamma_K, A\otimes_R A))} \\
\mathrm{Hom}_S(K\otimes L, A\otimes_R A) & \xrightarrow{\xi_{K,L}} & \mathrm{Hom}_S(L, \mathrm{Hom}(K, A\otimes_R A))
\end{array}$$

The significance of the theorem is that in later applications it will become necessary to compute the cohomology of graded algebras of the form $PA \otimes \wedge B$ with finitely generated free modules A and B and suitable differentials. This cohomology will then be known if we understand how to treat differential graded algebras defined over algebras of the form $P_R A \otimes_R \wedge_R A$. A substantial portion of the discussion in the next sections is devoted to the structure theory of such differential graded algebras.

In later applications we will need the following result, which belongs to the general context discussed in the current section:

Proposition 1.23. *Let S be a ring with identity, K a coalgebra (i. e. an abelian group with a morphism $\gamma_K\colon K \to K \otimes K$), and L with $\gamma_L\colon L \to L \otimes L$ a coalgebra over S. Let A be an arbitrary left S-module, and an R-module over some commutative ring R with identity whose operations commute with those of S. Let $K \otimes L$ (with the tensor product over $\mathbb{Z}$) be the tensor product of coalgebras with the S-module operation given by $s \cdot (k \otimes 1) = k \otimes s \cdot 1$. Then there is a natural isomorphism*

$$\xi_{K,L}\colon \mathrm{Hom}_S(K \otimes L, A) \to \mathrm{Hom}_S(L, \mathrm{Hom}(K, A))$$

and a commutative diagram (see the adjoining column). *If A itself is an R-algebra, then both* $\mathrm{Hom}_S(K\otimes L, A)$, $\mathrm{Hom}_S(L, \mathrm{Hom}(K, A))$ *are R-algebras as described in Lemma 1.3, ff. above, and ξ is a natural isomorphism of R-algebras.*

Proof. As usual ξ is defined by $\xi(f)(y)(x) = f(x \otimes y)$. It is well-known that this morphism is an isomorphism (Cartan-Eilenberg [11], p. 27). In the diagram, the commuting of the top rectangle is straightforward from the definitions; the commuting of the lower rectangle is a consequence of the naturality of ξ. The remainder is then clear.

Section 2

The arithmetic of certain spectral algebras

This section is a crucial one for any explicit statement about the cohomology of a finite abelian group, and this is its principal motivation. The basic topic is the following: Let R be a commutative ring with identity. If $\varphi: A \to B$ is a morphism of R-modules, then the bi-graded algebra $PB \otimes \wedge A$ is given a unique derivative and differential d_φ with $PB \otimes 1$ in its kernel and with $d_\varphi(1 \otimes a) = \varphi(a) \otimes 1$; the bi-degree of d_φ is $(2, -1)$, so that $E_2(\varphi) = PB \otimes \wedge A$ together with d_φ is the E_2 term of a spectral bi-graded algebra $E_r(\varphi)$ with trivial differential on $E_3(\varphi)$ and the following terms. The main effort is to describe $E_3(\varphi)$ for a given φ. Satisfactory results are obtainable for special morphisms $\varphi: A \to A$ of free modules A, which are adequate for our purpose of computing the cohomology of a finite group. Also the ring R has to be restricted, but a good deal of the theory covers the rings Z and $\mathsf{Z}/n\,\mathsf{Z}$. With considerable computational effort, we determine the terms $E_3^{p,q}(\varphi)$ for $p < 3$ or $q < 2$. Much more can be said in the case that R is a principal ideal domain, a situation which still covers the case of integral coefficients in cohomology — the most important case anyhow. Quite explicit, although technically somewhat involved results will be obtained in that case.

Definition 2.1. A *spectral algebra* over a commutative ring R with identity is a sequence of differential bi-graded R-algebras (E_r, d_r), $r = r_0, r_0 + 1, \ldots$, such that each differential d_r has total degree $+1$ and bi-degree $(r, 1-r)$, E_{r+1} is the cohomology algebra of E_r, and $(E_r^{p,q}, d_r)$ is a first quadrant spectral sequence. (For details we refer to Borel [3], p. 122, Leray [31], MacLane [32], p. 318ff.) We shall always assume that $E_r^{0,0} = R$. Note that for any bi-graded differential algebra (E_r, d_r) with bi-degree $(r, 1-r)$, one always obtains a spectral algebra $E_r, E_{r+1}, \ldots$ by defining $E_{r+k} = HE_r$ and $d_{r+k} = 0$ for $k > 0$. We shall be concerned mostly with the case $r = 2$.

For a spectral algebra E we define the edge algebras E^{I} and E^{II} to be the sequences $E_r^{\mathrm{I}} = \bigoplus_p E_r^{p,0}$ and $E_r^{\mathrm{II}} = \bigoplus_q E_r^{0,q}$, $r = 2, 3, \ldots$ There are surjective morphisms e_r^{I}: $E_r^{\mathrm{I}} \to E_{r+1}^{\mathrm{I}}$ and injective morphisms e_r^{II}: $E_{r+1}^{\mathrm{II}} \to E_r^{\mathrm{II}}$, $r = 2, 3, \ldots$, the so-called edge morphisms (see MacLane [32], p. 321). If the algebra E_r is commutative as a graded R-algebra, then the multiplication in E_r defines a morphism of graded commutative algebras μ_r: $E_r^{\mathrm{I}} \otimes E_r^{\mathrm{II}} \to E_r$.

Lemma. *Let*

$$\begin{array}{ccc} A & \xrightarrow{\;d\;} & A \\ \big\downarrow{\scriptstyle\varepsilon} & & \big\downarrow{\scriptstyle\varepsilon} \\ B & \xrightarrow{\;\delta\;} & B \end{array}$$

be a diagram, where ε is a morphism of graded algebras and d and δ are derivations. Let e_A be the automorphism associated with d and e_B that associated with δ. If $e_B\,\varepsilon = \varepsilon\, e_A$ and if the equalizer of εd and $\delta\varepsilon$ (as group morphisms) contains a generating set for A, then $\varepsilon d = \delta\varepsilon$.

Proof. We need only show that the equalizer E is a ring, since it is clearly an additive group. Suppose that $a, a' \in E$. Then

$$\begin{aligned} \delta\varepsilon(a\,a') &= \delta(\varepsilon(a)\cdot\varepsilon(a')) = \delta\varepsilon(a)\cdot\varepsilon(a') + e_B(\varepsilon(a))\cdot\delta\varepsilon(a') \\ &= \varepsilon\, d(a)\cdot\varepsilon(a') + \varepsilon(e_A(a))\cdot\varepsilon\, d(a') = \varepsilon\,(d(a)\,a' + e_A(a)\,d(a')) \\ &= \varepsilon\, d(a\,a'). \end{aligned}$$

Thus, $a\,a' \in E$.

Lemma 2.2. (a) *Let A be an R-module and M an R-algebra with identity, and suppose that φ': $A \to M$ is a morphism of modules. The endomorphism d of $M \otimes \wedge A$ defined by $d\,(M \otimes 1) = \{0\}$ and*

$$\begin{aligned} & d\,(m \otimes (a_1 \wedge \cdots \wedge a_q)) \\ & = \sum \{(-1)^{i-1}\, m\varphi'(a_i) \otimes (a_1 \mathbin{\hat{}} \cdots \mathbin{\hat{}} a_q)_i,\ i = 1, \ldots, q\} \end{aligned}$$

where $(\cdots)_i$ denotes the omission of the i-th factor, is a differential and a derivation. If $M \otimes \wedge A$ is considered a left M-module under $m \cdot (m' \otimes x) = m\,m' \otimes x$, then d is an M-module endomorphism.

(b) *If φ: $A \to B$ is a morphism of R-modules we let $M = PB$ and define φ': $A \to M$ by mapping A into the homogeneous component B of degree 2 in PB under φ. Then* (a) *applies and the morphism $d_\varphi = d$ is a morphism of the bigraded algebra $PB \otimes \wedge A$ of degree* $(2, -1)$. *We define $d^{i,q} = 0$ for $i = -2, -1$.*

(c) *Let $A^{(k)}$ be R-modules, $k = 1, 2$ and $M^{(k)}$ two R-algebras with identity, suppose that $\varphi^{(k)}\colon A^{(k)} \to M^{(k)}$ are morphisms, and let φ' be the morphism*

$$A = A^{(1)} \oplus A^{(2)} \xrightarrow{\varphi^{(1)} \oplus \varphi^{(2)}} M^{(1)} \oplus M^{(2)} \longrightarrow M^{(1)} \otimes 1 + 1 \otimes M^{(2)}$$

$$\xrightarrow{\subset} M = M^{(1)} \otimes M^{(2)}.$$

Let $d^{(k)}$ be the derivation on $M^{(k)} \otimes \wedge A^{(k)}$ and d the one on $M \otimes \wedge A$ associated with $\varphi^{(k)}$ and φ' respectively. Then $(M^{(1)} \otimes \wedge A^{(1)}) \otimes (M^{(2)} \otimes \wedge A^{(2)})$ is a differential graded algebra in the usual fashion relative to the derivation

$$d^{(1)} \otimes 1 + e \otimes d^{(2)},$$

where $e(x) = (-1)^p$ for a homogeneous element x of degree p in $M^{(1)} \otimes \wedge A^{(1)}$. Moreover the morphism

$$\psi\colon (M^{(1)} \otimes \wedge A^{(1)}) \otimes (M^{(2)} \otimes \wedge A^{(2)}) \to M \otimes \wedge A$$

given by $M^{(1)} \otimes M^{(2)} = M$ and $\wedge A^{(1)} \otimes \wedge A^{(2)} \to \wedge (A^{(1)} \oplus A^{(2)})$ is an isomorphism of differential graded algebras and of M-modules.

Proof. (a) The definition immediately implies

$$d(m \otimes x) = (m \otimes 1)\, d\,(1 \otimes x)$$

and

$$d\,(1 \otimes a)\,(1 \otimes a')$$
$$= (d\,(1 \otimes a))\,(1 \otimes a') + (1 \otimes a)\, d\,(1 \otimes a')$$

for $a' = a_1' \wedge \cdots \wedge a_{q'}'$, $a, a_i' \in A$. By induction, for any $a = a_1 \wedge \cdots \wedge a_q$, we derive

$$d\,(1 \otimes aa') = (d\,(1 \otimes a))\,(1 \otimes a') + (-1)^q\,(1 \otimes a)\, d\,(1 \otimes a').$$

Thus, with the previous remarks, it follows that d is a derivation and an M-module endomorphism. Since the zero set of dd for a derivation d is a subring, $dd = 0$ provided dd vanishes on a generating subset of $M \otimes \wedge A$ as an M-algebra. Such a set is $1 \otimes A$, and $dd\,(1 \otimes A)$ clearly is $\{0\}$.

(b) follows directly from (a).

(c) It is straightforward to check that ψ is an isomorphism of graded algebras and of M-modules. We check that $\psi\,(d^{(1)} \otimes 1 + e \otimes d^{(2)})$ agrees with $d\psi$ on the elements $(1 \otimes a) \otimes (1 \otimes 1)$ and $(1 \otimes 1) \otimes (1 \otimes b)$, $a \in A^{(1)}$ and $b \in A^{(2)}$ which generate the domain of ψ as an M-algebra. But

$$\psi\,(d^{(1)} \otimes 1 + e \otimes d^{(2)})\,((1 \otimes a) \otimes (1 \otimes 1))$$
$$= \psi\,((\varphi^{(1)}(a) \otimes 1) \otimes (1 \otimes 1)) = (\varphi^{(1)}(a) \otimes 1) \otimes 1 \in M \otimes \wedge A;$$

on the other hand

$$\begin{aligned} & d\psi\,((1 \otimes a) \otimes (1 \otimes 1)) \\ & = d\,(1 \otimes (a, 0)) \\ & = \varphi'(a, 0) \otimes 1 \in M \otimes \wedge A\,. \end{aligned}$$

But by the definition of φ' we have $\varphi'(a, 0) = \varphi^{(1)}(a) \otimes 1$. The case of the second type of generator is completely analogous.

Lemma 2.3. *The derivations d, resp., d_φ in Lemma 2.2 are uniquely determined by the requirements that $d(M \otimes 1) = 0$, resp., $d(B \otimes 1) = 0$ and that $d\,(1 \otimes a) = \varphi(a) \otimes 1$, resp., $d_\varphi\,(1 \otimes a) = \varphi a \otimes 1$ for all $a \in A$.*

Proof. Any derivation is uniquely determined on a generating set, since the subset on which two derivations agree must always be a subring. But again, $1 \otimes A$ generates $M \otimes \wedge A$ as an M-algebra, and $B \otimes 1 \oplus 1 \otimes A$ as an R-algebra.

Definition 2.4. Let $\mathfrak{M}$ be the category whose objects are R-module morphisms $\varphi\colon A \to B$, and whose morphisms are pairs (α, β) of maps making the diagram

$$\begin{array}{ccc} A & \xrightarrow{\ \varphi\ } & B \\ \big\downarrow{\scriptstyle\alpha} & & \big\downarrow{\scriptstyle\beta} \\ A' & \xrightarrow{\ \varphi'\ } & B' \end{array}$$

commute. To each morphism φ we associate the spectral algebra $(E_2(\varphi), d_\varphi)$, $(E_r, 0)$, $r > 2$, as follows: The bi-graded algebra $E_2(\varphi)$ is set to be $P_R B \otimes_R \wedge_R A$ and the differential d_φ is defined as

$$d_\varphi^{2p,q}(b^{2p} \otimes a_1 \wedge \cdots \wedge a_q) = \sum (-1)^{i-1} b^{2p}\, \varphi\, a_i \otimes (a_1 \wedge \cdots \wedge a_q)_i\,,$$

where $(\ldots)_i$ denotes the product inside the parentheses, where the factor with index i is omitted. We let $E_r(\varphi) = H(E_2(\varphi))$, $r > 2$.

Clearly the definition is functorial, i. e. for a morphism $(\alpha, \beta)\colon \varphi \to \varphi'$ in $\mathfrak{M}$, i. e. a pair of R-module morphisms giving a commutative diagram

$$\begin{array}{ccc} A & \xrightarrow{\ \varphi\ } & B \\ \big\downarrow{\scriptstyle\alpha} & & \big\downarrow{\scriptstyle\beta} \\ A' & \xrightarrow{\ \varphi'\ } & B' \end{array}$$

we obtain in the obvious fashion a morphism $E_r(\alpha, \beta)\colon E_r(\varphi) \to E_r(\varphi')$ of spectral algebras.

Lemma 2.5. *Let (A, d) be a differential R-algebra. Let $(A \otimes A, D)$ be the tensor product with $D = d \otimes A + e \otimes d$, where e is an involution of the algebra A with $ed + de = 0$. Let $\mu\colon A \otimes A \to A$ be the intrinsic multiplication of A (i. e. $\mu(a \otimes b) = ab$). Then the algebra morphism* $\ker d \otimes \ker d \to A \otimes A \to A$ *maps into* $\ker d$. *If $\psi\colon \ker d \to H(A)$ and $\Psi\colon \ker D \to H(A \otimes A)$ are the cohomology maps then there is a commuting diagram*

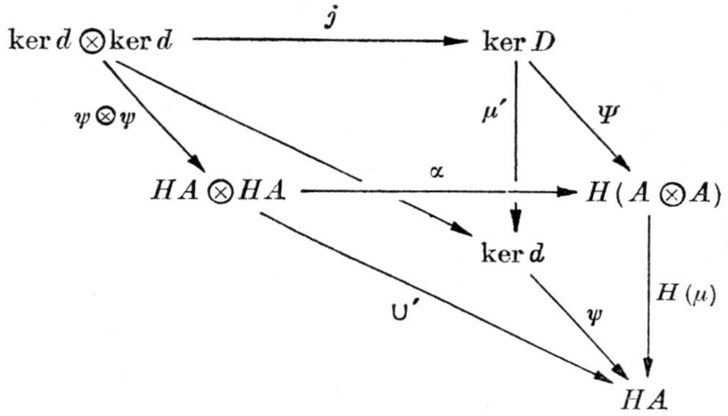

in which $\cup'$ is the intrinsic cup product, j the natural map, and μ' the restriction and co-restriction of μ. Moreover, ψ is a morphism of algebras. The assertions maintain for differential bi-graded algebras (A, d) with $\operatorname{bideg} d = (r, 1 - r)$. *In particular, the quotient map* $\ker d_\varphi \to E_3(\varphi)$ *is a morphism of bigraded algebras.*

Proof. If $da = db = 0$, then

$$\mu D(a \otimes b) = d\mu(a \otimes b) = d(ab) = (da)\, b + e(a)\, db = 0.$$

Further, $\alpha(\psi \otimes \psi)(a \otimes b) = \Psi j(a \otimes b)$; and if $Dc = 0$, then $H(\mu)\, \Psi c = \psi\mu' c$, because $\mu\colon A \otimes A \to A$ is a co-chain transformation due to the fact that $\mu D = d\mu$. Finally, we have

$$\begin{aligned}(\psi a) \cup (\psi b) &= \cup'(\psi a \otimes \psi b) = \cup'(\psi \otimes \psi)(a \otimes b)\\ &= H(\mu)\, \Psi j(a \otimes b) = \psi\mu' j(a \otimes b) = \psi(ab).\end{aligned}$$

That the results maintain in the presence of a bi-degree is straightforward.

Lemma 2.6. *In the notation of Definition 2.4, there is a morphism of bi-graded algebras ψ:* $\ker d_\varphi \to E_3$ *whose restriction to E_2^{I} is the edge morphism e_2^{I} of Definition 2.1.*

This follows immediately from Definition 2.4 and Lemma 2.5.

Definition 2.7. Under the circumstances of Definition 2.4, we denote with $B^{2p}(\varphi) \subset P^p B$ the submodule defined by $B^{2p} \otimes 1 = \operatorname{im} d^{2p-2,1}$. Note that $B^{2p}(\varphi)$ is generated by all $b\,\varphi(a)$, $b \in P^{p-1} B$, $a \in A$, and that

$$0 \longrightarrow B^{2p}(\varphi) \longrightarrow E_2^{\mathrm{I}2p} \xrightarrow{\;e_2^{\mathrm{I}2p}\;} E_3^{\mathrm{I}2p} \longrightarrow 0$$

is exact.

Definition 2.8. Let R be a ring with identity. An *integer* (or an *integral element*) of R is an element of the form $z \cdot 1$, where $z \in \mathsf{Z}$. We shall use z to denote the integral element of R as well as the integer in Z.

We shall say that the ring R is a *weakly principal ideal ring* if for each integer z, the ideal

$$\{x \in R \mid z\,x = 0\}$$

is a principal ideal. Note that in particular any homomorphic image of a principal ideal domain is a weakly principal ideal ring. More generally, we have the following:

Lemma 2.9. *Let R' be a commutative ring with identity whose additive group is torsion free. Let $a \neq 0$, 1 be a natural number and set $R = R'/a\,R'$. We have the following conclusions:*

(a) *If m is any natural number, then*

$$S(m\,R, R) = \{x \in R \colon m\,x = 0\} = (a/(a, m))\,R.$$

(b) *If b is any natural number, then $a\,R' + b\,R' = (a, b)\,R'$ and*

$$R/b\,R \cong R'/(a, b)\,R',$$

where (a, m) denotes the greatest common divisor of a and m.

Proof. There are integers p and q such that $(a, m) = p\,a + q\,m$. If $x + R' \in S(m\,R, R)$, then $m\,x = a\,y$ for some $y \in R'$. Then

$$a\,q\,y = q\,m\,x = (a, m)\,x - p\,a\,x,$$

so that $(a, m)\,x = a\,z$ for some $z \in R'$. Since $a = (a, m)\,(a/(a, m))$ and since the additive group of R' is torsion free, we obtain $x = (a/(a, m))\,z$. Thus,

$$S(m\,R, R) \subset (a/(a, m))\,R.$$

The other inclusion is trivial. Finally,

$$R/b\,R \cong (R'/a\,R')/[(b\,R' + a\,R')/a\,R] \cong R'/(b\,R' + a\,R').$$

But for suitable integers u, v, we have $u\,a + v\,b = (a, b)$, whence $a\,R' \subset a\,R' + b\,R'$. Conversely, $(a, b)\,R' \supset b\,R'$, so that $a\,R' + b\,R' = (a, b)\,R'$, which establishes the result.

Definition 2.10. We fix a well-ordered set I. For each natural number $q = 1, 2, \ldots$, we denote with $S(q)$ the set of all injective increasing functions $\{1, \ldots, q\} \to I$ and with $\Sigma(p)$ the set of all non-decreasing functions

$$\{1, \ldots, p\} \to I.$$

If $s \in S(q)$ and $i \in \{1, \ldots, q\}$, $q > 1$, then $s_i \in S(q-1)$ is defined by $s_i(x) = s(x)$ for $x < i$ and by $s_i(x) = s(x+1)$ for $q > x \geqq i$, i. e. s_i is the unique injective increasing function with domain $\{1, \ldots, q-1\}$ and image $\operatorname{im} s \setminus \{s(i)\}$. If $j \in \operatorname{im} s$ then s^j denotes the unique element of $S(q-1)$ with image $\operatorname{im} s \setminus \{j\}$. If $\sigma \in \Sigma(q-1)$, and $j \in I$, then σ^j is the unique element of $\Sigma(q)$ with image $\operatorname{im} \sigma \cup \{j\}$, and with

$$\operatorname{card}\{x : \sigma(x) = j\} + 1 = \operatorname{card}\{x : \sigma^j(x) = j\}.$$

If $\tau \in \Sigma(q)$, $q > 1$ and if $k \in \operatorname{im} \tau$, then $\tau_{[k]}$ denotes the unique element of $\Sigma(q-1)$ with $\operatorname{im} \tau \setminus \{k\} \subset \operatorname{im}_{\tau_{[k]}} \subset \operatorname{im} \tau$ and with

$$\operatorname{card}\{x : \tau(x) = k\} - 1 = \operatorname{card}\{x : \tau_{[k]}(x) = j\}.$$

For $p = 0$, we make the convention that $\Sigma(p) = \{\emptyset\} = S(p)$ and that $\emptyset^j$ is the constant map j, and that $j_{[j]} = \emptyset$. If A is a free R-module with a basis $\{a_i : i \in I\}$ for some well-ordered set I, then the elements

$$a_s = a_{s(1)} \wedge \cdots \wedge a_{s(q)}, \quad s \in S_{\mathrm{I}}(q)$$

constitute a basis of $\bigwedge^q A$ and the elements $a_\sigma = a_{\sigma(1)} \cdots a_{\sigma(p)}$, $\sigma \in \Sigma(p)$ constitute a basis of $P^{2p} A$. Occasionally we will write $a(s)$ in place of a_s etc.

It will be convenient to extend our prior conventions concerning 0:

$$\Sigma(0) = S(0) = \{\emptyset\}, \quad a_\emptyset = 1 = a_\emptyset \otimes a_\emptyset, \quad \Sigma(n) = S(n) = \emptyset \quad \text{if} \quad n < 0.$$

Definition 2.11. A morphism $\varphi: A \to B$ of R-modules will be called *elementary* if $A = B = R^{(I)}$ is free, where I is some well-ordered set, such that the family of elements $a_i \in A$ whose i-th component is 1 and whose other components are zero is a basis for A and that $\varphi(a_i) = z_i a_i$, where $z_i \in R$ and z_i divides z_{i+1}. (In later applications z_i will be a non-negative integer which will be identified with $z_i \cdot 1$ in R. This will not cause confusion since $n\,a_i = (n \cdot 1)\,a_i$ for any integer.)

A morphism φ is elementary if and only if its domain is free and

$$\varphi = \oplus \{z_j\,\varepsilon_j : j \in I\},$$

where I is a well-ordered set, ε_j is the identity on its domain, and $z_i | z_j$ if $i < j$.

The elementary morphisms define a full subcategory of the category $\mathfrak{M}$ introduced in Definition 2.4.

If R is a weakly principal ideal ring, and $S(q)$ and $\Sigma(q)$ are as in Definition 2.10, then for $s \in S(q)$, resp. $\sigma \in \Sigma(q)$, we pick a generator of

$$\{x \in R \mid z_{s(1)} x = 0\},$$

resp. of

$$\{x \in R \mid z_{\sigma(1)} x = 0\},$$

and denote it by w_s, resp. w_σ. For $i \in I$, w_i will then denote the element w_σ or w_s with $s \in S(1) = \Sigma(1)$ and $s(1) = i = \sigma(1)$.

The following proposition describes the edge terms of the spectral algebra $E_3(\varphi)$ for elementary φ. If M is any R-module, we call a family $\{m_i \colon i \in I\}$ of elements of M a basis of M, if M is the direct sum of the modules generated by the m_i.

Proposition 2.12. *Let A be a free module over a commutative ring R with a basis $\{a_i \colon i \in I\}$ with some well-ordered set I. Let $\varphi \colon A \to A$ be an elementary morphism. Then the following statements hold:*

(a) *$B^{2p}(\varphi)$ is a direct sum of the cyclic modules generated by*

$$z_{\sigma(1)} a_\sigma \in P^p A, \quad \sigma \in \Sigma(p).$$

(See Definition 2.7.)

(b) *$E_3^{2p,0}$ is a direct sum of the modules generated by $\bar{a}_\sigma = e_2^{\mathrm{I}}(a_\sigma)$, $\sigma \in \Sigma(p)$, and $R \cdot \bar{a}_\sigma$ is isomorphic to $R/z_{\sigma(1)} R$. The algebra $E_3^{\mathrm{I}}(\varphi)$ is isomorphic to P* coker φ.

(c) im $d^{0,q}$ *has a basis consisting of the elements $z_{s(1)} a_s^*$, $s \in S(q)$, where*

$$a_s^* = \Sigma(-1)^{i-1} (z_{s(i)}/z_{s(1)})\, a(s(i)) \otimes a(s_i)$$

summed over $i = 1, \ldots, q$. The a_s^* are independent in $E_2^{2,q-1}$, and $E_3^{0,q} \subset E_2^{0,q}$ has a basis of elements $1 \otimes w_s a_s$, $s \in S(q)$, provided that R is a weakly principal ideal ring. Define $a'_r = w_r a_r$, $r \in S(q)$. Then if J denotes the ideal of $\wedge A$ generated by all a'_i, $i \in I$, it follows that $\oplus E_3^{0,q} = E_3^{0,0} + 1 \otimes J$. Also, $J =$ ker $\wedge \varphi$. *(See Definition 2.11.)*

(d) *If R is a principal ideal ring, then there is an isomorphism of exact sequences*

$$\begin{array}{ccccccccc}
0 & \longrightarrow & E_3^{0,q}(\varphi) & \longrightarrow & \wedge^q A & \longrightarrow & \operatorname{im} d^{0,q} & \longrightarrow & 0 \\
 & & \downarrow & & \downarrow & & \downarrow & & \\
0 & \longrightarrow & \oplus\, w_s R \cdot a_s & \longrightarrow & \oplus\, R \cdot a_s & \longrightarrow & \oplus\, z_{s(1)} R \cdot a_s^* & \longrightarrow & 0
\end{array}$$

where the sums are extended over all $s \in S(q)$. *Finally*, $E_3^{\mathrm{II}}(\varphi) \cong \wedge \ker \varphi$ *as modules, but not as algebras.*

(e) *If* R *is a principal ideal ring then there is a commutative diagram of natural morphisms of bi-graded algebras*

$$\begin{array}{ccc} & P \operatorname{coker} \varphi \,\dot{\otimes} \wedge \ker \varphi & \\ \swarrow & & \searrow^{m} \\ P \operatorname{coker} \varphi \otimes (R + \ker \wedge \varphi) & \xrightarrow[n]{\qquad} & E_3(\varphi) \end{array}$$

such that the restriction of n *to the edge-terms is bijective.*

(f) *There are natural co-retractions*

$$E_3^{\mathrm{II}}(\varphi) \to E_3(\varphi)$$
$$E_3^{\mathrm{I}}(\varphi) \to E_3(\varphi)$$

whose inverses are given by the projections onto the edge terms.

Remark. Later (Corollary 2.28) we will show that in some cases the a_s^* in (c) are in fact in $\ker d^{2,q-1}$. Note that in general E_3^{II} is neither isomorphic to $\wedge \ker \varphi$ nor to $\ker \wedge \varphi$. More will be said about an intrinsic characterisation of E_3^{II} at a later point. The element $z_{s(i)}/z_{s(1)}$ is not uniquely determined in every ring, but anyhow in some rings of immediate interest such as Z, or in any unique factorisation ring, it is.

Proof. (a) By definition, $B^{2p}(\varphi)$ is generated by the elements $a_\sigma \varphi(a_i)$, $\sigma \in \Sigma(p)$, $i = 1, \ldots, p$. By hypothesis $z_i | z_{i+1}$. Therefore $B^{2p}(\varphi)$ is generated by the elements $z_{\sigma(1)} a_\sigma$, $\sigma \in \Sigma(p)$. This implies (a).

(b) Since $E_3^{2p,0} = E_2^{2p,0}/(B^{2p} \otimes 1)$, the first assertion is a consequence of (a). Now the algebra structure of E_3^{I}: Clearly $E_3^{2,0} \cong \operatorname{coker} \varphi$, since there is an exact sequence

$$0 \longrightarrow \bigoplus w_i R \cdot a_i \longrightarrow A \xrightarrow{\varphi} A \longrightarrow \bigoplus R \cdot \bar{a}_i = E_3^{2,0} \longrightarrow 0. \qquad (*)$$

The symmetric power $P^p E_3^{2,0}$ is the direct sum of the modules generated by the $\hat{a}_\sigma$, $\sigma \in \Sigma(p)$, the annihilator ideal of $\hat{a}_\sigma = \bar{a}_{\sigma(1)} \cdots \bar{a}_{\sigma(p)}$ being generated by $z_{\sigma(1)}$. This establishes a natural module isomorphism $E_3^{\mathrm{I}} \to PE_3^{2,0}$ sending $\bar{a}_\sigma$ onto $\hat{a}_\sigma$. That this module morphism preserves multiplication is most easily seen from the commuting diagram

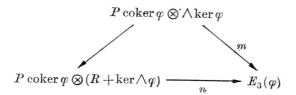

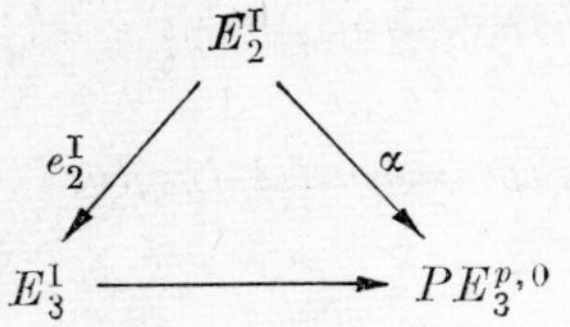

of epimorphisms of graded R-modules, in which the downward maps preserve multiplication, α being induced by the module morphism $e_2^{\mathrm{I}1}: E_2^{0,2} \to E_3^{0,2}$.

(c) Clearly im $d^{0,q}$ is generated by the elements $z_{s(1)} a_s^* = d(1 \otimes a_s)$, $s \in S(q)$. The independence of the a_s^* follows from the independence of the elements $a_{s(1)} \otimes a(s_1)$. From (d) below it will follow that im $d^{0,q}$ is a direct sum of the modules generated by $z_{s(1)} a_s^*$. An element $\Sigma\{1 \otimes n_s a_s \colon s \in S(q)\}$ is in

$$\ker d^{0,q} \quad \text{iff} \quad \Sigma\{(-1)^{i-1} n_s z_{s(i)} a_{s(i)} \otimes a(s_i)\} = 0 .$$

Because of the independence of the elements $a_{s(i)} \otimes a(s_i)$, $s \in S(q)$, $i = 1, \ldots, q$ this is equivalent to $n_s z_{s(i)} = 0$ for $i = 1, \ldots, q$, which, because $z_i | z_{i+1}$, again is equivalent to $n_s z_{s(1)} = 0$. By Definition 2.11 and the hypothesis about R, this means that $n_s = w_s t$ for some $t \in R$. Thus ker $d^{0,q}$ is generated by the elements $1 \otimes w_s a_s$, $s \in S(q)$. Moreover, ker $d^{0,q}$ is in fact the direct sum of the modules generated by these elements. We observe, that this module is exactly $\ker \wedge^q \varphi$.

(d) The exactness of both sequences is clear from the definitions. The commuting of the diagram is then straightforward in view of what has been said before.

We have to show that $w_s R \cdot a_s = \ker \wedge \varphi$ is isomorphic to $\wedge \ker \varphi$ as a module. But the module $\ker \varphi$ is spanned by the elements $w_i a_i$ for which $R \cdot w_i a_i \cong w_i R \cong R/z_i R$. Thus the module $\wedge^q \ker \varphi$ is the direct sum of the modules generated by the elements $(w_{s(1)} a_{s(1)}) \wedge \cdots \wedge (w_{s(q)} a_{s(q)})$, $s \in S(q)$ and the module generated by such an element is ismorphic to $w_{s(q)} R$, since $w_i R \subset w_j R$ for $i < j$. Thus there is an isomorphism of modules $\wedge^q \ker \varphi \to \bigoplus w_s R \cdot a_s$ which sends $(w_{s(1)} a_{s(1)}) \wedge \cdots \wedge (w_{s(q)} a_{s(q)})$ onto $w_s a_s$.

Note that this is not the morphism of graded algebras $\wedge \ker \varphi \to \ker \wedge \varphi$ which extends the inclusion $\ker \varphi \to \ker \wedge \varphi$, and that it is not in general a morphism of algebras.

(e) The natural morphism $\wedge \ker \varphi \to R + \ker \wedge \varphi$ defines the unlabelled morphism, and n is defined by the cup product in view of (a) and (c). Finally, m is defined so that the diagram commutes.

(f) Trivial.

In the next lemma, we obtain a system of linear equations describing the elements of $\ker d_\varphi$.

Lemma 2.13. *Let $\varphi: A \to A$ be an elementary morphism. Then* $\ker d^{2p,q}$ *is the module of all elements $x = \Sigma\, n(\sigma, s)\, a_\sigma \otimes a_s$, where $\sigma \in \Sigma(p)$, $s \in S(q)$, $n(\sigma, s) \in R$ and $n(\sigma, s) = 0$ with only finitely many exceptions, such that for each $\tau \in \Sigma(p+1)$, $t \in S(q-1)$,*

(Z) $\Sigma(-1)^{e(j,t)} n(\tau_{[j]}, t^j) z_j = 0$
where the sum is extended over all $j \in \operatorname{im} \tau \setminus (\operatorname{im} \tau \cap \operatorname{im} t)$ and where $e(j, t) = i - 1$ with $j = t^j(i)$.

Proof. We have $dx = 0$ iff $0 = \Sigma(-1)^{i-1} n(\sigma, s)\, z_{s(i)}\, a(\sigma^{s(i)}) \otimes a(s_i)$, where the sum is extended over all $\sigma \in \Sigma(p)$ and all $s \in S(q)$. Now fix an arbitrary $\tau \in \Sigma(q+1)$ and an arbitrary $t \in S(q-1)$. Then $dx = 0$ iff the coefficient of $a_\tau \otimes a_t$ in this sum vanishes. The term spanned by this element may be described as

$$\Sigma\{(-1)^{i-1} n(\sigma, s)\, z_{s(i)} a(\sigma^{s(i)}) \otimes a(s_i) : \sigma^{s(i)} = \tau \text{ and } s_i = t\}.$$

We set $s(i) = j$ and observe that then $j \in \operatorname{im} \tau$ and $j \notin \operatorname{im} t$. A simple transformation yields (Z).

The following lemma gives an explicit description of the elements of $\operatorname{im} d_\varphi$.

Lemma 2.14. *Under the hypotheses of Lemma* 2.13, $\operatorname{im} d^{2p-2,q+1}$ *is the module of all elements $x = \Sigma\, n(\sigma, s)\, a_\sigma \otimes a_s$, $\sigma \in \Sigma(p)$, $s \in S(q)$ satisfying the condition*

(B) $n(\sigma, s) = \Sigma\{(-1)^{e(j,s)} m(\sigma_{[j]}, s^j) z_j : j \in \operatorname{im} \sigma \setminus \operatorname{im} s\}$ *for suitable ring elements $m(\tau, t)$, $\tau \in \Sigma(p-1)$, $t \in S(q+1)$.*

Proof. An element x is a coboundary if it is of the form

$$\Sigma(-1)^{i-1} m(\tau, t)\, z_{t(i)} a(\tau^{t(i)}) \otimes a(t_i).$$

In order to compute the coefficient of $a_\sigma \otimes a_s$, we have to compute a sum extended over all $\tau \in \Sigma(p-1)$, $t \in S(q+1)$ with $t_i = s$ and $\tau^{t(i)} = \sigma$. We set $t(i) = j$ and observe that we have to sum over all $j \in \operatorname{im} \sigma \setminus \operatorname{im} s$. Writing out the coefficient of $a_\sigma \otimes a_s$, we get (B).

The following lemma is useful in the description of some features of the structure of $E_3(\varphi)$ as an $E_3^{\mathrm{I}}(\varphi)$-module as is discussed in Proposition 2.16 below.

Lemma 2.15. *Suppose that $x \in \operatorname{im} d^{2p-2,q+1} \cap P^p A \otimes a_s$, $s \in S(q)$. Then*

$$x = \Sigma\{(-1)^{e(j,s)} (z_j a(\tau^j)) \otimes (n(\tau, j)\, a_s) \mid \tau \in \Sigma(p-1), j < s(1)\}$$

and $z_{s(1)} n(\tau, j) = 0$ for all $\tau \in \Sigma(p-1)$, $j < s(1)$.

Proof. If $x \in \operatorname{im} d^{2p-2,q+1}$, then $x = dy$ with

$$y = \Sigma\{m(\tau, t)\, a_\tau \otimes a_t : \tau \in \Sigma(p-1),\ t \in S(q+1)\}$$

with suitable ring elements $m(\tau, t)$ which vanish for all but a finite number of arguments τ, t. Hence

$$x = \Sigma\{(-1)^{i-1} m(\tau, t)\, z_{t(i)} a(\tau^{t(i)}) \otimes a(t_i) \,|\, \tau \in \Sigma(p-1),\ t \in S(q+1),\ 1 \leqq i \leqq q+1\}.$$

Now suppose that x is also in $P^p A \otimes a_s$ for a fixed $s \in S(q)$. If $t \neq s^j$ for all $j \notin \operatorname{im} s$, then $t_i \neq s$ for all i. Thus necessarily $m(\tau, t) = 0$ for $t \neq s^j$, $j \notin \operatorname{im} s$. We abbreviate $m(\tau, s^j)$ with $n(\tau, j)$ and obtain the representation

$$x = \Sigma\{(-1)^{i-1} n(\tau, j)\, z(s^j(i))\, a(\tau^{s^j(i)}) \otimes a((s^j)_i) \,|\, \tau \in \Sigma(p-1),\ j \notin \operatorname{im} s,\ i \in \{1, \ldots, q+1\}\}.$$

This sum we split up into two sums in which the first summation is over all $\tau \in \Sigma(p-1)$ and all pairs (j, i) with $j \notin \operatorname{im} s$, $i \in \{1, \ldots, q+1\}$ with $s^j(i) = j$ and the second sum over all $\tau \in \Sigma(p-1)$ and all pairs (j, i) with $j \notin \operatorname{im} s$, $i \in \{1, \ldots, q+1\}$ with $s^j(i) \neq j$. If we write $e(j, s) = i - 1$ for $s^j(i) = j$, then the first sum may be written as

$$\Sigma\{(-1)^{e(j,s)} n(\tau, j)\, z_j a(\tau^j) \otimes a_s \,|\, j \notin \operatorname{im} s\}; \tag{1}$$

since this sum is the entire projection of x into $P^p A \otimes a_s$, the second sum must vanish. Suppose that $a((s'^{j'})_{i'}) = a((s^j)_i)$. Then it follows that $s'^{j'}(i') = j'$ and $s^j(i) = j$. Hence the factors $a((s^j)_i)$ occurring in the second sum are all different, hence independent. Thus the factor of $a((s^j)_i)$ in the second sum must vanish. This yields $\Sigma\{n(\tau, j)\, z(s^j(i))\, a(\tau^{s^j(i)}) \,|\, \tau \in \Sigma(p-1)\} = 0$ for all $j \notin \operatorname{im} s$, $i \in \{1, \ldots, q+1\}$ with $s^j(i) \neq j$. Under these circumstances, $s^j(i)$ ranges exactly through the set $\operatorname{im} s$ for each $j \notin \operatorname{im} s$. But $a(\tau^k) = a(\tau'^k)$ implies $\tau = \tau'$; so the elements $a(\tau^k)$ appearing in the last sum are all different, hence independent. Consequently their coefficients have to vanish. Hence $z_k n(\tau, j) = 0$ for all $\tau \in \Sigma(p-1)$, $j \notin \operatorname{im} s$ and all $k \in \operatorname{im} s$. Due to the fact that $z_k | z_{k'}$ for $k < k'$, this condition is equivalent to the condition

$$z_{s(1)} n(\tau, j) = 0 \quad \text{for all} \quad \tau \in \Sigma(p-1),\ \ j \notin \operatorname{im} s. \tag{2}$$

Returning to the first sum (1) we notice that in view of (2) we have $n(\tau, j)\, z_j = 0$ whenever $s(1) < j$. Hence it is only necessary to extend the sum over all j with $j < s(1)$. This proves the assertion.

Proposition 2.16. *Let R be a weakly principal ideal ring, A a free R-module, and let $\varphi: A \to A$ be an elementary morphism. Let $s \in S(q)$. Then the morphism*

$u \otimes a_s' \to u\, a_s'$ mapping $E_3^{2p,0} \otimes a_s' \subseteq E_3^{2p,0} \otimes E_3^{0,q}$ into $E_3^{2p,q}$ under the multiplication of E_3 is injective, where $a_s' = w_s a$. (See Proposition 2.12.)

Proof. We consider the commutative diagram

$$\begin{array}{ccc}
E_2^{2p,0} & \xrightarrow{\quad \psi^{2p,0} \quad} & E_3^{2p,0} \\
\downarrow & & \downarrow \\
E_2^{2p,0} \otimes E_3^{0,q} & \xrightarrow{\psi^{2p,0} \otimes \psi^{0,q}} & E_3^{2p,0} \otimes E_3^{0,q} \\
\downarrow & & \downarrow \\
\ker d^{2p,q} & \xrightarrow{\quad \psi^{2p,q} \quad} & E_3^{2p,q}
\end{array}$$

in which the left vertical maps all may be considered as inclusion maps between submodules of $E_2^{2p,0} \otimes E_2^{0,q}$ since $E_2^{2p,0}$ is free. If, for some $u \otimes a_s' \in E_3^{2p,0} \otimes E_3^{0,q}$, we have $u\, a_s' = 0$, then we write $u = \psi^{2p,0}(z \otimes 1)$ with $z \in P^p A$. Let

$$x = z \otimes w_{s(1)} a_s \in E_2^{2p,0} \otimes E_3^{0,q} \subset E_2^{2p,q}.$$

Then

$$\psi^{2p,q}(x) = \mu\,(\psi^{2p,0} \otimes \psi^{0,q})\,(z \otimes w_{s(1)} a_s) = (u \otimes a_s') = u\, a_s' = 0$$

with the intrinsic cup product by Lemma 2.5. Thus $x \in \operatorname{im} d^{2p-2,q+1}$. But also $x \in P^p A \otimes a_s$, so that Lemma 2.14 above is applicable. We obtain

$$\begin{aligned}
u \otimes a_s' &= (\psi^{2p,0} \otimes \psi^{0,q})\,(x) \\
&= \Sigma\{(-1)^{e(j,s)}\, \psi^{2p,0}\,(z_j\, a\,(\tau^j)) \\
&\qquad \otimes \psi^{0,q}\,(n\,(\tau, j)\, a_s) \mid \tau \in \Sigma(p-1), j < s\,(1)\}.
\end{aligned}$$

Note that here we have used that $n\,(\tau, j)\, a_s \in E_3^{0,q}$ because of $z_{s(1)} n\,(\tau, j) = 0$ by Lemma 2.15. But clearly $z_j a(\tau^j) \in B^{2p}$ (see Proposition 2.12). Hence the ψ-image of these elements vanishes. Thus we obtain $u \otimes a_s' = 0$ as asserted.

Next we compute the E_3-terms in the row and column next to the fringe of E_3, where in the following, R is a weakly principal ideal ring.

The module $E_2^{2p,q}$ is generated by the elements $a_\sigma \otimes a_s$, $\sigma \in \Sigma(p)$ and $s \in S(q)$. For each $\tau \in \Sigma(p+1)$ and $t \in S(q-1)$ we define the submodule $M_{\tau,t}$ of $E_2^{2p,q}$ to be the one generated by the elements $a(\tau_{[j]}) \otimes a(t^j)$, $j \in \operatorname{im} \tau \setminus (\operatorname{im} \tau \cap \operatorname{im} t)$. Then $E_2^{2p,q}$ is the sum over all $M_{\tau,t}$, $\tau \in \Sigma(p+1)$, $t \in S(q-1)$. For each

$\varrho \in \Sigma(p-1)$, $r \in S(q+1)$ let $M^{\varrho, r}$ be the submodule generated by all $a(\varrho^j) \otimes a(r_{[j]})$, $j \in \operatorname{im} r$, where $r_{[j]}$ is the unique function in $S(q)$ whose range is $\operatorname{im} r \backslash \{j\}$, i. e., $r_{[j]} = r_{r^{-1}(j)}$. Again $E_2^{2p,q}$ is the sum of all submodules $M^{\varrho r}$, $\varrho \in \Sigma(p-1), r \in S(q+1)$.

For $S(0) = \Sigma(0) = \{\emptyset\}$, we let $\emptyset^j$ denote the function $\{1\} \to I$ given by $\emptyset^j(1) = j$. We will also denote $M^{\emptyset r}$ by $M(r)$.

Case 1. $p = 1$.

$E_2^{2,q}$ is the direct sum of a subfamily of the family of the modules $M_{\tau t}$, where $\tau \in \Sigma(2)$ and $t \in S(q-1)$. The members of the subfamily are of one of two types:

Type 1: $M_{\tau t}$ is generated by all $a_i \otimes a(t_j)$, $t \in S(q-1)$, $i, j \notin \operatorname{im} t$ and $i \neq j$. For $i < j$, $i, j \notin \operatorname{im} t$, $t \in \Sigma(2)$ with $i, j \in \operatorname{im} \tau$ is uniquely determined and we denote this $M_{\tau t} = M(t, i, j)$.

Type 2: $M_{\tau t}$ is generated by all $a_j \otimes a(t^j)$, $t \in S(q-1)$, $j \notin \operatorname{im} t$. The module generated by $a_j \otimes a_s$, $s \in S(q)$, $j \in \operatorname{im} s$, is denoted by $M(s, j)$. Also,

$$\bigoplus \{M(t, i, j) : t \in S(q-1)\}$$

can be written as the direct sum of the modules $M(r) = M^{\emptyset r}$, $r \in S(q+1)$. Each of the latter is generated by the elements $a(j) \otimes a(r_{[j]})$, $j \in \operatorname{im} r$. Note that each $M(t, i, j)$ is contained in $M((t^i)^j)$. Thus

$$E_2^{2,q} = \bigoplus \{M(r) : r \in S(q+1)\} \oplus \bigoplus \{M(s, j) : s \in S(q), j \in \operatorname{im} s\}.$$

By Lemma 2.13,

$$\begin{aligned} \ker d^{2,q} &= \bigoplus \{\ker d^{2,q} \cap M(t, i, j) : t \in S(q-1), i, j \notin \operatorname{im} t, i \neq j\} \\ &= \bigoplus \{\ker d^{2,q} \cap M(r) : r \in S(q+1)\}. \end{aligned}$$

It then follows that

$$\begin{aligned} E_3^{2,q} \cong & \bigoplus \{[\ker d^{2,q} \cap M(r)]/R \cdot d(1 \otimes a_r) : r \in S(q+1)\} \\ & \oplus \bigoplus \{\ker d^{2,q} \cap M(s, j) : s \in S(q), j \in \operatorname{im} s\}. \end{aligned}$$

By Lemma 2.13, $\ker d^{2,q} \cap M(s, j)$ is generated by $w_{s(1)} a_j \otimes a_s = a_j \otimes a_s'$, with $a_s' = w_{s(1)} a_s$, $s \in S(q)$, $j \in \operatorname{im} s$. It remains to determine the structure of the modules $[\ker d^{2q} \cap M(r)]/R \cdot d(1 \otimes a_r)$. By Lemma 2.13, it follows that

$$x = \Sigma\, n_i a_{r(i)} \otimes a_{r_i} \in \ker d^{2q} \cap M(r) \text{ iff } z_{r(j)}\, n_i + (-1)^{j-i+1} z_{r(i)} n_j = 0$$

for $1 \leqq i < j \leqq q+1$. Now suppose that in fact only the conditions

$$z_{r(i)}\, n_1 + (-1)^i z_{r(1)}\, n_i = 0 \quad \text{for} \quad 1 < i \leqq q+1 \tag{*}$$

are satisfied. Let $i < j$ and define u, v, and $w \in R$ by $u\, z_{r(1)} = z_{r(i)}$,

$$v\, z_{r(i)} = z_{r(j)}, \quad w = z_{r(1)}\, v .$$

Then $u\, w = u\, z_{r(1)}\, v = z_{r(i)}\, v = z_{r(j)}$ and $w\, z_{r(i)} = z_{r(1)}\, v\, z_{r(i)} = z_{r(1)}\, z_{r(j)}$. Multiply (*) with v and obtain $v\, z_{r(i)}\, n_1 + (-1)^i\, v\, z_{r(1)}\, n_i = 0$ which is equivalent to $z_{r(j)}\, n_1 + (-1)^i\, w\, n_i = 0$. Together with (*), this yields

$$w\, n_i + (-1)^{j-i+1}\, z_{r(1)}\, n_j = 0 .$$

Now we multiply with u and obtain $z_{r(j)}\, n_i + (-1)^{j-i+1}\, z_{r(i)}\, n_j = 0$. Consequently, the elements x are characterized by the equations (*).

Let z^i be defined by $z_{r(i)} = z^i\, z_{r(1)}$. In view of the ring properties, $z_{r(1)}\, x = 0$ is equivalent to $x = w_{r(1)}\, t$ for some $t \in R$ (see Definition 2.11). Hence (*) is in fact equivalent to

$$n_i = w_r\, t_i + (-1)^{i-1}\, z^i\, n_1 \text{ for } i = 2, \ldots, q+1 \text{ with suitable } t_i \in R. \quad (**)$$

Now we define a morphism $\eta\colon R^{q+1} \to W(r)$ by

$$\eta(x_1, \ldots, x_{q+1}) = x_1\, a_{r(1)} \otimes a_{r_1} + (x_2\, w_r - x_1\, z^2)\, a_{r(2)} \otimes a_{r_2} + \cdots$$
$$+ (x_{q+1}\, w_r + (-1)^q\, x_1\, z^{q+1})\, a_{r(q+1)} \otimes a_{r_{q+1}} .$$

Then, because of (**), $\operatorname{im} \eta = \ker d^{2q} \cap M(r)$. Furthermore,

$$\eta(1, 0, \ldots, 0) = \Sigma\{(-1)^{i-1}\, z^i\, a_{r(i)} \otimes a_{r_i} \colon i = 1, \ldots, q+1\},$$

whence

$$z_{r(1)}\, \eta(R \times \cdots \times \{0\}) = R \cdot d(1 \otimes a_r) .$$

If we let $a_r^* = \Sigma(-1)^{i-1}\, z^i\, a_{r(i)} \otimes a(r_i)$ for each $r \in S(q+1)$, as in Proposition 2.12, then $\ker d^{2,q}$ is generated by the elements

$$\eta(1, 0, \ldots, 0) = a_r^*, \quad \psi(0, 1, 0, \ldots, 0) = a_{r(2)} \otimes a'(r_2), \ldots,$$
$$\eta(0, \ldots, 0, 1) = a_{r(q+1)} \otimes a'(r_{q+1}) \quad \text{with} \quad a'_s = w_s\, a_s \in E_3^{2,q}$$

for each $s \in S(q)$. Each of these generators generates a direct summand. By direct calculation using Lemma 2.14, one sees that the submodule $\operatorname{im} d^{2,q} \cap M(r)$ is generated by the element $z_{r(1)}\, a_r^*$. Thus, if $H_r = [\ker d^{2,q} \cap M(r)]/R \cdot d(1 \otimes a_r)$, then

$$H_r \cong R \cdot \tilde{a}_r \otimes \oplus \{R \cdot a_{r(i)} \otimes a'_{r_i} \colon i = 2, \ldots, q+1\}$$

where $\tilde{a}_r = \psi^{2,q}(a_r^*)$, $\psi^{2,q} = \psi \mid E_2^{2,q}$ being the co-cycle map. Notice that $R \cdot \tilde{a}_r \cong R/z_{r(1)}\, R$ and $R \cdot a_{r(i)} \otimes a'(r_i) \cong w_r\, R$. Note that a_r^* can be written as $(1/z_{r(1)})\, d(1 \otimes a_r)$.

Finally we have the following formula:

$$\begin{aligned} E_3^{2,q} &\cong \oplus \{R \cdot \tilde{a}_r : r \in S(q+1)\} \\ &\qquad \oplus \oplus \{R \cdot a_{r(i)} \otimes a'(r_i) : r \in S(q+1),\ i = 2, \ldots, q+1\} \\ &\qquad \oplus \oplus \{R \cdot a_{s(i)} \otimes a'_s : s \in S(q),\ i = 1, \ldots, q\} \\ &\cong \oplus \{R \cdot \tilde{a}_r : r \in S(q+1)\} \\ &\qquad \oplus \oplus \{R \cdot a_k \otimes a'_s : s \in S(q),\ k \in I,\ s(1) \leqq k\}. \end{aligned}$$

Case 2. $q = 1$.

$E_2^{2p,1}$ is a direct sum of the modules $M_{\tau t}$, $\tau \in \Sigma(p+1)$, $t \in S(0) = \{\emptyset\}$, each of which is generated by $a(\tau_{[j]}) \otimes a_j$, $j \in \operatorname{im} \tau$. We denote them by $M(\tau)$.

Also $E_2^{2p,1}$ is the sum of the modules $M^{\varrho r}$, $\varrho \in \Sigma(p-1)$, $r \in S(2)$. Each of these is generated by the elements $a(\varrho^j) \otimes a(r_{[j]})$, $j \in \operatorname{im} r$. One may describe these modules equivalently as being generated by $a(\tau_{[i]}) \otimes a_i$ and

$$a(\tau_{[j]}) \otimes a_j,\ \tau \in \Sigma(p+1) \quad \text{and} \quad i, j \in \operatorname{im} \tau,\ i < j.$$

We will label these modules by $M(\tau, i, j)$. It is then clear that each $M(\tau, i, j)$ is contained in $M(\tau)$. We observe now that

$$E_3^{2p,1} \cong \oplus \{[\ker d^{2p,1} \cap M(\tau)]/[\operatorname{im} d^{2p-2,2} \cap M(\tau)] : \tau \in \Sigma(p+1)\}.$$

By Lemma 2.13, an element $x = \Sigma\{n_i\, a(\tau_{[i]}) \otimes a_i :\ i \in \operatorname{im} \tau\}$ belongs to $\ker d^{2p,1}$ exactly when $\Sigma z_j n_j = 0$. Suppose that $k = \tau(1)$. For $j \in \operatorname{im} \tau$ define z^j by $z_j = z^j z_k$. Then x is in the kernel iff $n_k = w_k t - \Sigma\{z^j n_j : j \neq k\}$ for some $t \in R$. Moreover, if

$$y = \Sigma\{n(\sigma, i)\, a_\sigma \otimes a_i :\ \sigma \in \Sigma(p),\ i \in I\},$$

a simple calculation (as in Lemma 2.13) shows that $y \in \ker d^{2p,1}$ iff, for each $\tau \in \Sigma(p+1)$,

$$\Sigma\{n(\tau_{[i]}, i)\, z_i :\ i \in \operatorname{im}\ \tau\} = 0.$$

That is,

$$\ker d^{2p,1} = \oplus \{(\ker d^{2p,1} \cap M(\tau) : \tau \in \Sigma(p+1)\}.$$

On the other hand, for $\tau \in \Sigma(p+1)$, $\operatorname{im} d^{2p-2,2} \cap M(\tau)$ is generated by the elements

$$z_i\, a(\tau_{[j]}) \otimes a_j - z_j\, a(\tau_{[i]}) \otimes a_i,\quad i < j,\quad i, j \in \operatorname{im} \tau.$$

We assert that in fact it is already generated by the elements

$$b_i = z_k\, a(\tau_{[i]}) \otimes a_i - z_i\, a(\tau_{[k]}) \otimes a_k,\quad i \in \operatorname{im} \tau,\quad k < i.$$

In fact, suppose that $i < j$ in $\operatorname{im} \tau$. Define u and v by $z_j = u z_i$, $z_i = v z_k$. Then

$$v(b_j - u\, b_i) = z_i\, a(\tau_{[j]}) \otimes a_j - z_j\, a(\tau_{[i]}) \otimes a_i.$$

If we define

$$c_i = a(\tau_{[i]}) \otimes a_i - z^i\, a(\tau_{[k]}) \otimes a_k,$$

then $b_i = z_k\, c_i$.

We define a morphism $\eta\colon R^{p+1} \to M(\tau)$ by

$$\begin{aligned}\eta(x_1, \ldots, x_{p+1}) = {} & (w_k\, x_1 - \Sigma \{x_\nu\, z^{\tau(\nu)}\colon\ 1 < \nu \leqq p+1\})\, a(\tau_{[k]}) \otimes a_k \\ & + x_2\, a(\tau_{[\tau(2)]}) \otimes a_{\tau(2)} + \cdots \\ & + x_{p+1}\, a(\tau_{[\tau(p+1)]}) \otimes a_{\tau(p+1)}.\end{aligned}$$

Then, by the preceding $\operatorname{im} \eta = \ker d^{2p,1} \cap M(\tau)$. Further, $\psi(0, \ldots, 1, \ldots, 0)$ with 1 in place $i > 1$ is equal to $c_{\tau(i)}$. Thus η maps $\{0\} \times z_k R \times \cdots \times z_k R$ onto $\operatorname{im}\, d^{2p-2,2} \cap M(\tau)$. The module $\ker d^{2p,1} \cap M(\tau)$ is generated by the elements

$$\begin{aligned}\eta(1, 0, \ldots, 0) &= w_k\, a(\tau_{[k]}) \otimes a_k \\ \eta(0, \ldots, 1, \ldots, 0) &= -z^{\tau(\nu)}\, a(\tau_{[k]}) \otimes a_k + a\,(\tau_{[\tau(\nu)]}) \otimes a_{\tau(\nu)},\end{aligned}$$

where on the left side, the 1 is in the ν-th place, $1 < \nu$ and where $k = \tau(1)$. Each of these generators generates a direct summand. The module

$$\operatorname{im}\, d^{2p-2,2} \cap M(\tau)$$

is generated by the elements

$$\eta(0, \ldots, z_k, \ldots, 0) = z_k\, \eta(0, \ldots, 1, \ldots, 0),$$

where the only non-zero entry is in place $\nu > 1$. Thus

$$E_3^{2p,1} = \oplus \{[\ker d^{2p,1} \cap M(\tau)]/[\operatorname{im}\, d^{2p-2,2} \cap M(\tau)]\colon\ \tau \in \Sigma(p+1)\}$$

is the direct sum of all $R \cdot a_\sigma \otimes a_i'$ in $E_2^{2p,1}$ with $\sigma \in \Sigma(p)$, $i \leqq \sigma(1)$ and of all $R \cdot \tilde{a}_{ij}^\varrho$, where $\tilde{a}_{ij}^\varrho = \psi^{2p,1}(a_{ij}^\varrho)$, $a_{ij}^\varrho = a(\varrho^i) \otimes a_j - z_j/z_i\, a(\varrho^j) \otimes a_i$, $\varrho \in \Sigma(p-1)$, $i \leqq \varrho(1) < j$. Moreover, $R \cdot \tilde{a}_{ij}^\varrho \cong R/z_i R$. Note that the family of elements $\tilde{a}_{ij}^\varrho$ may also be described as the family of all

$$\begin{aligned}&\psi^{2p,q}((1/z_{r(1)})\, d(a_\varrho \otimes a_r)), \quad \varrho \in \Sigma(p-1), \quad r \in S(2), \\ &r(1) \leqq \varrho(1) < r(2).\end{aligned}$$

Before we collect the available information in unified form, we observe that for $s(1) \leqq \sigma(1)$ we have that $R \cdot a_\sigma \otimes a_s' \cong R \cdot \tilde{a}_\sigma \otimes a_s'$ with $\tilde{a}_\sigma = \psi^{2p,0}(a_\sigma)$, since the kernel of $R \cdot a_\sigma \to R \cdot \tilde{a}_\sigma$ is generated by $z_{\sigma(1)}\, a_\sigma$ according to Proposition 2.12 (a), but $z_{\sigma(1)}\, a_\sigma \otimes w_s\, a_s = 0$ if $z_{\sigma(1)} w_s = 0$, which is the case if $s(1) \leqq \sigma(1)$. We now write down the result of the computation of E_3 on the edge terms and the adjacent terms:

Proposition 2.17. *Let R be a weakly principal ideal ring, A a free R-module and $\varphi\colon A \to A$ an elementary morphism. If $p < 2$ or $q < 2$, then*

$$E_3^{2p,q}(\varphi) = \oplus \{R \cdot \psi^{2p-2,0}(a_\varrho \otimes 1)\, \psi^{2,q}((1/z_{r(1)})\, d(1 \otimes a_r)):$$
$$\varrho \in \Sigma(p-1),\ r \in S(q+1),\ r(1) \leqq \varrho(1) < r(2)\}$$
$$\oplus \oplus \{R \cdot \psi^{2p,0}(a_\sigma \otimes 1)\, \psi^{0,q}(1 \otimes w_s\, a_s) \colon \sigma \in \Sigma(p),\ s \in S(q),$$
$$s(1) \leqq \sigma(1)\}.$$

Further,

$$R \cdot \psi^{2p-2,0}(a_\varrho \otimes 1)\, \psi^{2,q}(1/z_{r(1)})\, d(1 \otimes a_\sigma) \cong R/z_{r(1)} R \quad \textit{for} \quad r(1) \leqq \varrho(1)$$

and

$$R \cdot \psi^{2p,0}(a_\sigma \otimes 1)\, \psi^{0,q}(1 \otimes w_s\, a_s) \cong w_s R \quad \textit{for} \quad s(1) \leqq \sigma(1).$$

Remark. Note that conditions of the form $r(1) \leqq \varrho(1) < r(2)$ are sometimes void, e. g. if $p = 1$, or to be interpreted accordingly, e. g. as $r(1) \leqq \varrho(1)$, if $q = 1$. Also, $w_\emptyset = 0$.

The term $(1/z_{r(1)})\, d(1 \otimes a_r)$ is defined by a_r^* as in Proposition 2.12 (c). Despite the non-uniqueness of the a_s^* for some rings R, the cyclic module $R \cdot a_s^*/R \cdot a_s$ is uniquely determined.

Proof. The preceding discussions show the assertion for the cases $p = 1$, q arbitrary and $q = 1$, p arbitrary. Proposition 2.12 (b) und (c) show that with appropriate interpretations of the summation requirements, the same formula applies to the case $p = 0$, q arbitrary and $q = 0$, p arbitrary.

Example. It may be appropriate to work out some more details in a very simple example. Let $R = \mathsf{Z}/z\mathsf{Z}$ and

$$z_1 | z_2 | z,\ w_1 = z/z_1,\ w_2 = z/z_2,\ \varphi\colon R^2 \to R^2.$$

Then $E_3^{2,2}(\varphi)$ is the direct sum of the cyclic subgroups spanned by the elements $h_{11} = \psi(a_1 \otimes w_1\, a_1)$, $h_{21} = \psi(a_2 \otimes w_1\, a_1)$, $h_{22} = \psi(a_2 \otimes w_2\, a_2)$, and

$$k = \psi(a_1 \otimes a_2 - (z_2/z_1)\, a_2 \otimes a_1).$$

Let $x_1 = \psi(a_1 \otimes 1)$, $x_2 = \psi(a_2 \otimes 1)$, $y_1 = \psi(1 \otimes w_1\, a_1)$, $y_2 = \psi(1 \otimes w_2\, a_2)$. Then $R \cdot x_1 \cong R/z_1 R$, $R \cdot x_2 \cong R/z_2 R$,

$$R \cdot y_1 \cong w_1 R \cong R/z_1 R, \quad R \cdot y_2 \cong w_2 R = R/z_2 R.$$

Consequently (using Proposition 2.16) we have

$$R \cdot x_1 \otimes y_1 \cong R \cdot x_1\, y_1 = R \cdot h_{11} \cong R/z_1 R,$$
$$R \cdot x_1 \otimes y_2 \cong R \cdot x_1\, y_2 \cong R/z_1 R,$$
$$R \cdot x_2 \otimes y_1 \cong R \cdot x_2\, y_1 = R \cdot h_{21} \cong R/z_1 R,$$
$$R \cdot x_2 \otimes y_2 \cong R \cdot x_2\, y_2 = R \cdot h_{22} \cong R/z_2 R.$$

Thus

$$X = E_3^{2,0}(\varphi) \otimes E_3^{0,2}(\varphi) \cong (R/z_1R)^3 \oplus R/z_2R.$$

Since $R \cdot k \cong R/z_1 R$ we also have

$$Y = E_3^{2,2}(\varphi) \cong (R/z_1R)^3 \oplus R/z_2R.$$

However, the morphism $\mu: X \to Y$ defined by the multiplication in E_3 is, in general, neither surjective nor injective: We compute easily that

$$x_1 y_2 = w_2 k + x_2 y_1 = w_2 k + h_{21}.$$

In order to compute the kernel of μ, take $x = \Sigma \alpha_{ij} x_i \otimes y_j \in X$. Then

$$\mu(x) = \alpha_{11} h_{11} + w_2 \alpha_{12} k + (\alpha_{21} + \alpha_{12}) h_{21} + \alpha_{22} h_{22}.$$

Now $\mu(x) = 0$ if and only if $\alpha_{11} \in z_1 R$, $(z/z_2)\alpha_{12} \in z_1 R$, $\alpha_{21} + \alpha_{12} \in z_1 R$ and $\alpha_{22} \in z_2 R$. It follows that the kernel is contained in the subgroup spanned by $x_1 \otimes y_2$ and $x_2 \otimes y_1$. The coefficients of these elements may be considered as elements of R/z_1R rather than R, in which case we over-bar them. Then an element $\bar{\alpha}_{12} x_1 \otimes y_2 + \bar{\alpha}_{21} x_2 \otimes y_1$ is in the kernel iff $(z/z_2)\bar{\alpha}_{12} = 0$ and

$$\bar{\alpha}_{21} = -\bar{\alpha}_{12}.$$

Since $(z/z_2)\bar{\alpha}_{12} = 0$ is equivalent to $\bar{\alpha}_{12} = t(((z/z_2), z_1) + z_1 R)$ for some t, the kernel is generated by the element $((z/z_2), z_1)(x_1 \otimes y_2 - x_2 \otimes y_1)$. Since the image of μ is spanned by $w_2 k, h_{11}, h_{21}$k, h_{22}, the co-kernel of μ is isomorphic to $R/(w_2, z_1) R = R/((z, z_2), z_1) R$.

If conditions on the ring R imply $w_i = 0$ for all $i \in I$, we obtain the following result:

Corollary 2.18. *Let R be a principal ideal domain, A a free R-module, and $\varphi: A \to A$ an elementary morphism such that all $z_i \neq 0$ in R (i. e. φ is injective). If $p < 2$ or $q < 2$, then*

$$E_3^{2p,q}(\varphi) = \oplus \{R \cdot \psi^{2p,q}((a_\varrho \otimes 1)((1/z_{r(1)}) d(1 \otimes a_r)): \varrho \in \Sigma(p-1),\ r \in S(q+1),\ r(1) \leqq \varrho(1) < r(2)\}$$

$$= \begin{cases} \{0\} \quad \textit{for} \quad p = 0,\ q > 0, \\ \oplus \{R \cdot \psi^{2,q}((1/z_{r(1)}) d(1 \otimes a_r)): r \in S(q+1)\} \quad \textit{for} \quad p = 1, \\ \oplus \{R \cdot \psi^{2p,1}(a(\varrho^i) \otimes a_j - (z_j/z_i)\, a(\varrho^j) \otimes a_i): \\ \qquad \varrho \in \Sigma(p-1), i, j \in I, i \leqq \varrho(1) < j\} \quad \textit{for} \quad q = 1, \\ \oplus \{R \cdot \psi^{2p,0}(a_\sigma): \sigma \in \Sigma(p)\} \quad \textit{for} \quad q = 0. \end{cases}$$

In particular, there is an exact sequence

$$0 \to \oplus \{R \cdot z_{r(1)}\, a_r \colon r \in S(q+1)\} \to E_2^{0,q}(\varphi) \to E_2^{2,q}(\varphi) \to 0$$

and $E_3^{2,q}(\varphi)$ *and* $\wedge^{q+1}$ coker φ *are isomorphic modules. In particular, if* $z_1 \neq 0$, *then* $E_3^{2,q}(\varphi) = \{0\}$ *if and only if* $E_2^{0,q+1}(\varphi) = 0$, *i. e. if and only if* $|I| \leqq 2$.

Proof. The first assertion follows immediately from Proposition 2.17.

We define a morphism $E_2^{0,q+1}(\varphi) \to E_3^{2,q}$ by assigning the element

$$\tilde{a}_r = \psi^{2,q}(a_r^*)$$

to the element a_r. The kernel of the map $x \to x \cdot \tilde{a}_r \colon R \to R \cdot \tilde{a}_r$ is $Rz_{r(1)}$ since $R \cdot \tilde{a}_r \cong R/z(1)\, R$. Moreover, $\Sigma\, x_r\, \tilde{a}_r = 0$ implies $x_r\, \tilde{a}_r = 0$ for each r. Hence there is an exact sequence

$$\sigma \to \oplus \{R \cdot (z_{r(1)} a_r) \colon r \in S(q+1)\} \to E_2^{0,q+1}(\varphi) \to E_3^{2,q}(\varphi) \to 0\,.$$

One can easily show that if $A \xrightarrow{\varphi} A \xrightarrow{\pi} B \longrightarrow 0$ is exact, and $b_i = \pi(a_i)$, $i \in I$, then $\wedge^{q+1} B = \oplus \{R \cdot b_s \colon s \in S(q+1)\}$ and $R \cdot b_s \cong R/z_{s(1)}R$. Since

$$\wedge^{q+1} A = \oplus \{R \cdot a_s \colon s \in S(q+1)\},$$

and clearly $\wedge^{q+1}(\pi)\,(a_s) = b_s$, we have an exact sequence

$$0 \to \oplus \{R \cdot (z_{r(1)}\, a_R) \colon r \in S(q+1)\} \to \wedge^{q+1} A \to \wedge^{q+1} \ \text{coker} \ \varphi \to 0.$$

The left hand map is equivalent to the left hand map in the exact sequence

$$0 \to \oplus \{R \cdot (z_{r(1)}\, a_r) \colon r \in S(q+1)\} \to E_2^{0,q+1}(\varphi) \to E_3^{2,q}(\varphi) \to 0$$

that was given above. Thus, we have $E_3^{2,q}(\varphi) \cong \wedge^{q+1}$ coker φ.

The results of Corollary 2.18 suggest the following:

Conjecture 2.19. Let R be a principal ideal domain, A a free R-module with basis $\{a_i \colon i \in I\}$, and $\varphi \colon A \to A$ an injective elementary morphism. Then there is an epimorphism $f^{2p-2,q+1} \colon E_2^{2p-2,q+1}(\varphi) \to E_3^{2p,q}(\varphi)$ which sends $a_\varrho \otimes a_r$ onto $\psi^{2p,q}((a_\varrho \otimes 1)\, a_r^*)$ with $a_r^* = (1/z_{r(1)})\, d(1 \otimes a_r)$ as in Proposition 2.12 (c), where a_r^* under the present circumstances is uniquely determined by a_r.

In Corollary 2.29, we will establish the existence of the morphism f, and in Proposition 2.38 prove its surjectivity, providing that I is finite. However, we must first provide some more background material — some of which is of independent interest.

Lemma 2.20. *Let* R *be an integral domain and* K *be its field of quotients. Let* A *be a free* R*-module,* $\varphi \colon A \to A$ *an elementary morphism, and* $\bar{A} = K \otimes A$, $\bar{\varphi} \colon \bar{A} \to \bar{A}$ *the ground ring extensions. Then the* R*-monomorphism* $A \to \bar{A}$ *induces an injection* $E_2(\varphi) \to E_2(\bar{\varphi})$ *of differential bi-graded algebras and* $E_2(\bar{\varphi})$ *is a* K*-algebra.*

Proof. There are natural isomorphisms $K \otimes \bigwedge_R A \cong \bigwedge_K \bar{A}$ and

$$K \otimes P_R A \cong P_K \bar{A}.$$

Hence $E_2(\bar{\varphi}) \cong K \otimes E_2(\varphi)$ naturally. Since $E_2(\varphi)$ is a free R-module, the natural map $E_2(\varphi) \to K \otimes E_2(\varphi)$ is injective.

Proposition 2.21. *If A and B are two R-modules and $\varphi\colon A \to A$ and $\psi\colon B \to B$ two morphisms, then there is a canonical isomorphism*

$$\varepsilon\colon E_2(\varphi \oplus \psi) \to E_2(\varphi) \otimes E_2(\psi)$$

and $\varepsilon\, d_{\varphi\oplus\psi} = (d_\varphi \otimes d_\psi)\,\varepsilon$, *where*

$$d_\varphi \otimes d_\psi\,(a^q \otimes a^p) = d_\varphi\, a^q \otimes a^p + (-1)^q\, a^q \otimes d_\psi\, a^p.$$

Proof. The isomorphisms of graded algebras $\bigwedge(A \oplus B) \to \bigwedge A \otimes \bigwedge B$ and $P(A \oplus B) \to PA \otimes PB$ induce a unique isomorphism

$$\varepsilon\colon E_2(\varphi \oplus \psi) \to E_2(\varphi) \otimes E_2(\psi)$$

of graded algebras. The remainder then follows from Lemma 2.2 (c).

Remark. In the terminology of Definition 1.1, $E_2(\cdot)$ is an exponential functor.

Lemma 2.22. *Let R be any commutative ring with identity, and let φ and ψ be elementary morphisms of free R-modules. If d_φ is exact, $\operatorname{Tor}^1_R(\operatorname{im} d_\varphi, \operatorname{im} d_\psi) = 0$ and $\operatorname{Tor}^1_R(\operatorname{im} d_q, \ker d_\psi) = 0$, then there is a natural isomorphism*

$$E_3(\psi) \to E_3(\varphi \oplus \psi).$$

Proof. We apply the Künneth theorem to the complexes $E_2(\varphi)$ and $E_2(\psi)$ with their total gradation and obtain a natural isomorphism

$$H(d_\varphi) \otimes H(d_\psi) \to H(d_\varphi \otimes d_\psi),$$

where $d_\varphi \otimes d_\psi$ is the differential on the tensor product of the complexes (see Cartan and Eilenberg [11], pp. 112, 113). By hypothesis, $H(d_\varphi) = R$, so $H(\psi) \cong H(d_\varphi \otimes d_\psi)$ naturally. Since $H(d_{\varphi\oplus\psi}) \cong H(d_\varphi \otimes d_\psi)$ by Proposition 2.21, the assertion follows.

Lemma 2.23. *Let R be a commutative ring with identity. If $\varphi\colon R \to R$ is an isomorphism, then d_φ on $E_2(\varphi)$ is exact.*

Proof. We have $\bigwedge R = R \oplus R$, whence $E_2(\varphi) = PR \otimes 1 \oplus PR \otimes R$. Clearly

$$\ker d_\varphi = PR \otimes 1 = \operatorname{im} d_\varphi \oplus (R \otimes 1).$$

Lemma 2.24. *Let R be a commutative ring. If A is a finitely generated free R-module, and φ is the identity, or if R is a field and φ is diagonalisable (i. e.*

a direct sum of homotheties, where a homothety is multiplication by a scalar), then d_φ is exact.

Proof. This follows immediately from Lemmas 2.22 and 2.23 by induction.

Proposition 2.25. *Let R be a commutative ring with identity and $\varphi\colon A \to A$ a morphism which is a direct limit of morphisms $\varphi_j\colon A_j \to A_j$ of finitely generated free modules. (If R is a principal ideal domain, every torsion free module is a direct limit of finitely generated free submodules.) If $\varphi\colon A \to A$ is the identity morphism, or if R is a field and all φ_j are diagonalisable, then $E_r^{q,p}(\varphi) = 0$ for $p + q \neq 0$, $r > 1$ (and $E_3^{0,0}(\varphi) = R$).*

Proof. The functor E_2 commutes with direct limits. The direct limit functor is exact. Since d_φ is the direct limit of the d_{φ_j} and the latter all are exact by our hypotheses and Lemma 2.24, then d_φ is exact. But this is the assertion.

Now we can take Lemma 2.20 together with Proposition 2.25 and obtain

Proposition 2.26. *Let R be an integral domain, and K its field of fractions. Let $\varphi\colon A \to A$ be an elementary morphism, and let $\bar{A} = K \otimes A$, and $\bar{\varphi}\colon \bar{A} \to \bar{A}$ be the ground field extensions. Then there is an injection of differential bi-graded R-algebras $E_2(\varphi) \to E_2(\bar{\varphi}) \cong K \otimes E_2(\varphi)$, and $E_3^{p,q}(\bar{\varphi}) = \{0\}$ for $p + q > 0$.*

Corollary 2.27. *Under the assumptions of Proposition 2.26, if we consider $E_2(\varphi)$ as an R-subalgebra of $E_2(\bar{\varphi})$, $\ker d_{\bar{\varphi}}^{2p,q}$ is spanned over K by $\operatorname{im} d_\varphi^{2p,q}$. If $x \in \ker d_\varphi^{2p,q}$, then $\{t' \in R\colon t'x \in \operatorname{im} d_\varphi^{2p,q}\}$ is a non-zero ideal.*

Corollary 2.28. *Under the assumptions of Corollary 2.27,*

$$a_r^* = (1/z_{r(1)})\, d_\varphi(1 \otimes a_r) \in \ker d_\varphi \quad \textit{for all} \quad r \in S(q+1).$$

Proof. Any element of the form $k\, d_\varphi\, (1 \otimes a_r)$ is in $\ker d_{\bar{\varphi}}$. But $a_r^* \in E_2(\varphi)$, and so is in $\ker d_\varphi$.

Remark. When R is a weakly principal ideal ring, as we have shown in the discussion after Proposition 2.16, the $\psi(a_r^*)$, $r \in \Sigma(q+1)$, generate $E_3^{2,q}(\varphi)$.

Corollary 2.29. *Under the same assumptions, there is a morphism of R-modules*

$$f\colon E_2^{2p-2,q+1}(\varphi) = P^{p-1} A \otimes \textstyle\bigwedge^{q+1} A \to E_3^{2p,q}(\varphi)$$

which is defined by

$$f(a_\varrho \otimes a_r) = \psi^{2p,q}((a_\varrho \otimes 1)\, a_r^*) = \psi^{2p-2,0}(a_\varrho \otimes 1)\, \psi^{2,q}(a_r^*)$$

for $\varrho \in \Sigma(p-1)$, $r \in S(r+1)$.

The direct computational confirmation of Conjecture 2.19 saying that the morphism of R-modules $f\colon E_2(\varphi) \to E_3(\varphi)$ of bi-degree $(2, -1)$ is an epimorphism requires further considerations. If $R = \mathsf{Z}$ then Corollary 2.27 expresses the fact that the smallest pure sub-group of $\ker d_\varphi$ generated by the elements $(1/z_{r(1)})\, d_\varphi(a_\varrho \otimes a_r)$ is $\ker d_\varphi$ itself. In order to exhibit more clearly what will have to be established to show that these elements in fact generate $\ker d_\varphi$, we list the following equivalent statements:

Lemma 2.30. *Let R be a principal ideal domain and let $\varphi\colon A \to A$ be an elementary morphism. Then the following statements are equivalent:*

(a) *The morphism $f\colon E_2(\varphi) \to E_3(\varphi)$ of R-modules of bi-degree $(2, -1)$ is surjective (except for bi-degree $(0, 0)$).*

(b) *The R-algebra $E_3(\varphi)$ is generated by $E_3^{0,0}(\varphi) \oplus \bigoplus \{E_3^{2,q}(\varphi)\colon q = 0, 1, \ldots\}$.*

(c) *If E_3' is the bi-graded module which coincides with $E_3(\varphi)$ for $p < 3$ and is zero otherwise, then the morphism of bi-graded modules P coker $\varphi \otimes E_3' \to E_3(\varphi)$ defined by the isomorphism P coker $\varphi \to E_3^{\mathrm{I}}(\varphi)$ of Proposition 2.12 (b) and the internal cup product $E_3(\varphi) \otimes E_3(\varphi) \to E_3(\varphi)$ of $E_3(\varphi)$ is surjective.*

(d) $\ker d_\varphi$ *is generated by $R = E_2^{0,0}(\varphi)$ and $\Sigma\{\ker d_\varphi^{2,q}\colon q = 1, 2, \ldots\}$, hence by $R = E_2^{0,0}(\varphi)$ and all a_r^*, $r \in S(q+1)$, $q = 0, 1, 2, \ldots$*

Proof. By the definition of f, (a) and (c) are equivalent. (See the remark after Corollary 2.28.) Trivially, (c) implies (b), but likewise, (b) implies (c). Clearly, (d) implies (a). But (a) implies that $\ker d_\varphi$ is generated by $R = E_2^{0,0}(\varphi)$ and $\Sigma\{\ker d_\varphi^{2,q}\colon q = 1, 2, \ldots\} \oplus \operatorname{im} d_\varphi$. But $d_\varphi(E_2^{\mathrm{II}}(\varphi)) \subseteqq \Sigma \ker d_\varphi^{2,q}$ already generates $\operatorname{im} d_\varphi$. Hence (a) implies (d).

It is perhaps instructive to discuss some special cases.

Lemma 2.31. *Let R be an arbitrary commutative ring with identity and let $\varphi, \varphi^*\colon A \to A$ be elementary morphisms of free R-modules. Suppose that $z_i^* = u_i z_i$ with units u_i in R. Then $E_3(\varphi) \cong E_3(\varphi^*)$.*

Proof. We define an automorphism of the graded algebras $E_2(\varphi)$ by

$$\alpha = P\beta \otimes \wedge 1(A), \quad \text{where} \quad \beta(a_i) = u_i a_i.$$

We write $u_\sigma = u_{\sigma(1)} \cdots u_{\sigma(p)}$ for $\sigma \in \Sigma(p)$. Then

$$\begin{aligned} d_{\varphi^*} \alpha(a_\sigma \otimes a_s) &= \Sigma\{(-1)^{i-1} z_{s(i)}^* u_\sigma a_\sigma a_{s(i)} \otimes a(s_i)\} \\ &= \Sigma\{(-1)^{i-1} z_{s(i)} u(\sigma^{s(i)}) a(\sigma^{s(i)}) \otimes a(s_i)\} \\ &= \alpha d_\varphi (a_\sigma \otimes a_s). \end{aligned}$$

Thus $d_{\varphi^*} \alpha = \alpha d_\varphi$. But this implies the assertion.

Proposition 2.32. *Let R be a commutative ring with identity and let $z \in R$ be an element such that $z x = 0$ implies $x = 0$. Let A be a finitely generated free*

R-module and let $\varphi\colon A \to A$ *be defined by* $\varphi(a) = z \cdot a$. *Let*

$$\nu, \varepsilon\colon A/zA \to A/zA$$

be the zero, resp. the identity, morphism and let $\pi'\colon \varphi \to \nu$ *be the morphism in* $\mathfrak{M}$ *(see Definition 2.4) defined by the commutative diagram*

$$\begin{array}{ccc} A & \xrightarrow{\varphi} & A \\ \downarrow{\scriptstyle\pi} & & \downarrow{\scriptstyle\pi} \\ A/zA & \xrightarrow{\nu} & A/zA \end{array}$$

Then the morphism of bi-graded algebras

$$E_3(\pi')\colon E_3(\varphi) \to E_3(\nu) = PA/zA \otimes \wedge A/zA$$

maps $E_3(\varphi)$ *bijectively onto* $R \oplus \operatorname{im} d_\varepsilon$ *in* $E_3(\nu) = E_2(\varepsilon)$, *where* $R = E_2^{0,0}(\varepsilon)$. *(As usual,* $\pi = \operatorname{coker} \varphi$.)

Proof. Let $\varepsilon'\colon A \to A$ be the identity and $d = d_{\varepsilon'}$. By the hypothesis about z and the freeness of A, we have $z\,a = 0$ for $a \in A$ iff $a = 0$. Suppose that $z\,a \in \ker d$. Then $0 = d\,(z\,a) = z \cdot da$, whence $a \in \ker d$. Thus

$$zE_2(\varepsilon') \cap \ker d = z \cdot \ker d.$$

But $\ker d = E_2^{0,0}(\varepsilon') + \operatorname{im} d$ by Lemma 2.24. Hence $zE_2(\varphi) \cap \operatorname{im} d = z \cdot \operatorname{im} d$. Further, by the properties of z, we have $\ker d_\varphi = \ker d$ and $\operatorname{im} d_\varphi = z \cdot \operatorname{im} d$. The morphism $\pi\colon A \to A/zA$ induces a surjective morphism

$$\pi|\operatorname{im} d\colon \operatorname{im} d \to \operatorname{im} d_\varepsilon.$$

Its kernel is

$$\ker \pi \cap \operatorname{im} d = zE_2(\varphi) \cap \operatorname{im} d = z \cdot \operatorname{im} d = \operatorname{im} d_\varphi.$$

But $\operatorname{im} d^{2p,q} = \ker d_\varphi^{2p,q}$. Hence, since $\ker d_\varphi^{2p,q}/\operatorname{im} d_\varphi = E_3(\varphi)$, the induced morphism $E_3(\pi)\colon E_3(\varphi) \to E_3(\nu)$ maps $E_3(\varphi)$ isomorphically onto $\operatorname{im} d_\varepsilon$.

This situation can be described in slightly more general terms.

Proposition 2.33. *Let* R *be a commutative ring with identity and* $z \in R$ *an element with* $z\,x = 0$ *iff* $x = 0$. *Let* A *be a finitely generated free R-module and let* $\varphi, \varphi^*\colon A \to A$ *be elementary morphisms with* $\varphi = z\,\varphi^*$. *Consider the commutative diagrams*

$$\begin{array}{ccc} A & \xrightarrow{\varphi,\,\varphi^*} & A \\ \downarrow{\scriptstyle\pi} & & \downarrow{\scriptstyle\pi} \\ A/zA & \xrightarrow{\varphi',\,\varphi^{*\prime}} & A/zA \end{array}$$

Then there is an exact sequence

$$0 \to \operatorname{im} d_{\varphi^{*\prime}} \to E_3(\varphi) \to E_3(\varphi^*) \to 0 .$$

Proof. Again we have $\ker d_\varphi = \ker d_{\varphi^*}$ and $\operatorname{im} d_\varphi = z \cdot \operatorname{im} d_{\varphi^*}$, and

$$zE_2(\varphi^*) \cap \operatorname{im} d_{\varphi^*} = z \cdot \operatorname{im} d_{\varphi^*} .$$

Thus there is a surjective morphism

$$\begin{aligned} E_3(\varphi) &= \ker d_\varphi / \operatorname{im} d_\varphi = \ker d_{\varphi^*}/z \cdot \operatorname{im} d_{\varphi^*} \\ &= \ker d_{\varphi^*}/zE_2(\varphi^*) \cap \operatorname{im} d_{\varphi^*} \to \ker d_{\varphi^*}/\operatorname{im} d_{\varphi^*} = E_3(\varphi') \end{aligned}$$

with kernel $\operatorname{im} d_{\varphi^*}/zA \cap \operatorname{im} d_{\varphi^*}$. There is a surjective morphism

$$\pi \,|\, \operatorname{im} d_{\varphi^*} : \operatorname{im} d_{\varphi^*} \to \operatorname{im} d_{\varphi^{*\prime}}$$

with kernel $\operatorname{im} d_{\varphi^*} \cap zE_2(\varphi^*)$. This finishes the proof.

Note that for $\varphi^* =$ identity, we have $E_3(\varphi^*) = \{0\}$ except in bi-degree $(0, 0)$.

Proposition 2.34. *Let R be a commutative ring with identity, A a finitely generated free module over R, and $\varphi = \varphi_1 \oplus \varphi_2 : A \to A$ with $A = A_1 \oplus A_2$ an elementary morphism. If $\varphi_1 : A_1 \to A_1$ is the identity and if $\operatorname{im} d(\varphi_1)$ is flat, then $E_3(\varphi) \cong E_3(\varphi_2)$ in a natural fashion.*

Proof. By Lemma 2.24, $d(\varphi_1)$ is exact. Thus Lemma 2.22 is applicable and yields the assertion.

Note that $\operatorname{im} d(\varphi_1)$ is automatically flat if R is a principal ideal domain, since then any submodule of a free module is free.

The preceding results allow the build-up of an inductive procedure to compute $E_3(\varphi)$ for an elementary morphism of a finitely generated free module over a principal ideal domain. The technical description is somewhat involved, although not basically difficult.

Proposition 2.35. *Suppose that R is a principal ideal domain and that the following conditions are satisfied:*

(a) *Let A^i, $i = 1, \ldots, n$ be finitely generated free R-modules.*

(b) *Let $\varphi : A \to A = \oplus \{A^i : i = 1, \ldots, n\}$ be the elementary morphism defined by $\varphi(a^{(i)}) = z^{(1)} \cdots z^{(i)} a^{(i)}$. We define $A_i = \oplus \{A^j : i < j\}$ and define inductively morphisms $\varphi_i : A_i \to A_i$ by $\varphi_n = z^{(n)} 1(A^n)$ and $\varphi_i = z^{(i)}(1(A^i) \oplus \varphi_{i+1})$, $i = n-1, \ldots, 1$. Note that $\varphi = \varphi_1$. Further we define the endomorphism Φ_i of $A_i/z^{(i)}A_i$ by the commutative diagram*

$$\begin{array}{ccc} A_i & \xrightarrow{\;1(A^i) \oplus \varphi_{i+1}\;} & A_i \\ \downarrow & & \downarrow \\ A_i/z^{(i)}A_i & \xrightarrow{\;\Phi_i\;} & A_i/z^{(i)}A_i \end{array} \qquad \Phi_{n+1} = 0 .$$

We abbreviate $d(\Phi_i)$ with d_i. Then there are exact sequences of morphisms of bigraded R-modules:

$$
\begin{aligned}
&0 \to \operatorname{im} d_1 \to E_3(\varphi) \to E_3(\varphi_2) \to 0,\\
&0 \to \operatorname{im} d_2 \to E_3(\varphi_2) \to E_3(\varphi_3) \to 0,\\
&\cdot\\
&\cdot\\
&\cdot\\
&0 \to \operatorname{im} d_{n-1} \to E_3(\varphi_{n-1}) \to E_3(\varphi_n) \to 0,\\
&0 \to \operatorname{im} d_n \to E_3(\varphi_n) \to R \to 0.
\end{aligned}
$$

Proof. By Proposition 2.33, we have the exact sequences

$$0 \to \operatorname{im} d_i \to E_3(\varphi_i) \to E_3(1(A^i) \oplus \varphi_{i+1}) \to 0.$$

By Proposition 2.34, the last term is isomorphic to $E_3(\varphi_{i+1})$, and in particular $E_3(1(A_n)) = R$.

Lemma 2.36. *Let I be an ideal of an R-algebra A, and let $\pi: A \to B$ be the quotient mapping onto $B = A/I$. Suppose that M is a subset of A such that $\pi(M)$ generates B as an R-algebra. Suppose that $N \subset I$ is a subset generating I as an R-algebra. Then $N \cup M$ generates A as an R-algebra.*

Proof. Let A' be the smallest sub-algebra containing $N \cup M$. Since N generates I, $I \subset A'$. Hence $\pi(M) \subset A'/I = A/I = B$. Since $\pi(M)$ generates B, we have $A'/I = A/I$, whence $A' = A$.

Notation. For any φ, let

$$E_3^2(\varphi) = E_3^{0,0}(\varphi) + \oplus \{E_3^{2,q}(\varphi) : q = 0, 1, \ldots\}.$$

Lemma 2.37. *Let the hypotheses be the ones of Proposition 2.33: If $E_3(\varphi^*)$ is generated by $E_3^2(\varphi^*)$, then $E_3(\varphi)$ is generated by $E_3^2(\varphi)$.*

Proof. We apply Lemma 2.36 with $I = \operatorname{im} d_{\varphi^{*\prime}}$,

$$N = R + \oplus \{\operatorname{im} d_{\varphi^{*\prime}}^{2,q} : q = 0, 1, \ldots\}$$

(with R in bi-degree $(0, 0)$), $A = E_3(\varphi)$, $M = E_3^2(\varphi)$, $B = E_3(\varphi^*)$, $\pi(M) = E_3^2(\varphi^*)$. The fact that N generates I follows from the fact that $E_2(\varphi^{*\prime})$ is a differential $E_2^{\mathrm{I}}(\varphi^{*\prime})$-module, whence

$$d_{\varphi^{*\prime}}(E_2^{\mathrm{I}}(\varphi^{*\prime}) \otimes E_2^{\mathrm{II}}(\varphi^{*\prime})) = (E_2^{\mathrm{I}}(\varphi^{*\prime}) \otimes 1)\, d_{\varphi^{*\prime}}(1 \otimes e_2^{\mathrm{II}}(\varphi^{*\prime})).$$

Proposition 2.38. *Let R be a principal ideal domain and A a finitely generated free R-module. If $\varphi: A \to A$ is an injective elementary morphism, then* *the module morphism* $f\colon E_2(\varphi) \to E_3(\varphi)$ *of bi-degree* $(2, 1)$ *defined in Corollary* 2.29 *is surjective. The four equivalent statements of Lemma* 2.30 *then hold.*

Proof. Since φ is elementary, the hypotheses of Proposition 2.35 are satisfied. The assertion then follows from Lemma 2.37 by induction over the exact sequences in the conclusion of Proposition 2.35, where we proceed from the last to the first sequence.

Thus, Conjecture 2.19 is established for finitely generated A. We now collect the results obtained from the above consideration (beginning after Corollary 2.18) in the following theorem:

Theorem I. *Let R be a principal ideal domain and $\varphi\colon A \to A$ an injective elementary morphism of finitely generated free R-modules. Let $\theta\colon E_2(\varphi) \to E_2(\varphi)$ be the PA-module endomorphism defined by $\theta(1) = 0$,*

$$\theta(1 \otimes a_s) = (1/z_{s(1)})\, d_\varphi(1 \otimes a_s) \quad \textit{for} \quad s \in S(q), \quad q = 1, 2, \ldots$$

Then $\ker d_\varphi^{2p,q} = \operatorname{im} \theta^{2p-2,q+1}$; *in short,* $\ker d_\varphi = \operatorname{im} \theta$. *The R-algebra $E_3(\varphi)$ is generated by the R-sub-module* $M = E_3^{0,0}(\varphi) \oplus \bigoplus \{E_3^{2,q}(\varphi)\colon q = 0, 1, \ldots\}$. *There is a commutative diagram*

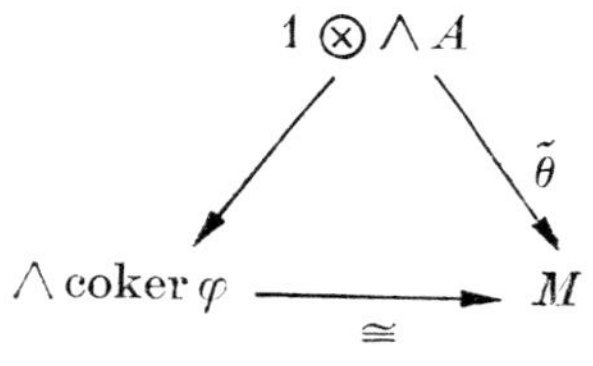

with $\tilde{\theta}(1) = 1$ *and*

$$\tilde{\theta}(1 \otimes a_s) = \psi^{2,q}\, \theta^{0,q+1}(1 \otimes a_s) \quad \textit{for} \quad s \in S(q+1), \quad q = 0, 1, \ldots,$$

and the horizontal map is a morphism of modules. The cup product maps $E_3^{\mathrm{I}}(\varphi) \otimes M$ *onto* $E_3(\varphi)$, *and* $E_3^{\mathrm{I}}(\varphi) \cong P \operatorname{coker} \varphi$. *In particular,* $E_3(\varphi)$ *is generated as a* (P *coker* φ)*-module by* M.

Definition 2.39. Let R be a principal ideal domain and A a finitely generated free R-module. Let $\varphi\colon A \to A$ be an injective elementary morphism, K the ring of quotients of R. By Proposition 2.26, we may assume that the differential bi-graded R-algebra $E_2(\varphi)$ is a subalgebra of the differential bi-graded R-algebra $E_2(\bar{\varphi})$, which is also a K-algebra. Let $\lambda\colon E_2(\bar{\varphi}) \to E_2(\bar{\varphi})$ be the morphism defined by $\lambda(a_\sigma \otimes a_s) = z_{s(1)} a_\sigma \otimes a_s$ for $\sigma \in \Sigma(p)$, $s \in S(q)$, $p, q = 1, 2, \ldots$, and $\lambda(a_\sigma \otimes 1) = a_\sigma \otimes 1$, $\sigma \in \Sigma(p)$, $p = 0, 1, \ldots$ There will be little confusion if we denote various restrictions and corestrictions of λ with the same letter. Let $E_2^*(\varphi) = \lambda^{-1}E_2(\varphi)$. We abbreviate $d_{\bar{\varphi}}$ with d, and $d|E_2^*(\varphi)$ with d^*; finally we set $E_3^*(\varphi) = \ker d^*/\operatorname{im} d^*$.

Lemma 2.40. *The differential bigraded module $(E_2^*(\varphi), d^*)$ is isomorphic to the differential bi-graded module $(E_2(\varphi), d')$, where $d' = \lambda\, d\lambda^{-1}|E_2(\varphi)$, and*

$$\begin{aligned} d'(a_\sigma \otimes a_s) &= z_{s(2)} s\, a_\sigma a_{s(1)} \otimes a(s_1) - z_{s(2)} a_\sigma a_{s(2)} \otimes a(s_2) \\ &\quad + z_{s(3)} a_\sigma a_{s(3)} \otimes a(s_3) + \cdots \\ &= (z_{s(2)} - z_{s(1)}) a_\sigma a_{s(1)} \otimes a(s_1) + d(a_\sigma \otimes a_s) \end{aligned}$$

for

$$s \in \Sigma(p), \quad s \in S(q), \quad p = 0, 1, \ldots, \quad q = 2, 3, \ldots,$$
$$d'(a_\sigma \otimes a_i) = a_\sigma a_i \otimes 1, \quad \textit{and} \quad d'(x \otimes 1) = 0.$$

The bi-graded module $E_3'(\varphi) = \ker d'/\operatorname{im} d'$ is isomorphic to $E_3^*(\varphi)$.

Proof. The isomorphism is $\lambda: E_2^*(\varphi) \to E_2(\varphi)$ and all computations are straightforward.

Lemma 2.41. *There is an isomorphism of exact sequences of bi-graded R-modules*

$$\begin{array}{ccccccccc} 0 & \longrightarrow & E_2(\varphi) & \xrightarrow{\subset} & E_2^*(\varphi) & \longrightarrow & PA \otimes \wedge^+ \operatorname{coker} \varphi & \longrightarrow & 0 \\ & & \| & & \downarrow{\scriptstyle\lambda} & & \downarrow{\scriptstyle\varkappa} & & \\ 0 & \longrightarrow & E_2(\varphi) & \xrightarrow{\lambda} & E_2(\varphi) & \longrightarrow & E_2(\varphi)/\lambda E_2(\varphi) & \longrightarrow & 0 \end{array}$$

where $\wedge^+ M$ denotes the bi-graded algebra $\{0\} + \wedge^1 M + \cdots$. Moreover, $\varkappa$ is a morphism of algebras.

Proof. We recall that coker φ is a direct sum of cyclic modules generated by elements b_i with $R \cdot b_i \cong R/z_i R$. Thus $\wedge^q$ coker φ is a direct sum of submodules generated by the elements $b_s = n_{s(1)} \wedge \cdots \wedge b_{s(q)}$, $s \in S(q)$, $q = 1$, $2, \ldots$ Because $z_{s(1)}|z_{s(2)}|\ldots$, we have $R \cdot b_s \cong R/z_{s(1)} R$. Clearly $E_2(\varphi)/\lambda E_2(\varphi)$ is a direct sum of cyclic submodules generated by $a_\sigma \otimes a_s + \lambda E_2(\varphi)$, $\sigma \in \Sigma(p)$, $s \in S(q)$, $p = 0, 1, \ldots, q = 1, 2, \ldots$, and the submodule generated by such an element is isomorphic to $R/z_{s(1)} R$. Thus there is an isomorphism of bi-graded modules

$$E_2(\varphi)/\lambda E_2(\varphi) \to PA \otimes \wedge^+ \operatorname{coker} \varphi$$

which maps $a_\sigma \otimes a_s + \lambda E_2(\varphi)$ onto $a_\sigma \otimes b_s$. Moreover, this isomorphism is an isomorphism of algebras.

Proposition 2.42. *If $\varphi: A \to A$ is an injective elementary morphism of finitely generated free R-modules over a principal ideal domain, then there is a commutative diagram*

$$\begin{array}{ccccccc}
E_2^{2p-2,q+1}(\varphi) & & & & & & \\
\downarrow \lambda & & & & & & \\
E_2^{2p-2,q+1}(\varphi) & \xrightarrow{d'^{2p-2,q+1}} & \operatorname{im} d'^{2p-2,q+1} \cong \ker d^{2p,q} & \overset{\subset}{\rightarrowtail} & E_2^{2p,q}(\varphi) & \overset{d_\varphi^{2p,q}}{\twoheadrightarrow} & \operatorname{im} d^{2p,q} \\
 & & \downarrow \cap & & \downarrow \lambda & & \downarrow \lambda \\
 & & \ker d'^{2p,q} & \overset{\subset}{\rightarrowtail} & E_2^{2p,q}(\varphi) & \twoheadrightarrow & \ker d^{2p+2,q-1} \cong \operatorname{im} d'^{2p,q} \\
 & & \downarrow & & \downarrow & & \downarrow \\
 & & E_3^{*2p,q}(\varphi) & \rightarrowtail & E_2^{2p,q}(\varphi)/\lambda E_2^{2p,q}(\varphi) \cong P^p A \otimes \Lambda^q \operatorname{coker} \varphi & \twoheadrightarrow & E_3^{2p+2,q-1}(\varphi)
\end{array}$$

in which the three by three diagram has exact rows and columns. There is an exact sequence of bi-graded modules

$$0 \longrightarrow \lambda E_2(\varphi) + \ker d' \overset{\subset}{\longrightarrow} E_2(\varphi) \overset{p}{\longrightarrow} E_3(\varphi) \longrightarrow 0$$

in which the map $f: E_2(\varphi) \to E_3(\varphi)$ has bi-degree $(2, -1)$. Note that

$$\ker d' = \ker d_\varphi \lambda^{-1}.$$

Proof. The commuting of the diagram and the exactness of the rows and columns is most easily deduced from the equalities $d\,(E_2^{*2p,q}) = \ker d^{2p+2,q-1}$ and the equivalent diagram

$$\begin{array}{ccccc}
\ker d = \operatorname{im} d^* & \overset{\subset}{\rightarrowtail} & E_2(\varphi) & \twoheadrightarrow & \operatorname{im} d \\
\downarrow \cap & & \downarrow \cap & & \downarrow \cap \\
\ker d^* & \rightarrowtail & E_2^*(\varphi) & \twoheadrightarrow & \operatorname{im} d^* = \ker d \\
\downarrow & & \downarrow & & \downarrow \\
E_3^*(\varphi) & \rightarrowtail & E_2^*(\varphi)/E_2(\varphi) & \longrightarrow & E_3(\varphi).
\end{array}$$

The exactness of the last row follows from diagram chasing. The map f is defined by $f = \psi\, d\lambda^{-1}$ and sends $a_\sigma \otimes a_s$ onto

$$\psi^{2p+2,q-1}\left((1/z_{s(1)})\,(a_\sigma \otimes 1)\, d^{2p,q}(1 \otimes a_s)\right)$$

for $\sigma \in \Sigma(p)$, $s \in S(q)$ (see Corollary 2.29). The kernel of the equivalent map $\psi\, d: E_2^*(\varphi) \to E_3(\varphi)$ is $E_2(q) + \ker d^*$. Application of λ yields the last assertion.

Remark. Since $E_3^{*0,q}(\varphi) = \{0\}$ for $q = 1, 2, \ldots$, we have retrieved the isomorphism $\wedge^{q+1} \operatorname{coker} \varphi \to E_3^{2,q}(\varphi)$ of Corollary 2.18.

It is a natural question to ask to what extent the elements $d^*(a_\sigma \otimes a_s)$ in the notation of Definition 2.39, which are crucial in the computation of $E_3(\varphi)$, are linearly dependent. This question has a fairly satisfactory answer. But first we elaborate somewhat on an earlier observation, namely that $E_2(\varphi)$ is a free $E_2^{\mathrm{I}}(\varphi)$-module, that $E_3(\varphi)$ is an $E_3^{\mathrm{I}}(\varphi)$-module, and finally that these module structures are compatible with the morphism $\psi: E_2 \to E_3$. The following observation is clear:

Proposition 2.43. *Let R be a principal ideal domain, $\varphi: A \to A$ an elementary morphism of finitely generated free R-modules. Let $\psi': PA \to P$ coker φ be the natural morphism (namely the one which makes the diagram*

$$\begin{array}{ccc} PA & \xrightarrow{\psi'} & P \operatorname{coker} \varphi \\ \downarrow & & \downarrow \\ E_2^{\mathrm{I}}(\varphi) & \xrightarrow{\psi^{\mathrm{I}}} & E_3^{\mathrm{I}}(\varphi) \end{array}$$

commutative). Then $E_2(\varphi)$ is a free differential graded (PA)-module generated by $E_2^{\mathrm{II}}(\varphi) \cong \wedge A$ and as such, may be considered as the ground ring extension of $\wedge A$ by PA. Moreover, $E_3(\varphi)$ is a graded (P coker φ)-module such that

$$\psi(p \cdot x) = \psi'(p) \cdot \psi(x) \quad \textit{for} \quad p \in PA, \quad x \in E_2(\varphi).$$

(The degree of an element $x \in E_2(\varphi)$ as a (PA)-module of bi-degree (m, n) is n, and the degree of the differential d_φ relative to the structure of $E_2(\varphi)$ as differential graded (PA)-module is -1. Consequently $E_3(\varphi)$ is the *homology* module of $E_2(\varphi)$ when looked at from this point of view.)

Proposition 2.44. *Under the conditions of Proposition 2.43, let $n_s \in PA$, $s \in S(q)$, $q \in \{1, 2, \ldots\}$. Then $\Sigma\{n_s \cdot d(1 \otimes a_s): s \in S(q)\} = 0$ if and only if there are elements $m(t) \in PA$, $t \in S(q+1)$ such that*

$$n_s = \Sigma\{m(s^i) \cdot a_i : i < s(1)\} + (1/z_{s(1)})\, \Sigma\{(-1)^{e(i,s)-1} z_i\, a_i\, m(s^i) : s(1) < i\}$$

with $s^i(e(i,s)) = i$.

Proof. The condition $\Sigma\{n_s \cdot d(1 \otimes a_s) \colon s \in S(q)\} = 0$ is equivalent to

$$\Sigma\{n_s \cdot (1 \otimes a_s) \colon s \in S(q)\} = \Sigma\, n_s \otimes a_s \in \ker d_\varphi .$$

This, by Proposition 2.38 and Lemma 2.30, is equivalent to the existence of elements $m(t) \in PA$, $t \in S(q+1)$ with

$$\Sigma\, n_s \otimes a_s = \Sigma\{(1/z_{t(1)})\, m(t) \cdot d(1 \otimes a_t) \colon t \in S(q+1)\}.$$

This means

$$n_s = \Sigma\{(1/z_{t(1)})\, m(t)\, (-1)^{i-1} z_{t(i)} a_{t(i)} \colon t_i = s\}.$$

We may extend this sum over all $j \notin \operatorname{im} s$ and put $t = s^j$, then $t(i) = j$ and $i = e(j, s)$. For $j < s(1)$ we have $e(j, s) = 0$, and $t(1) = j$. The assertion now follows.

Corollary 2.45. *If, in Proposition 2.44, the elements n_s in PA all have degree $2p$, then all $m(t)$ in PA have degree $2p - 2$.*

In particular this again confirms the independence of the $d(1 \otimes a_s)$ over R.

Lemma 2.46. *If $n \in PA$ and $n \cdot d(1 \otimes a_s) \in z_{s(1)} \operatorname{im} d$, then $n \in z_{s(1)} PA$.*

Proof. We have

$$\Sigma\{(-1)^{i-1} z_{s(i)}\, n\, a_{s(i)} \otimes a(s_i) \colon i = 1, \ldots, q\}$$
$$= z_{s(1)}\, \Sigma\{m_t \cdot d(1 \otimes a_t) \colon t \in S(q),\ q = 1, 2, \ldots\}.$$

Since $E_2(\varphi)$ is a free (PA)-module, we sum at most over those t for which $t_j = s_i$ for suitable j and i. However, if $t \neq s$, then there is always a $j \in \operatorname{im} t \setminus \operatorname{im} s$ such that t_j differs from all s_i. Hence $m_t = 0$ for those t. Thus we have

$$n \cdot d(1 \otimes a_s) = z_{s(1)} m \cdot d(1 \otimes a_s).$$

Comparison of the coefficient of $1 \otimes a(s_1)$ yields $z_{s(1)} n = z_{s(1)}^2 \cdot m$, hence $n = z_{s(1)} m$.

Corollary 2.47. *Under the conditions of Proposition 2.42, let*

$$\Psi \colon PA \otimes \wedge \operatorname{coker} \varphi \to E_3(\varphi)$$

be the (PA)-algebra morphism defined by $\Psi(1) = 0$ and

$$\Psi(1 \otimes b_s) = \psi^{2,q}(d\lambda^{-1}(1 \otimes a_s)).$$

Then $PA \cdot \Psi(1 \otimes b_s) \cong PA/z_{s(1)} PA$.

Proof. The relation $x \cdot \Psi(1 \otimes b_s) = 0$ is equivalent to

$$(1/z_{s(1)})\, x \cdot d(1 \otimes a_s) \in \operatorname{im} d .$$

By Lemma 2.46, this implies $x \in z_{s(1)} PA$. Conversely every such x annihilates $\Psi(1 \otimes b_s)$.

The preceding considerations explicitly excluded the case of a non-injective φ. This case is discussed in the following.

Proposition 2.48. *Let R be a principal ideal domain and $\varphi = \varphi_1 \oplus \varphi_2$ an elementary morphism such that $\varphi_2 = 0$. Then there is a natural isomorphism*

$$E_3(\varphi) \cong E_3(\varphi_1) \otimes E_2(\varphi_2) = E_3(\varphi_1) \otimes (PA_2 \otimes \wedge A_2)$$

with $A = \operatorname{dom} \varphi$, $A_i = \operatorname{dom} \varphi_i$, $i = 1, 2$.

Proof. We apply Proposition 2.21 and observe that

$$E_2(\varphi) \cong E_2(\varphi_1) \otimes E_2(\varphi_2).$$

But $E_2(\varphi_2) = PA_2 \otimes \wedge A_2$ is a free trivial complex, whence

$$E_3(\varphi) = H(E_2(\varphi)) \cong H(E_2(\varphi_1)) \otimes E_2(\varphi_2) = E_3(\varphi_1) \otimes E_2(\varphi_2).$$

Note that Proposition 2.47 is a generalisation of Proposition 2.32, and that it can be easily combined with Lemma 2.24 to yield the following result:

Proposition 2.49. *Let R be a principal ideal domain and φ an elementary morphism of a finitely generated free R-module. Suppose that $\varphi = \varphi_1 \oplus \varphi_2 \oplus \varphi_3$ with φ_1 being the identity of its domain A_1, φ_3 being zero on its domain, and that φ_2 and $1 - \varphi_2$ are injective. Then $E_3(\varphi) \cong E_3(\varphi_2) \otimes (PA_1 \otimes \wedge A_1)$, and φ_2 is minimal relative to this decomposition.*

Proof. The differential d_{φ_1} is exact by Lemma 2.24, so that Lemma 2.22 applies. The preceding proposition then proves the remainder.

Section 3

Some analogues of the results about spectral algebras with dual derivations

In this section we complement our results of Section 2 by considering on the bi-graded algebra $PB \otimes \wedge A$, for any morphism $\varphi\colon B \to A$, the unique derivation and differential ∂_φ which has $1 \otimes \wedge A$ in its kernel and satisfies $\partial_\varphi(b \otimes 1) = 1 \otimes \varphi(b)$. Many observations will parallel the corresponding ones in Section 2. In the latter part of the section, we will consider differential bi-graded Hopf algebras with two differentials. This part refers back to some of the results in Section 1.

We conclude with a development of a duality theory for certain spectral algebras.

Lemma 3.1. *Let R be a commutative ring with identity and $\varphi: B \to A$ a morphism of R-modules. There is a unique derivation and differential ∂_φ on $PB \otimes \wedge A$ of bi-degree $(-2, 1)$ satisfying the conditions*

(i) $\partial_\varphi(1 \otimes \wedge A) = 0$.

(ii) $\partial_\varphi(b \otimes 1) = 1 \otimes \varphi(b)$, *for* $b \in B$.

The explicit formula for the derivation ∂_φ is as follows:

(iii) $\partial_\varphi(b_1 \cdots b_p \otimes x) = \Sigma\{(b_1 \cdots b_p)_i \otimes \varphi(b_i) \wedge x \colon i = 1, \ldots, p\}$, *for* $b_i \in B$, $x \in \wedge A$, *where $(b_1 \cdots b_p)_i$ is the product of the b_j with the exception of the factor b_i.*

(iv) *If $PB \otimes \wedge A$ is considered as a $\wedge A$-right-module under $(u \otimes v) \cdot w = u \otimes (v \wedge w)$, then $PB \otimes \wedge A$ is a differential $\wedge A$-module relative to ∂_φ.*

Proof. The proofs of these facts are analogous to the proofs of Lemmas 2.2 and 2.3.

Definition 3.2. Under the assumptions of Lemma 3.1, we denote the differential bi-graded algebra $PB \otimes \wedge A$ with $E_2[\varphi]$. The bi-graded algebra $\ker \partial_\varphi / \operatorname{im} \partial_\varphi$ will be called $E_r[\varphi]$ with $r > 2$ and given the zero differential. Then the (two term) sequence $E_r[\varphi]$ is a spectral algebra. The total degree of the differential ∂_φ on $E_2[\varphi]$ is -1. If $B = A$, then $E_2\{\varphi\}$ is the bi-graded algebra $PA \otimes \wedge A$ together with the differentials d_φ and ∂_φ.

Lemma 3.3. *If $\mathfrak{M}$ denotes again (see Definition 2.4) the category of morphisms of the category of R-modules, then $\varphi \to E_r[\varphi]$ is a functor.*

Proof. Clear.

Lemma 3.4. *If A and B are two R-modules and $\varphi: A \to A$ and $\psi: B \to B$ are morphisms, then there is a canonical isomorphism*

$$\varepsilon_{\varphi,\psi} \colon E_2[\varphi \oplus \psi] \to E_2[\varphi] \otimes E_2[\psi]$$

and $\partial_{\varphi \oplus \psi} = (\partial_\varphi \otimes \partial_\psi)$, *where*

$$(\partial_\varphi \otimes \partial_\psi)(a^q \otimes a^p) = \partial_\varphi a^q \otimes a^p + (-1)^q a^q \otimes \partial_\psi a^p.$$

Proof. Analogous to the proof of Proposition 2.21.

Remark. In the terminology of Section 1, $E_2[\cdot]$ is an exponential functor.

The following lemma is straightforward:

Lemma. *If E is any differential algebra relative to two differentials d and ∂ relative to commuting involutions e, f (i. e.*

$$d(a\,b) = (da)\,b + e(a)\,db, \quad \partial(a\,b) = (\partial a)\,b + f(a)\,\partial b$$

with $de + e\,d = \partial e + e\,\partial = 0 = df + f\,d = \partial f + f\,\partial$) then $D = \partial d + d\partial$ satisfies $D(a\,b) = (Da)\,b + e\,f\,(a)\,(Db)$.

With the aid of this lemma it is easy to establish

Lemma 3.5. *Let $\varphi: A \to A$ be a morphism of free R-modules. We set*

$$D_\varphi = \partial_\varphi d_\varphi + d_\varphi \partial_\varphi .$$

Then for $x \in E_2^{2p,q}\{\varphi\}$, we have $D_\varphi x = (p+q)\,x$, provided that φ is the identity.

Proof. By straightforward calculation, it follows that

$$D(1 \otimes a_s) = \partial_\varphi d_\varphi (1 \otimes a_s) = q(1 \otimes a_s) \quad \text{for} \quad s \in S(q)$$

and

$$D(a_\sigma \otimes 1) = d_\varphi \partial_\varphi (a_\sigma \otimes 1) = p(a_\sigma \otimes 1) \quad \text{for} \quad \sigma \in \Sigma(p).$$

Then, with the preceeding lemma it follows that

$$D(a_\sigma \otimes a_s) = (p+q)\,(a_\sigma \otimes a_s).$$

Remark. The hypothesis of freeness is not really relevant; we used it only to fall back on our old notation involving the basis elements of $E_2(\varphi)$.

Proposition 3.6. *Let A be a free module over a commutative ring with identity. Let $\varphi: A \to A$ be an elementary morphism (see Definition 2.11). Then the edge algebra $E_3^{\mathrm{II}}[\varphi]$ is naturally isomorphic to $\wedge$ coker φ.*

Proof. The proof of this fact is analogous to the proof of the fact that the edge algebra $E_3^{\mathrm{I}}(\varphi)$ is naturally isomorphic to P coker φ (see Proposition 2.12 and the proof of Lemma 2.41).

Definition 3.6a. An *R-coalgebra A* is an R-module together with a comultiplication $\gamma: A \to A \otimes A$. It is then clear how a graded or bigraded coalgebra is to be defined. A coderivation d of a coalgebra is an endomorphism of A with $\gamma\,d = (d \otimes 1(A) \oplus e \otimes d)\,\gamma$, where e is a fixed involution of A with $de + e\,d = 0$. If A is a graded coalgebra, then it is called a *differential graded coalgebra* with differential d if d has a fixed degree r and e is defined by $e\,(x) = (-1)^p x$ for a homogeneous element x of degree p. An R-algebra is a *Hopf algebra* if it is both an algebra and a coalgebra. It is a *differential graded Hopf algebra* relative to a derivation d if it is both a differential graded algebra and a differential graded coalgebra relative to d.

Proposition 3.7. *Let $\varphi: A \to A$ be an elementary morphism of R-modules. Then $E_2\{\varphi\}$ is a differential bi-graded Hopf algebra relative to d_φ and ∂_φ.*

Proof. By the remarks following Definition 1.1, clearly $E_2\{\varphi\}$ is a Hopf algebra. The following diagram will be shown to commute:

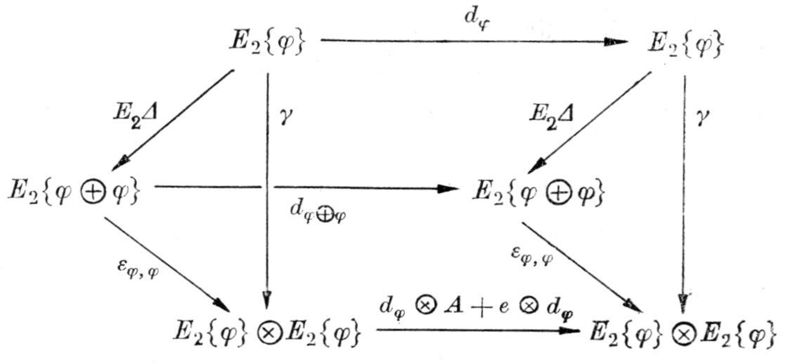

where $\Delta\colon \varphi \to \varphi \oplus \varphi$ is the diagonal morphism and

$$\varepsilon_{\varphi,\psi}\colon E_2\{\varphi \oplus \psi\} \to E_2\{\varphi\} \otimes E_2\{\psi\}$$

the natural isomorphism defining the exponentiality of E_2. Indeed, the commuting of the triangles is the definition of the comultiplication according to the remarks following Definition 1.1 of Section 1. The parallelogram with the sides $\varepsilon_{\varphi,\varphi}$ commutes by Proposition 2.21. The parallelogram with sides $E_2\Delta$ commutes by the naturality of d_φ. Thus the rectangle with sides γ commutes and this finally shows that d is a codifferential. The case of ∂_φ is treated similarly.

Lemma 3.8. *Let F be a finitely generated free S-module and $f\colon F \to F$ an elementary morphism. If R is a commutative ring with identity and an S-algebra, then $\varphi\colon A \to A$ with $A = \mathrm{Hom}_S(F, R)$, $\varphi = \mathrm{Hom}_S(f, R)$ is an elementary morphism of finitely generated free R-modules. The assignment $f \to E_2\{\varphi\}$ is a cofunctor from the category of elementary morphisms of finitely generated free abelian groups to the category of bi-graded differential Hopf algebras over R.*

Proof. Since $\mathrm{Hom}_S(-, R)$ is additive, we have

$$\mathrm{Hom}_S(S^n, R) \cong \mathrm{Hom}(S, R)^n;$$

but $\mathrm{Hom}(S, R) \cong R$ as an R-module. Thus A is a finitely generated free module. If $f = z_1 f_1 \oplus \ldots \oplus z_m f_m$ with $z_1 | z_2 | \ldots$, then

$$\varphi = z_1 \varphi_1 \oplus \ldots \oplus z_m \varphi_m \quad \text{with} \quad \varphi_i = \mathrm{Hom}_S(f_i, R).$$

In particular, if f is elementary, so is φ. The remaining statements are clear.

We are now in position to show that for a weakly principal ideal ring R and an elementary morphism f of finitely generated abelian groups, the edge term $E_3^{\mathrm{I}}(\mathrm{Hom}(f, R))$ is isomorphic to $P\,\mathrm{Ext}(G, R)$ and the edge term $E_3^{\mathrm{II}}(\mathrm{Hom}(f, R))$ is isomorphic to $\mathrm{Hom}(\wedge G, R)$. However, in order to make it transparent to what extent these isomorphisms are functorial in G, we make the following considerations first. The categories which we are about to introduce are important for our later discussion of finite abelian groups.

Lemma 3.9. *Let G be a finite abelian group. There exists a free resolution*

$$(f, \pi) = 0 \longrightarrow F \xrightarrow{f} F \xrightarrow{\pi} G \longrightarrow 0$$

with the following properties:

(i) *In the category of short exact sequences of abelian groups, (f, π) is the direct sum of exact sequences*

$$(f_i, \pi_i) = 0 \longrightarrow F_i \xrightarrow{f_i} F_i \xrightarrow{\pi_i} G_i \longrightarrow 0, \quad i = 1, \ldots, n,$$

such that f_i is multiplication with the natural number z_i, $F_i \cong \mathbf{Z}$, and G_i is cyclic of the order z_i.

(ii) *In the decomposition in* (i), *we have $z_i | z_{i+1}$, $i = 1, \ldots, n-1$.*

(iii) *The resolution with the properties* (i) *and* (ii) *is unique in the following sense: If*

$$(f', \pi') = 0 \to F' \to F' \to G \to 0$$

is another resolution of G with the same properties, then there is an isomorphism $(\alpha, \beta): f \to f'$ in the category of morphisms of abelian groups such that the diagram

$$\begin{array}{ccccccccc} 0 & \longrightarrow & F & \xrightarrow{f} & F & \longrightarrow & G & \longrightarrow & 0 \\ & & \downarrow{\scriptstyle\alpha} & & \downarrow{\scriptstyle\beta} & & \| & & \\ 0 & \longrightarrow & F' & \xrightarrow{f'} & F' & \longrightarrow & G & \longrightarrow & 0 \end{array}$$

commutes.

Proof. All this in one form or another is well known. See e. g. Bourbaki [6], p. 91.

Definition 3.10. The free resolution of G in Lemma 3.9 will be called a *standard resolution* of G.

Lemma 3.11. *Let $\gamma: G_1 \to G_2$ be a morphism of finite abelian groups and let*

$$0 \longrightarrow F_i \xrightarrow{f_i} F_i \xrightarrow{\pi_i} G_i \longrightarrow 0$$

be any standard resolutions of G_i, $i = 1, 2$. *Then there is a morphism* $(\alpha, \beta): f_1 \to f_2$ *in the category of morphisms of abelian groups such that the diagram*

$$\begin{array}{ccccccccc} 0 & \longrightarrow & F_1 & \xrightarrow{f_1} & F_1 & \xrightarrow{\pi_1} & G_1 & \longrightarrow & 0 \\ & & \downarrow{\scriptstyle\alpha} & & \downarrow{\scriptstyle\beta} & & \downarrow{\scriptstyle\gamma} & & \\ 0 & \longrightarrow & F_2 & \xrightarrow{f_2} & F_2 & \xrightarrow{\pi_2} & G_2 & \longrightarrow & 0 \end{array}$$

commutes. If $(\alpha', \beta'): f \to f'$ *is a second morphism with the same property, then* (α, β) *and* (α', β') *are homotopic, which is to say there is a morphism* $\xi: F_1 \to F_2$ *such that* $\alpha - \alpha' = \xi f_1$ *and* $\beta - \beta' = f_2 \xi$.

Proof. This is standard homological algebra. Observe that $\mathrm{Hom}(G, F_2) = 0$.

Lemma 3.12. *Let* A *be an* R*-module over a commutative ring* R *with identity. Then with each morphism* $f: F \to F$ *there is functorially associated an exact sequence of* R*-modules*

$$0 \longrightarrow \mathrm{Hom}(\mathrm{coker}\, f, A) \longrightarrow \mathrm{Hom}(F, A) \xrightarrow{\mathrm{Hom}(f, A)} \mathrm{Hom}(F, A)$$
$$\longrightarrow \mathrm{Ext}(\mathrm{coker}\, f, A) \longrightarrow 0.$$

Proof. Well known (see e. g. MacLane [32], p. 74).

Lemma 3.13. *Let* Z *be an integral domain (such as* $\mathbb{Z}$*) and let* $f: F \to F$ *be an elementary morphism of finitely generated free* Z*-modules* F. *Let*

$$F = \mathrm{Hom}(F, Z) \quad \text{and} \quad \hat{f} = \mathrm{Hom}(f, Z).$$

Let A *be any* Z*-module. Then we have for each positive natural number* p *the following isomorphisms which are natural isomorphisms of functors from the category of elementary morphisms to the category of* Z*-modules:*

$$\begin{aligned} &\mathrm{Hom}(\textstyle\bigwedge^p \mathrm{coker}\, f, A) \cong \ker \textstyle\bigwedge^p \mathrm{Hom}(f, A) \\ &\cong \ker \mathrm{Hom}(\textstyle\bigwedge^p f, A) \cong \ker(A \otimes \mathrm{Hom}(\textstyle\bigwedge^p f, Z)) \\ &\cong \ker(A \otimes \textstyle\bigwedge^p \hat{f}) \cong \mathrm{Tor}(A, \textstyle\bigwedge^p \mathrm{coker}\, \hat{f}). \end{aligned}$$

Proof. The sequence $0 \longrightarrow F \xrightarrow{f} F \xrightarrow{\pi} G \longrightarrow 0$ with $G = \mathrm{coker}\, f$ is exact. Since $\bigwedge$ is a left adjoint functor, it preserves colimits, hence epics, so $\bigwedge \pi$ is surjective; this could be checked directly without difficulty. If e_i is a basis element of F with $f(e_i) = z_i \cdot e_i$ so that $z_1 | z_2 | \ldots | z_n$, then $(\bigwedge f)(e_s) = z_{s(1)} \cdots z_{s(q)} \cdot e_s$ $s \in S(q)$. Since Z is an integral domain, this element is never zero. Hence $\bigwedge f$

is injective, so that

$$0 \longrightarrow \bigwedge^p F \xrightarrow{\bigwedge^p f} \bigwedge^p F \xrightarrow{\bigwedge^p \pi} \bigwedge^p G \longrightarrow 0 \tag{1}$$

is exact. Since F is finitely generated free we can produce a commuting diagram in which all vertical maps are isomorphisms:

$$\begin{array}{ccc} \bigwedge^p \operatorname{Hom}(F, A) & \xrightarrow{\bigwedge^p \operatorname{Hom}(f, A)} & \bigwedge^p \operatorname{Hom}(F, A) \\ \downarrow & & \downarrow \\ \operatorname{Hom}(\bigwedge^p F, A) & \xrightarrow{\operatorname{Hom}(\bigwedge^p f, A)} & \operatorname{Hom}(\bigwedge^p F, A) \\ \uparrow & & \uparrow \\ A \otimes \operatorname{Hom}(\bigwedge^p F, Z) & \xrightarrow{A \otimes \operatorname{Hom}(\bigwedge^p f, Z)} & A \otimes \operatorname{Hom}(\bigwedge^p F, A) \\ \downarrow & & \uparrow \\ A \otimes \bigwedge^p \hat{F} & \xrightarrow{A \otimes \bigwedge^p \hat{f}} & A \otimes \hat{F} \end{array} \tag{2}$$

It follows that

$$\ker \bigwedge^p \operatorname{Hom}(f, A) = \ker \operatorname{Hom}(\bigwedge^p f, A) \cong \ker (A \otimes \operatorname{Hom}(\bigwedge^p F, Z) \cong \ker (A \otimes \bigwedge^p \hat{f}).$$

By applying $\operatorname{Hom}(-, A)$ and $A \otimes -$ to the exact sequence (1), we obtain the exact sequences

$$0 \longrightarrow \operatorname{Hom}(\bigwedge^p G, A) \xrightarrow{\operatorname{Hom}(\bigwedge^p \pi, A)} \operatorname{Hom}(\bigwedge^p F, A) \xrightarrow{\operatorname{Hom}(\bigwedge^p f, A)} \operatorname{Hom}(\bigwedge^p F, A), \tag{3}$$

$$\operatorname{Tor}(A, \bigwedge^p F) \xrightarrow{\operatorname{Tor}(A, \bigwedge^p \pi)} \operatorname{Tor}(A, \bigwedge^p G) \longrightarrow A \otimes \bigwedge^p F \xrightarrow{A \otimes \bigwedge^p f} A \otimes \bigwedge^p F. \tag{4}$$

Since F is free, we have

$$\operatorname{Tor}(A, \bigwedge^p F) = 0. \tag{5}$$

From (3) we deduce readily that $\operatorname{Hom}(\bigwedge^p G, A) \cong \ker \operatorname{Hom}(\bigwedge^p f, A)$. We apply (4) and (5) to $\hat{F}$ and $\hat{f}$ in place of F and f and derive

$$\operatorname{Tor}(A, \bigwedge^p G') = \ker A \otimes \bigwedge^p \hat{f},$$

where $G' = \operatorname{coker} \hat{f}$.

Recall that in a category $\mathfrak{A}$, a *coretraction* is a morphism which has a left inverse and a *retraction* is a morphism that has a right inverse. Now we formulate the proposition which has previously announced:

Proposition 3.14. *Let R be a commutative ring with identity, and let $f: F \to F$ be an elementary morphism of finitely generated free abelian groups. Then:*

(a) *There is a natural coretraction of R-algebras*

$$P\,\mathrm{Ext}(\mathrm{coker}\, f, R) \to E_3(\mathrm{Hom}(f, R))$$

whose image is $E_3^{\mathrm{I}}(\mathrm{Hom}(f, R))$.

(b) *There is a natural coretraction of R-algebras*

$$\mathrm{Hom}(\wedge\, \mathrm{coker}\, f, R) \to E_3(\mathrm{Hom}(f, R))$$

whose image is $E_3^{\mathrm{II}}(\mathrm{Hom}(f, R))$. *Also*

$$\mathrm{Hom}(\wedge\, \mathrm{coker}\, f, R) \cong R \oplus \ker \wedge \mathrm{Hom}(f, R) \cong \ker \wedge^p \mathrm{Hom}(f, R).$$

(b′) *If A is an abelian group then there is a coretraction of* $\mathrm{Hom}(\wedge\, \mathrm{coker}\, f, A)$ *into the cohomology group of $A \otimes E_2(\mathrm{Hom}(f, \mathsf{Z}))$ whose image is the vertical edge term in the obvious bi-gradation of the cohomology group. Moreover*

$$\begin{aligned}\mathrm{Hom}(\wedge\, \mathrm{coker}\, f, A) &\cong \ker(A \otimes \wedge^p \mathrm{Hom}(f, \mathsf{Z}))\\ &\cong \mathrm{Tor}(A, \wedge^p\, \mathrm{coker}\, \mathrm{Hom}(f, \mathsf{Z})).\end{aligned}$$

(c) *If*

$$\begin{array}{ccccccccc} 0 & \longrightarrow & F & \xrightarrow{f} & F & \xrightarrow{\pi} & G & \longrightarrow & 0\\ & & \downarrow{\scriptstyle\alpha} & & \downarrow{\scriptstyle\beta} & & \downarrow{\scriptstyle\gamma} & & \\ 0 & \longrightarrow & F' & \xrightarrow{f'} & F' & \xrightarrow{\pi'} & G' & \longrightarrow & 0 \end{array}$$

is a commutative diagram, then there are commutative diagrams involving the injections of (a) *and* (b):

$$\begin{array}{ccc} P\,\mathrm{Ext}(G, R) & \longrightarrow & E_3(\mathrm{Hom}(f, R))\\ \downarrow{\scriptstyle P\,\mathrm{Ext}(\gamma, R)} & & \downarrow{\scriptstyle E_3(\mathrm{Hom}(\beta, R),\, \mathrm{Hom}(\alpha, R))}\\ P\,\mathrm{Ext}(G', R) & \longrightarrow & E_3(\mathrm{Hom}(f', R)), \end{array}$$

$$\begin{array}{ccc} \mathrm{Hom}(\wedge G, R) & \longrightarrow & E_3(\mathrm{Hom}(f, R))\\ \downarrow{\scriptstyle \mathrm{Hom}(\wedge\gamma, R)} & & \downarrow{\scriptstyle E_3(\mathrm{Hom}(\beta, R),\, \mathrm{Hom}(\alpha, R))}\\ \mathrm{Hom}(\wedge G', R) & \longrightarrow & E_3(\mathrm{Hom}(f', R)) \end{array}$$

(d) *There is a module, but not in general a ring isomorphism*

$$\operatorname{Hom}(\bigwedge G, R) \to \bigwedge \operatorname{Hom}(G, R).$$

Proof. By Lemma 3.12, we have $\operatorname{coker} \operatorname{Hom}(f, R) \cong \operatorname{Ext}(\operatorname{coker} f, R)$ and $\ker \operatorname{Hom}(f, R) \cong \operatorname{Hom}(\operatorname{coker} f, R)$. Then (a) follows from Proposition 2.12. (b) follows from 2.12 and Lemma 3.13. For (d) we apply Proposition 2.12 (d). (c) just expresses the functoriality of our construction in f.

We abbreviate $\hat{f} = \operatorname{Hom}(f, \mathsf{Z})$ and $\hat{F} = \operatorname{Hom}(F, \mathsf{Z})$. There is a commutative diagram, in which all vertical maps are natural isomorphisms:

$$\begin{array}{ccc}
A \otimes E_2^{0,p}(f) = A \otimes \mathsf{Z} \otimes \bigwedge^p \hat{F} & \xrightarrow{A \otimes d_{\hat{f}}^{0,p}} & E_2^{2,p-1}(\hat{f}) = A \otimes F \otimes \bigwedge^{p-1} \hat{F} \\
\downarrow & & \downarrow \\
A \otimes \mathsf{Z} \otimes \operatorname{Hom}(\bigwedge^p F, \mathsf{Z}) & \longrightarrow & A \otimes \hat{F} \otimes \operatorname{Hom}(\bigwedge^{p-1} F, \mathsf{Z}) \\
\downarrow & & \downarrow \\
\operatorname{Hom}(\mathsf{Z} \otimes \bigwedge^p F, A) & \xrightarrow{\operatorname{Hom}(\partial, A)} & \operatorname{Hom}(F \otimes \bigwedge^{p-1} F, A)
\end{array}$$

where ∂ is determined by $\partial(1 \otimes \bigwedge^{p-1} F) = \{0\}$ and

$$\partial(e_i \otimes a) = z_i\,(1 \otimes (e_i \wedge a))$$

for the canonical generators of e_i of F. The image of ∂, when identified with a subgroup of $\bigwedge^p F$ via the isomorphisms $\mathsf{Z} \otimes \bigwedge^p F \to \bigwedge^p F$, is spanned by the elements $z_i e_i \wedge a$ with degree $a = p - 1$; therefore it is equal to $\operatorname{im} \bigwedge^p f$. By the exactness of the sequence (1) in the proof of Lemma 3.13, this equals $\ker \bigwedge^p \pi$ where $F \xrightarrow{\pi} \operatorname{coker} f$ is the quotient map.

Thus

$$\begin{array}{ccccc}
F \otimes \bigwedge^{p-1} F & \longrightarrow & \mathsf{Z} \otimes \bigwedge^p F & & \\
 & & \downarrow \cong & & \\
 & & \bigwedge^p F & \xrightarrow{\bigwedge^p \pi} & \bigwedge^p \operatorname{coker} f \longrightarrow 0
\end{array}$$

is exact. Applying $\operatorname{Hom}(-, A)$ to this sequence yields an exact sequence

$$0 \longrightarrow \operatorname{Hom}(\bigwedge^p \operatorname{coker} f, A) \longrightarrow \operatorname{Hom}(\mathsf{Z} \otimes \bigwedge^p F, A)$$

$$\xrightarrow{\operatorname{Hom}(\partial, A)} \operatorname{Hom}(F \otimes \bigwedge^{p-1} F, A).$$

In view of the commutative diagram at the beginning of the proof this shows that $\operatorname{Hom}(\wedge^p \operatorname{coker} f, A) = \ker A \otimes d_f^{0,p}$. Together with Lemma 3.13, this finishes the proof of (b′).

Corollary 3.15. *Under the hypotheses of Proposition* 3.14 *there is a morphism of bi-graded algebras*

$$P_R \operatorname{Ext}(\operatorname{coker} f, R) \otimes_R \operatorname{Hom}(\wedge \operatorname{coker} f, R) \to E_3(\operatorname{Hom}(f, R)),$$

which is bijective on the edge terms.

Proof. See Proposition 2.12 (e).

Corollary 3.16. *Let R be a commutative ring with identity and let G be a finite group. If (f, π) and (f', π') are two standard resolutions of G, then there is a commutative diagram*

$$\begin{array}{ccc} & & E_3(\operatorname{Hom}(f, R)) \\ & \nearrow & \Big\downarrow {\scriptstyle E_3(\operatorname{Hom}(\beta, R), \operatorname{Hom}(\alpha, R))} \\ P_R \operatorname{Ext}(G, R) \otimes_R \operatorname{Hom}(\wedge G, R) & & \\ & \searrow & \\ & & E_3(\operatorname{Hom}(f, R)) \end{array}$$

with the morphisms introduced in Corollary 3.15 *and suitable α, β as in Lemma* 3.9 (iii).

It is in this sense that we can say that the inclusion maps of Proposition 3.14 and the morphism of Corollary 3.15 are functorial in G. At a later stage we will see that $E_3(\operatorname{Hom}(f, R))$ as a graded algebra with its total degree also depends only on G, so that the morphism of Corollary 3.15 is in fact a morphism of graded algebras which is in fact natural on the category of finite abelian groups. Whether or not this prevails as a statement about morphisms of bi-graded algebras we do not know.

We conclude this section with a series of Propositions concerning R/Z as coefficient module.

Proposition 3.17. *Let $f\colon F \to F$ be a morphism of finitely generated free abelian groups. There is then a natural morphism of bi-graded abelian groups of bi-degree* $(2, -1)$

$$H(\mathsf{R}/\mathsf{Z} \otimes E_2(f)) \to E_3(f),$$

which is in fact an isomorphism off the origin.

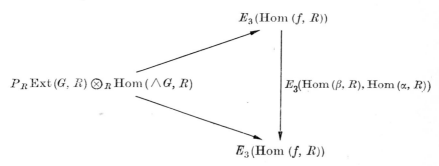

Proof. Since $E_2(f)$ is free, there is an exact sequence of complexes

$$0 \to E_2(f) \to \mathsf{R} \otimes E_2(f) \to \mathsf{R}/\mathsf{Z} \otimes E_2(f) \to 0.$$

From the long exact cohomology sequence, there is then a morphism k of bi-degree $(2, -1)$ such that the following triangle is exact:

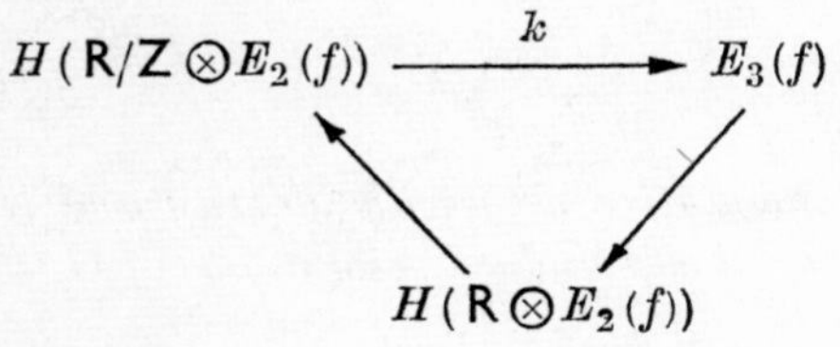

But we have

$$\mathsf{R} \otimes E_2(f) = \mathsf{R} \otimes PF \otimes \wedge F = P_{\mathsf{R}}\,(\mathsf{R} \otimes F) \otimes_{\mathsf{R}} \wedge_{\mathsf{R}}(\mathsf{R} \otimes F)$$
$$= E_2(\mathsf{R} \otimes f).$$

Thus

$$H(\mathsf{R} \otimes E_2(f)) = E_3(\mathsf{R} \otimes f).$$

Since R is a field, $d_{\mathsf{R}\otimes f}$ is exact by Proposition 2.24. Hence $E_3^{pq}(\mathsf{R} \otimes f) = 0$ if $p + q > 0$. This proves the assertion.

Remark. The proposition maintains with the rational field Q in place of R.

Proposition 3.18. *Under the conditions of Proposition* 3.17, *there is a natural isomorphism of bi-graded abelian groups*

$$H(\mathrm{Hom}\,(E_2(f), \mathsf{R}/\mathsf{Z})) \to \mathrm{Hom}\,(E_3(f), \mathsf{R}/\mathsf{Z}).$$

Hence, if $\hat{} = \mathrm{Hom}(-, \mathsf{R}/\mathsf{Z})$, *then there is an isomorphism of abelian bi-graded groups*

$$E_3(f) \cong H(E_2(f)\hat{}\,)\hat{}\,.$$

There is a natural morphism of bi-graded abelian groups of bi-degree $(2, -1)$, *which is an isomorphism off the origin:*

$$H(\mathsf{R}/\mathsf{Z} \otimes E_2(f)) \to H(E_2(f)\hat{}\,)\hat{}\,.$$

Proof. The second isomorphism follows from the first by duality, and the third follows from the second via Proposition 3.17. In order to prove the first isomorphism, we use the universal coefficient theorem connecting homology and cohomology (see Cartan-Eilenberg [11], p. 114, or MacLane [32],

p. 77); we obtain the exact sequence

$$0 \to \mathrm{Ext}\,(H\,(E_2\,(f)),\ \mathsf{R}/\mathsf{Z}) \to H\,(\mathrm{Hom}\,(E_2\,(f),\ \mathsf{R}/\mathsf{Z}))$$
$$\to \mathrm{Hom}\,(H\,(E_2\,(f)),\ \mathsf{R}/\mathsf{Z}) \to 0\,.$$

By definition, $H\,(E_2(f)) = E_3(f)$, and $\mathrm{Ext}(-, D) = 0$ if D is divisible.

Section 4

The Bockstein formalism

In this section we discuss some homological algebra which we will use in changing coefficient rings.

Lemma 4.1. *Suppose that A^*, B^*, C^*, R_1^*, R_2^* are graded R-modules over a commutative ring R, and $d\colon B^* \to A^*$ and $D\colon B^* \to R_1^*$ are morphisms of degree 1 such that in the following diagram, triangles which can be exact on the basis of their configuration are exact, the other triangles are commutative, and the square is commutative:*

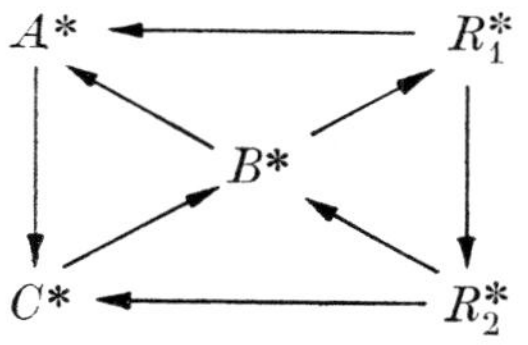

Then

(a) $\ker(C^* \leftarrow A^*) = \mathrm{im}\, d \subset \mathrm{im}(A^* \leftarrow R_1^*)$.

(b) $\ker D = \mathrm{im}(B^* \leftarrow R_2^*) \subset \mathrm{im}(B^* \leftarrow C^*) = \ker d$.

(c) $\ker d/\ker D \cong \mathrm{im}(R_1^* \leftarrow C^*)$
$\subset \ker(R_2^* \leftarrow R_1^*) \cap \ker(A^* \leftarrow R_1^*) = \mathrm{im}\, D \cap \ker(A^* \leftarrow R_1^*)$,
where $R_1^* \leftarrow C^* = D(B^* \leftarrow C^*)$.

(d) $\mathrm{im}\,(C^* \leftarrow A^*)\,(A^* \leftarrow R_1^*)$
$\cong \mathrm{im}\,(A^* \leftarrow R_1^*)/[\mathrm{im}(A^* \leftarrow R_1^*) \cap \ker(C^* \leftarrow A^*)]$
$= \mathrm{im}\,(A^* \leftarrow R_1^*)/\mathrm{im}\, d$
$= \mathrm{im}\,(C^* \leftarrow R_2^*)\,(R_2^* \leftarrow R_1^*)$
$\cong \mathrm{im}\,(R_2^* \leftarrow R_1^*)/[\ker(B^* \leftarrow R_2^*) \cap \ker(C^* \leftarrow R_2^*)]$.

Proof by diagram chasing.

Lemma 4.2. *If, in Lemma 4.1, we have* $R_1^* = R_2^*$, $A^* = B^*$, *and*

$$A^* \leftarrow R_1^* = B^* \leftarrow R_2^*,$$

then

(a) $d^2 = 0$ (*i. e. d is a differential on A*).

(b) $H(d) = \ker d/\operatorname{im} d \cong \operatorname{im}(R^* \leftarrow C^*)$
$\subset \operatorname{im} D \cap \ker(B^* \leftarrow R_2^*) = \operatorname{im} D \cap \operatorname{im}(R_2^* \leftarrow R_1^*)$.

In particular, d is exact, if $R_2^* \leftarrow R_1^* = 0$ *in positive dimensions.*

Proof from Lemma 4.1 and by diagram chasing.

Lemma 4.3. *Let* A^*, C^*, *and* R_i^*, $i = 1, 2, 3$ *be graded R-modules, and suppose* D_A *and* D_C *are morphisms of degree 1 such that in the following diagram, triangles are exact whenever they can be on the basis of their configuration, while the other triangles commute and the square commutes:*

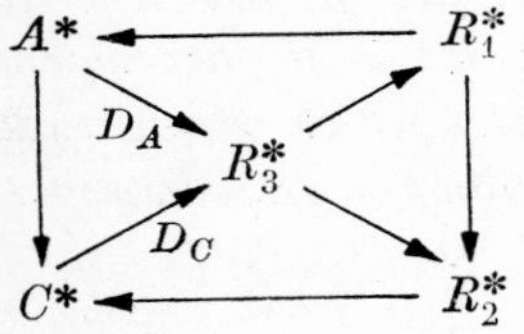

Then

(a) $\ker(R_1^* \leftarrow R_3^*) = \operatorname{im} D_A \subset \operatorname{im} D_C = \ker(R_2^* \leftarrow R_3^*)$.

(b) $\operatorname{im} D_C/\operatorname{im} D_A \cong \operatorname{im} D_C/\ker(R_1^* \leftarrow R_3^*) \cong \operatorname{im}(R_1^* \leftarrow C^*)$
$\subset \ker(A^* \leftarrow R_1^*) \cap \ker(R_1^* \leftarrow R_2^*) = \operatorname{im}(R_1^* \leftarrow R_3^*) \cap \ker(R_2^* \leftarrow R_1^*)$.

If, in addition, $R_1^* = R_2^* = R_3^*$ *and* $R_2^* \leftarrow R_1^* = R_1^* \leftarrow R_3^*$, *then*

$$(R_2^* \leftarrow R_1^*)^2 D_A = 0.$$

Proof by diagram chasing.

Lemma 4.4. *Suppose that the commutative diagram at the top of p. 87 is a diagram of complexes over a commutative ring R with exact rows and columns.*

Then the two rows give rise to a diagram of cohomology modules as described in Lemma 4.1 (the star denoting cohomology) and the two columns to a cohomology diagram as described in Lemma 4.3. If $A = B$, $R_1 = R_2$ *and* $\pi_A = \pi_B$ *we have the situation of Lemma 4.2.*

Proof. The fundamental theorem about the cohomology exact sequence of an exact sequence of complexes.

$$\begin{array}{ccccccccc}
 & & 0 & & 0 & & & & \\
 & & \downarrow & & \downarrow & & & & \\
 & & R_3 & = & R_3 & & & & \\
 & & \downarrow{\scriptstyle\alpha} & & \downarrow{\scriptstyle\gamma} & & & & \\
0 & \longrightarrow & R_1 & \xrightarrow{\beta} & R_2 & \xrightarrow{\pi_B} & B & \longrightarrow & 0 \\
 & & \downarrow{\scriptstyle\pi_A} & & \downarrow{\scriptstyle\pi_C} & & \| & & \\
0 & \longrightarrow & A & \xrightarrow{\varphi} & C & \xrightarrow{\psi} & B & \longrightarrow & 0 \\
 & & \downarrow & & \downarrow & & & & \\
 & & 0 & & 0 & & & &
\end{array}$$

Definition 4.5. A diagram of complexes of the form

$$\begin{array}{ccc}
R_1 & = & R_1 \\
\downarrow{\scriptstyle\alpha} & & \downarrow{\scriptstyle\alpha^2} \\
R_1 & \overset{\alpha}{\rightarrowtail} & R_1
\end{array}$$

will be called a *pre-Bockstein diagram*. It can be completed to a diagram as given above with $\beta = \alpha$, $\gamma = \alpha^2$, $R_1 = R_2 = R_3$, $\pi_A = \pi_B$. The diagram

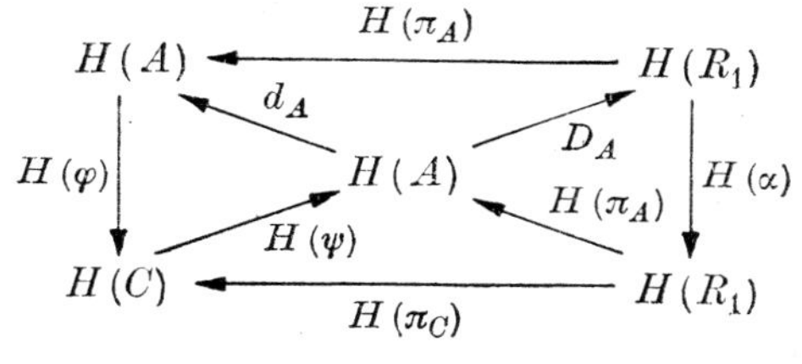

which arises on applying Lemma 4 to the two rows will be called the *standard Bockstein diagram.*

We recall from the earlier lemmas

Lemma 4.6. *In the standard Bockstein diagram, d_A is a differential with*

$$H(d_A) \cong \operatorname{im} D_A\, H(\psi) \subset \ker H(\alpha) \cap \operatorname{im} H(\alpha).$$

If $H^i(\alpha) = 0$ for $i > 0$, then d_A is exact and $H^i(\pi_A)$ is injective for $i > 0$.

Notation. We will write H^+ for $H^1 \oplus H^2 \oplus H^3 \oplus \ldots$

The following lemma is clear from the functorial nature of the assignment of the standard Bockstein diagram to the pre-Bockstein diagram:

Lemma 4.7. *A morphism of pre-Bockstein diagrams gives rise to a morphism of Bockstein diagrams.*

Lemma 4.8. *The standard Bockstein diagram defines a commutative diagram of exact sequences*

$$
\begin{array}{cccccccc}
 & & 0 & & & & 0 & \\
 & & \downarrow & & & & \downarrow & \\
 & & \operatorname{im} H(\alpha) & & & & \operatorname{im} H(\alpha) \cap \ker(H(\alpha)) & \\
 & & \downarrow & & & & \downarrow & \\
0 \longrightarrow \operatorname{im} H(\alpha) & \xrightarrow{\text{incl}} & H(R_1) & \xrightarrow{H(\pi_A)} & H(A) & \xrightarrow{\operatorname{coim} D_A} & \operatorname{im} D_A & \longrightarrow 0 \\
 & & \downarrow{\scriptstyle H(\pi_A)'} & & \| & & \downarrow{\scriptstyle H(\pi_A)''} & \\
0 & \longrightarrow & \ker d_A & \xrightarrow{\text{incl}} & H(A) & \xrightarrow{\operatorname{coim} d_A} & \operatorname{im} d_A & \longrightarrow 0 \\
 & & \downarrow & & & & \downarrow & \\
 & & \operatorname{im} D_A H(\psi) & & & & 0 & \\
 & & \downarrow & & & & & \\
 & & 0 & & & & &
\end{array}
$$

where $H(\pi_A)'$ and $H(\pi_A)''$ are corestrictions and restrictions of $H(\pi_A)$. Moreover, every morphism of pre-Bockstein diagrams gives rise to a morphism of configurations of the type above.

The proof follows straightforwardly from the previous results. We thus have the following immediate consequence.

Proposition 4.9. *If, in the standard Bockstein diagram, we have $H^+(\alpha) = 0$, then there is an isomorphism of exact sequences*

$$
\begin{array}{ccccccc}
0 \longrightarrow H(R_1) & \xrightarrow{H(\pi_A)} & H(A) & \xrightarrow{\operatorname{coim} D_A} & H^+(R_1) & \longrightarrow 0 \\
\downarrow{\scriptstyle H(\pi_A)'} & & \| & & \downarrow{\scriptstyle H(\pi_A)''} & \\
0 \longrightarrow \ker d_A & \longrightarrow & H(A) & \xrightarrow{\operatorname{coim} d_A} & \operatorname{im} d_A & \longrightarrow 0
\end{array}
$$

Lemma 4.10. *Every endomorphism $\alpha\colon R_1 \to R_1$ defines uniquely a pre-Bockstein diagram*

$$P(R_1, \alpha) = \begin{array}{ccc} R_1 & = & R_1 \\ \downarrow{\scriptstyle \alpha} & & \downarrow{\scriptstyle \alpha^2} \\ R_1 & \xrightarrow{\alpha} & R_1 \end{array}$$

and each triple of morphisms $\lambda, \mu, \nu \colon R_1 \to R_1'$ *such that*

$$\begin{array}{ccc} R_1 & \xrightarrow{\lambda} & R_1' \\ \downarrow{\scriptstyle \alpha} & & \downarrow{\scriptstyle \alpha'} \\ R_1 & \xrightarrow{\mu} & R_1' \\ \downarrow{\scriptstyle \alpha} & & \downarrow{\scriptstyle \alpha'} \\ R_1 & \xrightarrow{\nu} & R_1' \end{array}$$

commutes, defines a unique morphism of pre-Bockstein diagrams

$$(\lambda, \mu, \nu) \colon P(R_1, \alpha) \to P(R_1', \alpha').$$

Consequently each such triple defines a morphism

$$S(\lambda, \mu, \nu) \colon S(R_1, \alpha) \to S(R_1', \alpha')$$

of standard Bockstein diagrams in cohomology. More specifically such a morphism is as follows:

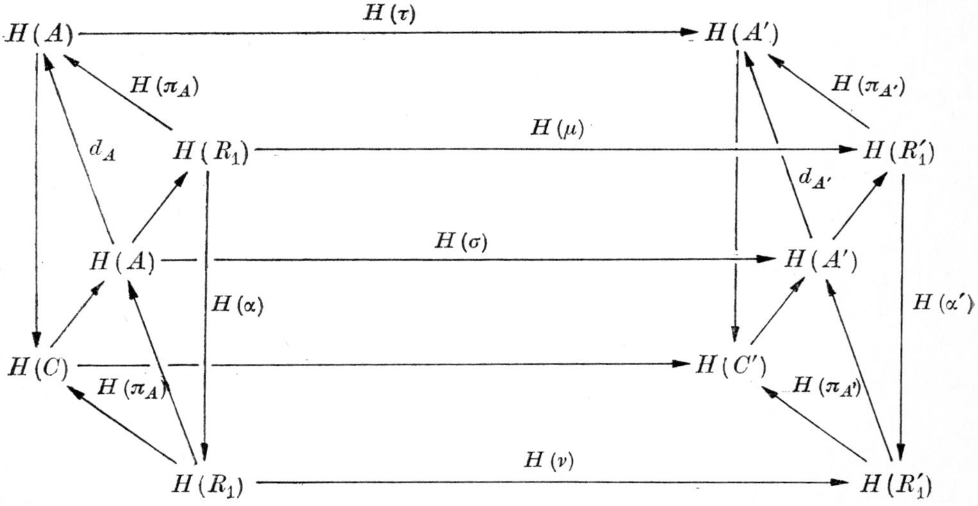

In particular, there is a commutative diagram

$$\begin{array}{ccc}
H(A) & \xrightarrow{d_A} & H(A) \\
\Big\downarrow{\scriptstyle H(\sigma)} & & \Big\downarrow{\scriptstyle H(\tau)} \\
H(A') & \xrightarrow{d_{A'}} & H(A')
\end{array}$$

where the morphisms σ *and* τ *are defined by*

$$\begin{array}{ccccccccc}
0 & \longrightarrow & R_1 & \xrightarrow{\alpha} & R_1 & \longrightarrow & A & \longrightarrow & 0 \\
 & & \Big\downarrow{\scriptstyle \mu} & & \Big\downarrow{\scriptstyle \nu} & & \Big\downarrow{\scriptstyle \sigma} & & \\
0 & \longrightarrow & R_1' & \xrightarrow{\alpha'} & R_1' & \longrightarrow & A' & \longrightarrow & 0
\end{array}$$

and

$$\begin{array}{ccccccccc}
0 & \longrightarrow & R_1 & \longrightarrow & R_1 & \longrightarrow & A & \longrightarrow & 0 \\
 & & \Big\downarrow{\scriptstyle \lambda} & & \Big\downarrow{\scriptstyle \mu} & & \Big\downarrow{\scriptstyle \tau} & & \\
0 & \longrightarrow & R_1' & \longrightarrow & R_1' & \longrightarrow & A' & \longrightarrow & 0.
\end{array}$$

Proposition 4.11. *Suppose that in the diagram of complexes*

$$\begin{array}{ccccccccc}
 & & 0 & & 0 & & & & \\
 & & \Big\downarrow & & \Big\downarrow & & & & \\
 & & R_1 & = & R_1 & & & & \\
 & & \Big\downarrow{\scriptstyle \alpha} & & \Big\downarrow & & & & \\
0 & \longrightarrow & R_1 & \xrightarrow{\alpha} & R_1 & \xrightarrow{\pi} & A & \longrightarrow & 0 \\
 & & \Big\downarrow{\scriptstyle \pi} & & \Big\downarrow{\scriptstyle \pi'} & & \Big\| & & \\
0 & \longrightarrow & A & \xrightarrow{\varphi} & C & \xrightarrow{\psi} & A & \longrightarrow & 0 \\
 & & \Big\downarrow & & \Big\downarrow & & & & \\
 & & 0 & & 0 & & & &
\end{array}$$

all rows and columns are exact as morphisms of groups (or modules), all complexes are differential graded rings and π and π' are morphisms of rings. Then condition (a) *implies condition* (b):

(a) $\alpha(x\,y) = (\alpha\,x)\,y = x(\alpha\,y)$ *for all* $x, y \in R_1$.

(b) $\varphi\,(x(\psi\,y)) = (\varphi\,x)\,y$ *and* $\varphi\,((\psi\,y)\,x) = y(\varphi\,x)$ *for all* $x \in A$, $y \in C$.

Proof. To the first equation in (a) we apply π', and with $\varphi\,\pi = \pi'\,\alpha$, we obtain $\varphi\,\pi\,x\,y = (\varphi\,\pi\,x)\,\pi'\,y$. Since π is a morphism of rings we have

$$\varphi\,\pi\,x\,y = \varphi(\pi\,x)\,(\pi\,y).$$

Now let $\pi\,x = u$ and $\pi'\,y = v$. Then $\psi\,v = \pi\,y$ and we have

$$\varphi\,(u(\psi\,v)) = (\varphi\,u)\,v.$$

Since π and π' are surjective, we have proved the first equality of (b) for all $u \in A$ and $v \in C$. The remainder of the proof is analogous.

Proposition 4.12. *Suppose that* $0 \longrightarrow A \xrightarrow{\varphi} C \xrightarrow{\psi} A \longrightarrow 0$ *is a sequence of differential graded R-algebras in which φ and ψ are morphisms of graded R-modules such that*

(a) $\varphi\,(a(\psi\,c)) = (\varphi\,a)\,c$ *and* $\varphi\,((\psi\,c)\,a) = c(\varphi\,a)$ *for all* $a \in A$ *and* $c \in C$.

Then the connecting morphism $d\colon HA \to HA$ *is a derivation.*

Proof. Let $B = \psi^{-1}Z(A)$ and take x, y homogeneous elements in B of degree i and j respectively. Then, as is well known and easily checked using (a), the elements $\delta(x\,y)$, $(\delta x)\,y$, $x(\delta y)$ are in fact in $\varphi(A)$ and satisfy the equation $\delta(x\,y) = (\delta x)\;y + (-1)^i\,x(dy)$. We apply φ^{-1} to this equation, observe condition (a) and obtain

$$\begin{aligned}\varphi^{-1}\,\delta(x\,y) &= \varphi^{-1}\,((\delta x)\,y) + (-1)^i\,\varphi^{-1}\,(x(\delta y))\\ &= \varphi^{-1}(\delta x)\,(\psi\,y) + (-1)^i\,(\psi\,x)\,\varphi^{-1}(\delta y).\end{aligned}$$

Now let $h\colon ZA \to HA$ be the quotient map and let $a = h\,\psi\,x$, $b = h\,\psi\,y$, and $a\,b = h\,\psi(x\,y)$ be the corresponding cohomology classes. By the definition of d we have $da = h\,\varphi^{-1}(\delta x)$ etc. Hence, by applying h to the equation which we derived previously, we get

$$d(a\,b) = (da)\,b + (-1)^i\,a(db).$$

This proves the assertion.

Corollary 4.13. *Let R be a commutative ring with identity and let $a \in R$ be an element such that $a\,x = 0$ implies $x = 0$. Let R_1 be a differential graded algebra over R and define $\alpha\colon R_1 \to R_1$ by $\alpha(x) = a \cdot x$. Then α defines a diagram*

$$\begin{array}{ccccccccc}
 & & 0 & & 0 & & & & \\
 & & \downarrow & & \downarrow & & & & \\
 & & R_1 & = & R_1 & & & & \\
 & & \downarrow{\scriptstyle\alpha} & & \downarrow & & & & \\
0 & \longrightarrow & R_1 & \xrightarrow{\alpha} & R_1 & \xrightarrow{\pi} & A & \longrightarrow & 0\\
 & & \downarrow{\scriptstyle\pi} & & \downarrow & & \| & & \\
0 & \longrightarrow & A & \xrightarrow{\varphi} & C & \xrightarrow{\psi} & A & \longrightarrow & 0\\
 & & \downarrow & & \downarrow & & & & \\
 & & 0 & & 0 & & & &
\end{array}$$

of differential graded R-modules such that π is a morphism of graded algebras, all rows and columns are exact and conditions (a) *and therefore* (b) *of Proposition* 4.11 *are satisfied. The Bockstein operator $d_A\colon HA \to HA$ is a derivation and a differential.*

The proof follows directly from our previous results.

Corollary 4.14. *Suppose that X is a projective complex of S-modules for a not necessarily commutative ring S, and suppose that R is a commutative ring which is considered to be a trivial-S-module with identity. Then every element $a \in R$ such that $a\,x = 0$ implies $x = 0$, defines a diagram*

$$\begin{array}{ccccccccc}
 & & 0 & & 0 & & & & \\
 & & \downarrow & & \downarrow & & & & \\
 & & \mathrm{Hom}_S(X, R) & = & \mathrm{Hom}_S(X, R) & & & & \\
 & & \downarrow{\scriptstyle a} & & \downarrow{\scriptstyle a^2} & & & & \\
0 & \longrightarrow & \mathrm{Hom}_S(X, R) & \xrightarrow{a} & \mathrm{Hom}_S(X, R) & \longrightarrow & \mathrm{Hom}_S(X, R/aR) & \longrightarrow & 0\\
 & & \downarrow & & \downarrow & & \downarrow & & \\
0 & \longrightarrow & \mathrm{Hom}_S(X, R/aR) & \longrightarrow & \mathrm{Hom}_S(X, R/a^2R) & \longrightarrow & \mathrm{Hom}_S(X, R/aR) & \longrightarrow & 0\\
 & & \downarrow & & \downarrow & & & & \\
 & & 0 & & 0 & & & &
\end{array}$$

in which the maps marked a, resp. a^2, are multiplication by a, resp. a^2, and which satisfies the conditions of Corollary 4.13. *Thus the connecting morphism $d\colon H(\mathrm{Hom}_S(X, R/aR)) \to H(\mathrm{Hom}_S(X, R/aR))$ is a differential and derivation.*

In the formulation of the next result, we use the concept of the *I-socle* $S(I, M)$ of an R-module M, where R is a commutative ring with identity, I an ideal of R. The R-module $S(I, M)$ is defined by

$$S(I, M) = \{m \in M : Im = 0\}.$$

Proposition 4.15. *Let G be any group and R any commutative ring which is regarded as a trivial G-module. Suppose that $a \in R$ is an element such that $a\,x = 0$ implies $x = 0$. Then there is a derivation and differential*

$$d_a : H(G, R/aR) \to H(G, R/aR)$$

and the standard Bockstein diagram

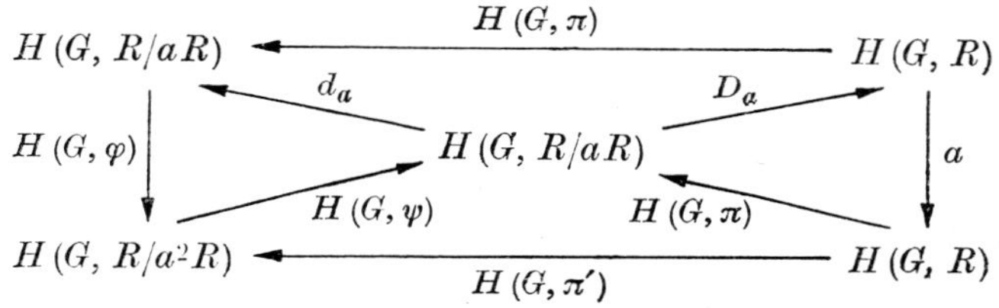

where

$$\begin{array}{ccccccccc}
 & & 0 & & 0 & & & & \\
 & & \downarrow & & \downarrow & & & & \\
 & & R & = & R & & & & \\
 & & \downarrow{\scriptstyle a} & & \downarrow{\scriptstyle a^2} & & & & \\
0 & \longrightarrow & R & \xrightarrow{a} & R & \xrightarrow{\pi} & R/aR & \longrightarrow & 0 \\
 & & \downarrow{\scriptstyle \pi} & & \downarrow{\scriptstyle \pi'} & & \| & & \\
0 & \longrightarrow & R/aR & \xrightarrow{\varphi} & R/a^2R & \xrightarrow{\psi} & R/aR & \longrightarrow & 0 \\
 & & \downarrow & & \downarrow & & \downarrow & & \\
 & & 0 & & 0 & & 0 & &
\end{array}$$

is a commutative diagram with exact rows and columns. The left and right triangles are exact, the top and bottom triangle commute. The cohomology module of the differential d_a is isomorphic to

$$\operatorname{im} D_a H(G, \psi) \subset S(aR, H(G, R)) \cap aH(G, R).$$

We have $d_a\, H(G, \pi) = 0$ *and* $\ker d_a / \operatorname{im} H(G, \pi) \cong H(d_a)$. *In particular, if* $aH(G, R) = \{0\}$, *then*

(i) d_a *is exact,*

(ii) $H^+(G, \pi)$ *is injective* $(H^+ = H^1 \oplus H^2 \oplus \ldots)$,

(iii) *The diagonal sequences are exact.*

There is a commutative diagram of exact sequences

$$\begin{array}{ccccccccc}
 & & 0 & & & & 0 & & \\
 & & \downarrow & & & & \downarrow & & \\
 & & a\,H(G, R) & & & & aH^+(G, R) \cap S\,(aR, H(G, R)) & & \\
 & & \downarrow & & & & \downarrow & & \\
0 \to a\,H(G, R) & \xrightarrow{\text{incl}} & H(G, R) & \xrightarrow{H(G,\pi)} & H(G, R/aR) & \xrightarrow{\text{coim}\, D_a} & S\,(aR, H^+(G, R)) & \to & 0 \\
 & & \downarrow{\scriptstyle H(G,\pi)'} & & & & \downarrow{\scriptstyle H(G,\pi)''} & & \\
0 & \longrightarrow & \ker d_a & \xrightarrow{\text{incl}} & H(G, R/aR) & \xrightarrow{\text{coim}\, d_a} & \operatorname{im} d & \longrightarrow & 0 \\
 & & \downarrow & & & & \downarrow & & \\
 & & \operatorname{im} D_a\, H(G, \psi) & & & & 0 & & \\
 & & \downarrow & & & & & & \\
 & & 0 & & & & & &
\end{array}$$

in which $H(G, \pi)'$ *and* $H(G, \pi)''$ *are restrictions and corestrictions of* $H(G, \pi)$. *If* $aH(G, R) = 0$, *we have in addition*

$$\text{(iv)}\quad \begin{array}{ccccccccc}
0 & \longrightarrow & H(G, R) & \longrightarrow & H(G, R/aR) & \longrightarrow & H^+(G, R) & \longrightarrow & 0 \\
 & & \downarrow & & \downarrow & & \downarrow & & \\
0 & \longrightarrow & \ker d_a & \longrightarrow & H(G, R/aR) & \longrightarrow & \operatorname{im} d_a & \longrightarrow & 0
\end{array}$$

is an isomorphism of exact sequences. If R *is a principal ideal domain, then they split. The entire set of conclusions remains valid if* $H(G, M)$ *is replaced by* $H(Y, M)$, *where* Y *is an arbitrary complex, and* $M = R,\ R/aR,\ R/a^2 R$ *etc.*

All these assertions follow from our results. In condition (iv) we observe that the top sequence is the sequence of the universal coefficient theorem, which splits.

The previous results are now applied to the spectral algebras of Section 2.

Let R be commutative ring with identity and $a \in R$ an element with the property that $a\,x = 0$ and $x = 0$ are equivalent for all $x \in R$. If $\varphi\colon A \to A$ is an elementary morphism of R-modules, then there is an exact sequence of morphisms

$$
\begin{array}{ccccccccc}
0 & \longrightarrow & A & \xrightarrow{a} & A & \xrightarrow{\pi} & A' & \longrightarrow & 0 \\
 & & \downarrow{\scriptstyle\varphi} & & \downarrow{\scriptstyle\varphi} & & \downarrow{\scriptstyle\varphi'} & & \\
0 & \longrightarrow & A & \xrightarrow{a} & A & \xrightarrow{\pi} & A' & \longrightarrow & 0
\end{array}
$$

In this diagram φ is an elementary morphism of R-modules and φ' is an elementary morphism of R/aR-modules. The sequence induces an exact sequence of differential bi-graded R-algebras

$$0 \longrightarrow E_2(\varphi) \xrightarrow{a} E_2(\varphi) \xrightarrow{\bar{\pi} = E_2(\pi)} E_2(\varphi') \longrightarrow 0$$

and $E_2(\varphi')$ is also a free differential bi-graded R/aR-algebra. This can easily be seen by observing that $\ker \pi$ generates the kernel of the morphisms

$$PA \xrightarrow{P(\pi)} PA',$$

$$\wedge A \xrightarrow{\wedge(\pi)} \wedge A',$$

and thus $\ker \pi \otimes 1 + 1 \otimes \ker \pi$ generates the kernel of

$$PA \otimes \wedge A \to PA \otimes \wedge A .$$

But this is just the ideal generated by $\operatorname{im} a \otimes 1 + 1 \otimes \operatorname{im} a$, which is $aE_2(\varphi)$. In the usual manner we derive a commuting diagram with exact rows and columns

$$
\begin{array}{ccccccccc}
 & & 0 & & 0 & & & & \\
 & & \downarrow & & \downarrow & & & & \\
 & & E_2(\varphi) & = & E_2(\varphi) & & & & \\
 & & \downarrow{\scriptstyle a} & & \downarrow{\scriptstyle a^2} & & & & \\
0 & \longrightarrow & E_2(\varphi) & \xrightarrow{a} & E_2(\varphi) & \xrightarrow{\bar{\pi}} & E_2(\varphi') & \longrightarrow & 0 \\
 & & \downarrow{\scriptstyle \bar{\pi}} & & \downarrow & & \| & & \\
0 & \longrightarrow & E_2(\varphi') & \longrightarrow & E_2(\varphi') & \longrightarrow & E_2(\varphi') & \longrightarrow & 0 \\
 & & \downarrow & & \downarrow & & & & \\
 & & 0 & & 0 & & & &
\end{array}
$$

A direct application of the results of this section then yields the following

Proposition 4.16. *Let $\varphi: A \to A$ be an elementary morphism of R-modules for some commutative ring R with identity. If $A \in R$ does not annihilate any non-zero element of R, then there are elementary morphisms $\varphi': A' \to A'$ and $\varphi'': A'' \to A''$ of R/aR-, resp. R/a^2R-modules and a standard Bockstein diagram of bi-graded R-algebras*

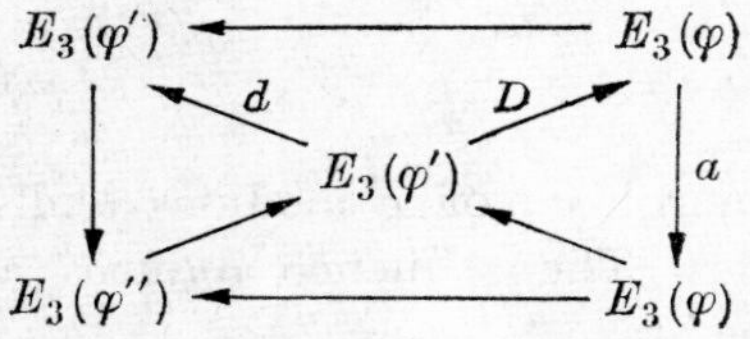

such that d is a derivation and differential of bi-degree $(2, -1)$. If $a\,E_3^+(\varphi) = \{0\}$, then d is exact, $E_3(\varphi) \to E_3(\varphi')$ is injective and the diagonal sequences are exact. Moreover, we then have an isomorphism of exact sequences

$$\begin{array}{ccccccccc} 0 & \longrightarrow & E_3(\varphi) & \longrightarrow & E_3(\varphi') & \longrightarrow & E_3^+(\varphi) & \longrightarrow & 0 \\ & & \downarrow & & \| & & \downarrow & & \\ 0 & \longrightarrow & \ker d & \longrightarrow & E_3(\varphi') & \longrightarrow & \operatorname{im} d & \longrightarrow & 0 \end{array}$$

where the module E_3^+ is obtained from E_3 by omission of the direct summand $E_3^{0,0}$.

Remark. Suppose that R is a principal ideal domain. If z_i, $i \in I$ are the ring elements appearing as the eigenvalues of the morphism φ and if $z_i | a$ for all i, then the hypothesis $a\,E_3^+(\varphi) = 0$ is satisfied. Then Proposition 4.15 allows one to compute $E_3(\varphi)$ provided that $E_3(\varphi')$ and d are known. It may also be worthwhile to recall explicitly in the present situation how d is computed. Denote with

$$E_2^{2p,q}(\pi) : E_2^{2p,q}(\varphi) \to E_2^{2p,q}(\varphi')$$

the reduction modulo a and with

$$\psi^{2p,q}: \ker d_{\varphi'}^{2p,q} \to E_2^{2p,q}(\varphi')$$

the cohomology map. Let $M^{2p,q} \subset E_2^{2p,q}(\varphi)$ be the submodule of all x with $d_\varphi x \in a\,E_2^{2p+2,q+1}(\varphi)$. Then, for $m \in M^{2p,q}$, the element $a^{-1} d_\varphi^{2p,q}\, m$ is well defined in $\ker d_\varphi^{2p+2,q-1}$ and we have

$$d^{2p,q}\, \psi^{2p,q}\,(E_2^{2p,q}(\pi)\,(m)) = \psi^{2p+2,q-1}\, E_2^{2p+2,q-1}(\pi)\,(a^{-1} d_\varphi^{2p,q}\, m);$$

in short, $d\psi^* E_2(\pi)\,(m) = \psi^* E_2(\pi)\, a^{-1} d_\varphi(m)$.

Corollary 4.17. *If, with the notation of Proposition* 4.16, *we have* $\varphi = a\,\varphi^*$ *for some endomorphism* φ^* *of* A, *then* $d = d_{\varphi^*}$, *where* $\varphi^{*\prime}\colon A' \to A'$ *is induced by* φ^* *as* φ' *is by* φ. *In particular, in Propositon* 2.33, $d_{\varphi^{*\prime}}$ *is the Bockstein morphism on* $E_3(\varphi') = E_2(\varphi')$.

Proof. Since $\operatorname{im}\varphi \subset a \cdot A$, we have $M^{2p,q} = E_2^{2p,q}(\varphi)$, and $d_{\varphi'} = 0$, whence $E_3(\varphi') = E_2(\varphi')$, and ψ is the identity on $E_2(\varphi')$. Thus by the remark above, we have a commutative diagram

$$\begin{array}{ccc} E_2(\varphi) & \xrightarrow{d_{\varphi^*}} & E_2(\varphi) = E_2(\varphi^*) \\ \downarrow{\scriptstyle E_2(\pi)} & & \downarrow{\scriptstyle E_2(\pi)} \\ E_2(\varphi') & \xrightarrow{d} & E_2(\varphi') \end{array}$$

This diagram also commutes with $d_{\varphi^{*\prime}}$ in place of d; since $E_2(\pi)$ is epic, $d = d_{\varphi^{*\prime}}$.

In later applications it is helpful to have the following general lemma concerning the Bockstein formalism:

Lemma 4.18. *Let the notation be as in Lemma* 4.4ff. *Suppose that there is a natural transformation* $n\colon F \to H^2$ *of functors from the category of complexes over* R *to the category of modules over* R. *If* $F\pi_A$ *is surjective, then* $d_a^2\, n_A = 0$.

Proof. We have a commutative diagram in which the rows are exact:

$$\begin{array}{ccccc} FR_1 & \xrightarrow{F\pi_A} & FA & & \\ \downarrow{\scriptstyle n_{R_1}} & & \downarrow{\scriptstyle n_A} & & \\ H^2R_1 & \xrightarrow{H^2(\pi_A)} & H^2A & \xrightarrow{D_a^2} & H^3R_1 \\ \downarrow{\scriptstyle H^2(\pi_C)} & & \| & & \downarrow{\scriptstyle H^3(\pi_A)} \\ H^2C & \longrightarrow & H^2A & \xrightarrow{d_a^2} & H^3A \end{array}$$

If $F\pi_A$ is surjective, then

$$\operatorname{im} n_A = \operatorname{im} H^2(\pi_A)\, n_{R_1} \subset \operatorname{im} H^2(\pi_A).$$

Since the second row is exact,

$$\operatorname{im} H^2(\pi_A) \subset \ker D_a^2 \subset \ker H^3(\pi_A)\, D_a^2 = \ker d_a^2.$$

Thus $d_a^2\, n_A = 0$.

Chapter II

The cohomology of finite abelian groups

Section 1

Products

In this section we discuss some homological algebra which we specifically apply to the cohomology of finite groups. Most of our statements in the beginning parts are inspired by the pertinent sections in Cartan and Eilenberg [11], and are frequently contained there so that we have a convenient source of reference. If tensor products are not marked, they are understood to be taken over Z. If S is a ring with identity, A a right S-module and B a left S-module, then $A \otimes_S B$ denotes the tensor product over S in the usual sense.

Let S_i, $i = 1, 2, 3$, $S_3 = S_1 \otimes S_2$ be rings with identity which are projective over Z, and suppose that Z is an S_i-module such that $(s_1 z_1)(s_2 z_2) = (s_1 \otimes s_2)(z_1 z_2)$. Let $0 \longleftarrow \mathsf{Z} \overset{\varepsilon_i}{\longleftarrow} X_i$, $i = 1, 2$, be resolutions of Z by left S_i-modules. Then if $X_3 = X_1 \otimes X_2$, $0 \longleftarrow \mathsf{Z} \overset{\varepsilon_3}{\longleftarrow} X_3$ is a projective resolution of Z by projective left S_3-modules, where $(s_1 \otimes s_2)(x_1 \otimes x_2) = s_1 x_1 \otimes s_2 x_2$ and where $\varepsilon_3 = m(\varepsilon_1 \otimes \varepsilon_3)$ with the multiplication $m\colon \mathsf{Z} \otimes \mathsf{Z} \to \mathsf{Z}$ (see Cartan-Eilenberg [11], p. 203, 1.1). Thus, if $0 \longleftarrow \mathsf{Z} \overset{\varepsilon_3'}{\longleftarrow} X_3'$ is any projective resolution of Z by left S_3-modules, then there is a morphism $f\colon X_3' \to X_3$ of complexes such that

$$\begin{array}{ccc} \mathsf{Z} & \overset{\varepsilon_3'}{\longleftarrow} & X_3' \\ \| & & \downarrow f \\ \mathsf{Z} & \overset{\varepsilon_3}{\longleftarrow} & X_3 \end{array}$$

commutes, and f is unique up to homotopy. (This is just the fundamental theorem of homological algebra.)

Let R_i, $i = 1, 2$, be left S_i-modules, suppose that R is a commutative ring with identity, and that R_i, $i = 1, 2$, is an R-module such that the module operations of R and S_i commute (i. e. $s_i(r\, x_i) = r(s_i x_i)$). The tensor product $R_3 = R_1 \otimes_R R_2$ is an R-module and a left $S_3 = S_1 \otimes S_2$-module under

$$(s_1 \otimes_{\mathsf{Z}} s_2)(r_1 \otimes_R r_2) = s_1 r_1 \otimes_R s_2 r_2,$$

and the operations of S_3 and R commute.

For any left S_i-module A_i, $i = 1, 2$, there is a natural morphism of R-modules

$$F\colon \mathrm{Hom}_{S_1}(A_1, R_1) \otimes_R \mathrm{Hom}_{S_2}(A_2, R_2) \to \mathrm{Hom}_{S_3}(A_1 \otimes A_2,\ R_1 \otimes_R R_2)$$

defined by $F(\varphi_1 \otimes \varphi_2)(r_1 \otimes_R r_2) = \varphi_1(r_1) \otimes_R \varphi_2(r_2)$. Thus there is a natural morphism

$$\mathrm{Hom}_{S_3}(f, R_1 \otimes R_2)F\colon \mathrm{Hom}_{S_1}(X_1, R_1) \otimes_R \mathrm{Hom}_{S_2}(X_2, R_2) \to \mathrm{Hom}_{S_3}(X_3', R_1 \otimes_R R_2).$$

Definition 1.1. The morphism constructed above will be denoted with $\varkappa$.

Note that the construction determines $\varkappa$ uniquely up to a cochain homotopy. Thus, on the cohomology level, the construction determines $H(\varkappa)$ uniquely. The question, of course, arises as to what extent the construction is functorial in the rings S_i.

Suppose that $\varphi\colon S \to T$ is a morphism of augmented rings with identity over Z; i. e. there is a commutative diagram of rings

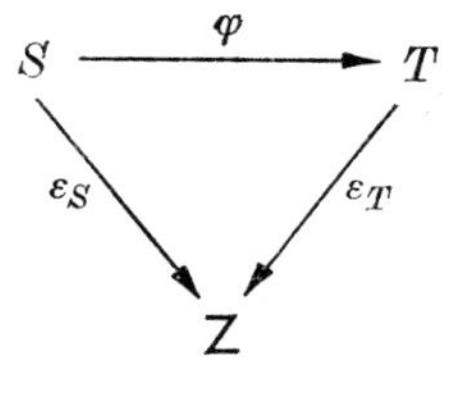

Assume that Z is a left T-module. Clearly every right (left) T-module is a right (left) S-module under $s \cdot x = \varphi(s)\, x$; in particular T is a right S-module and Z a left S-module.

Let

$$\begin{array}{l} 0 \leftarrow \mathsf{Z} \xleftarrow{\varepsilon_S} X, \quad X^0 = S, \\ \qquad\ \ \| \\ 0 \leftarrow \mathsf{Z} \xleftarrow{\varepsilon_T} Y, \quad Y^0 = T, \end{array}$$

be projective resolutions of Z by left S- (resp. T-)modules. Then

$$0 \longleftarrow T \otimes_S \mathsf{Z} \xleftarrow{T \otimes_S \varepsilon_S} T \otimes_S X$$

is a complex over $T \otimes_S \mathsf{Z}$ and this complex is in fact projective (see Cartan-Eilenberg [11], p. 30). There is a morphism of T-modules $T \otimes_S \mathsf{Z} \to \mathsf{Z}$ defined by $t \otimes_S z \to t\,z$. Thus there is a morphism of complexes of left T-modules $g\colon T \otimes_S X \to Y$ such that

$$\begin{array}{ccc} T \otimes_S \mathsf{Z} & \longleftarrow & T \otimes_S X \\ \downarrow & & \downarrow g \\ \mathsf{Z} & \longleftarrow & Y \end{array}$$

commutes, and this morphism is then uniquely determined up to homotopy. Let R' be a left T-module. Then there is a morphism of cochain complexes

$$\mathrm{Hom}_T(g, R')\colon \mathrm{Hom}_T(Y, R') \to \mathrm{Hom}_T(T \otimes_S X, R').$$

But also, there is a natural isomorphism

$$\mathrm{Hom}_T(T \otimes_S X, R') \to \mathrm{Hom}_S(X, \mathrm{Hom}_T(T, R'))$$

(see Cartan-Eilenberg [11], p. 28), and $\mathrm{Hom}_T(T, R')$ of course is naturally isomorphic to R'. Thus we have a morphism of cochain complexes

$$\mathrm{Hom}_T(Y, R') \to \mathrm{Hom}_S(X, R')$$

which is natural up to homotopy.

Definition 1.2. The morphism so constructed will be denoted with λ.

Lemma 1.3. *Let $S_i \xrightarrow{\varphi_i} T_i$, $i = 1, 2$, be a morphism of rings with augmentation, i. e. suppose that*

$$\begin{array}{ccc} S_i & \longrightarrow & T_i \\ & {\scriptstyle\varepsilon_i}\searrow \quad \swarrow{\scriptstyle\varepsilon_i'} & \\ & \mathsf{Z} & \end{array}$$

commutes and assume that Z is a left T_i-module. If we identify $\mathsf{Z} \otimes \mathsf{Z}$ with Z under the multiplication isomorphism m given by $m(z_1 \otimes z_2) = z_1 z_2$, then

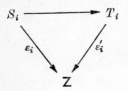

$$\begin{array}{ccc} S_3 & \longrightarrow & T_3 \\ & \varepsilon_3 \searrow \quad \swarrow \varepsilon_3' & \\ & \mathbf{Z} & \end{array}$$

is a morphism of augmented S_3 *modules, where* $S_3 = S_1 \otimes S_2$, $T_3 = T_1 \otimes T_2$, $\varepsilon_3 = \varepsilon_1 \otimes \varepsilon_2$, $\varepsilon_3' = \varepsilon_1' \otimes \varepsilon_2'$.

Let R_i, $i = 1, 2$ *be left* S_i*-modules and modules over a commutative ring* R *such that the operations from* S_i *and* R *commute. Let* $R_3 = R_1 \otimes R_2$. *Let* $0 \leftarrow \mathbf{Z} \leftarrow X_i$ *be projective resolutions of* $\mathbf{Z}$ *by left* S_i*-modules,* $i = 1, 2, 3$ *and* $0 \leftarrow \mathbf{Z} \leftarrow Y_i$ *projective resolutions of* $\mathbf{Z}$ *by left* T_i*-modules,* $i = 1, 2, 3$. *Then there is a diagram of* R*-modules*

$$\begin{array}{ccc} \operatorname{Hom}_{T_1}(Y_1, R_1) \otimes_R \operatorname{Hom}_{T_2}(Y_2, R_2) & \xrightarrow{\varkappa_{T_1, T_2}} & \operatorname{Hom}_{T_3}(Y_3, R_3) \\ \downarrow \lambda_1 \otimes \lambda_2 & & \downarrow \lambda_3 \\ \operatorname{Hom}_{S_1}(X_1, R_2) \otimes_R \operatorname{Hom}_{S_2}(X_2, R_2) & \xrightarrow{\varkappa_{S_1, S_2}} & \operatorname{Hom}_{S_3}(X_3, R_3) \end{array}$$

which commutes up to homotopy.

Proof. We consider the diagram of projective T_3-complexes

$$\begin{array}{cccc} \text{exact} \quad Y_1 \otimes Y_2 & \longleftarrow & Y_3 & \text{exact} \\ \uparrow & & & \\ (T_1 \otimes_{S_1} X_1) \otimes (T_2 \otimes_{S_2} X_2) & & \uparrow & \\ \uparrow \chi & & & \\ T_3 \otimes_{S_3} (X_1 \otimes X_2) & \longleftarrow & T_3 \otimes_{S_3} X_3 & \end{array}$$

in which the isomorphism χ is defined by

$$\chi((t_1 \otimes t_2) \otimes_{S_3} (x_1 \otimes x_2)) = (t_1 \otimes_{S_1} x_1) \otimes (t_2 \otimes_{S_2} x_2).$$

This diagram commutes up to homotopy by the uniqueness part in the fundamental theorem of homological algebra, since all the morphisms in the diagram are compatible with the augmentations. Now apply $\operatorname{Hom}_{T_3}(-, R_3)$ and observe some of the natural transformations used before to obtain the assertion.

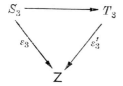

Corollary 1.4. *Under the conditions of Lemma* 1.3, *there is a commutative diagram of R-modules* (see the adjoining column) *in which* α *is the standard natural transformation* (see Cartan-Eilenberg [11], p. 64).

$$\begin{array}{ccccc} H(\mathrm{Hom}_{T_1}(Y_1, R_1)) \otimes_R H(\mathrm{Hom}_{T_2}(Y_2, R_2)) & \xrightarrow{\alpha_T} & H(\mathrm{Hom}_{T_1}(Y_1, R_1) \otimes_R \mathrm{Hom}_{T_2}(Y_2, R_2)) & \xrightarrow{H(\varkappa_T)} & H(\mathrm{Hom}_{T_3}(Y_3, R_3)) \\ \downarrow{\scriptstyle H(\lambda_1) \otimes H(\lambda_2)} & & \downarrow{\scriptstyle H(\lambda_1 \otimes \lambda_2)} & & \downarrow{\scriptstyle H(\lambda_3)} \\ H(\mathrm{Hom}_{S_1}(X_1, R_1)) \otimes_R H(\mathrm{Hom}_{S_2}(X_2, R_2)) & \xrightarrow{\alpha_S} & H(\mathrm{Hom}_{S_1}(X_1, R_1) \otimes_R \mathrm{Hom}_{S_2}(X_2, R_2)) & \xrightarrow{H(\varkappa_S)} & H(\mathrm{Hom}_{S_3}(X_3, R_3)) \end{array}$$

Now we suppose that S_i, $i = 1, 2$ are augmented Hopf algebras, i. e. that we have a commutative diagram

$$\begin{array}{ccccc} S_i & \xrightarrow{\gamma_i} & S_i \otimes S_i & \xrightarrow{\mu_i} & S_i \\ \downarrow{\scriptstyle \varepsilon_i} & & \downarrow{\scriptstyle \varepsilon_i \otimes \varepsilon_i} & & \downarrow{\scriptstyle \varepsilon_i} \\ \mathsf{Z} & \longrightarrow & \mathsf{Z} & \longrightarrow & \mathsf{Z} \\ & & \| & & \\ & & \mathsf{Z} \otimes \mathsf{Z} & & \end{array}$$

where μ_i denotes multiplication (i. e. $\mu_i(s \otimes s') = s\,s'$). Then $S_3 = S_1 \otimes S_2$ is a Hopf algebra with comultiplication

$$S_1 \otimes S_2 \xrightarrow{\gamma_1 \otimes \gamma_2} (S_1 \otimes S_1) \otimes (S_2 \otimes S_2) \xrightarrow{\cong} (S_1 \otimes S_2) \otimes (S_1 \otimes S_2).$$

The previous considerations, notably Lemma 1.3 and Corollary 1.4, then apply with the comultiplications as φ_i, and with the resolutions

$$0 \leftarrow \mathsf{Z} \leftarrow X_i \otimes X_i$$

of Z by left $S_i \otimes S_i$-modules as the resolutions

$$0 \leftarrow \mathsf{Z} \leftarrow Y_i, \quad i = 1, 2, 3, \quad X_3 = X_1 \otimes X_2.$$

Now assume that R_i, $i = 1, 2$, are algebras over the commutative ring R. Consider $R_i \otimes_R R_i$ as a left S_i-module under $s(r \otimes r') = s\,r \otimes r'$. Then the multiplication $R_i \otimes_R R_i \to R_i$ induces natural morphisms of R-modules $\mathrm{Hom}_{S_i}(X, R_i \otimes_R R_i) \to \mathrm{Hom}_{S_i}(X, R_i)$. Consequently the sequence of morphisms (p. 104)

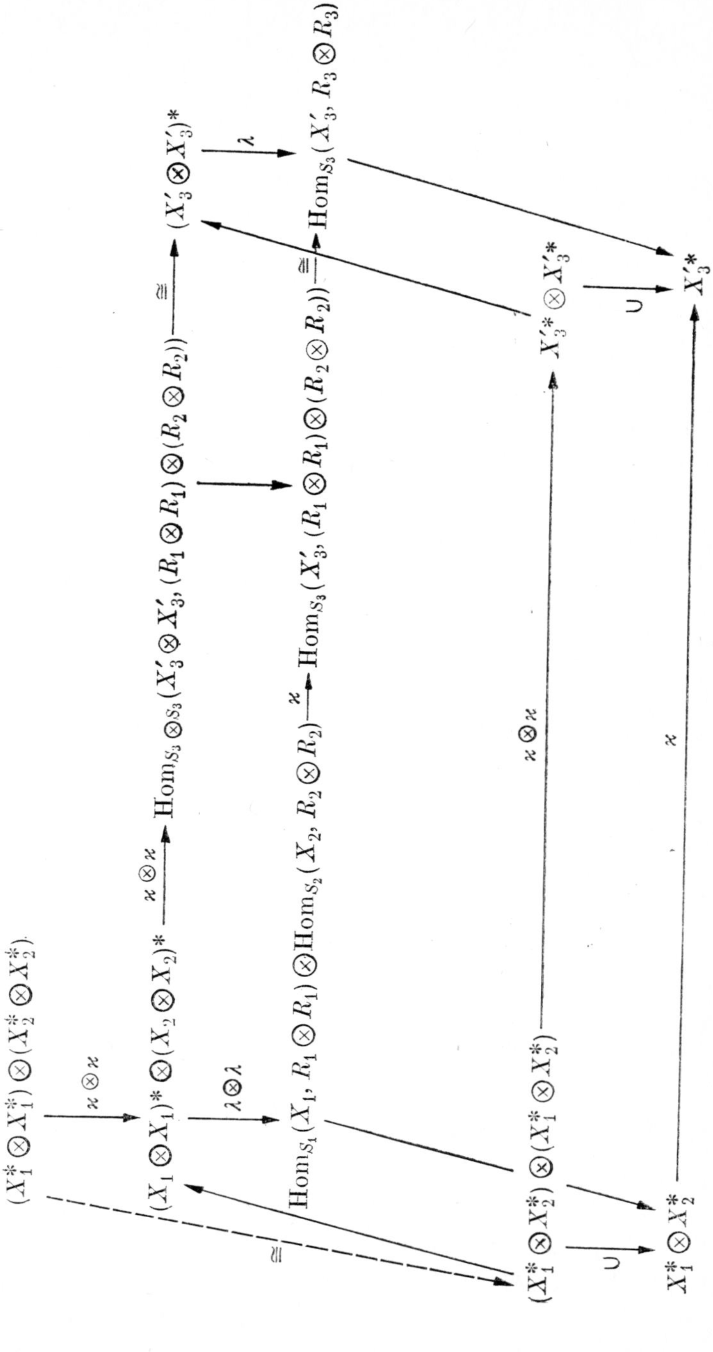
$(X_1^*\otimes X_1^*)\otimes(X_2^*\otimes X_2^*)$
$\varkappa\otimes\varkappa$
$(X_1\otimes X_1)^*\otimes(X_2\otimes X_2)^*$
$\varkappa\otimes\varkappa$
$\mathrm{Hom}_{S_3\otimes S_3}(X_3'\otimes X_3', (R_1\otimes R_1)\otimes(R_2\otimes R_2))$
$\cong$
$(X_3'\otimes X_3')^*$
$\lambda\otimes\lambda$
$\mathrm{Hom}_{S_1}(X_1, R_1\otimes R_1)\otimes\mathrm{Hom}_{S_2}(X_2, R_2\otimes R_2)$
$\varkappa$
$\mathrm{Hom}_{S_3}(X_3', (R_1\otimes R_1)\otimes(R_2\otimes R_2))$
$\cong$
$\mathrm{Hom}_{S_3}(X_3', R_3\otimes R_3)$
λ
$\cong$
$(X_1^*\otimes X_2^*)\otimes(X_1^*\otimes X_2^*)$
$\varkappa\otimes\varkappa$
$X_3'^*\otimes X_3'^*$
$\cup$
$X_3'^*$
$\cup$
$X_1^*\otimes X_2^*$
$\varkappa$

$$\mathrm{Hom}_{S_i}(X_i, R_i) \otimes_R \mathrm{Hom}_{S_i}(X_i, R_i) \xrightarrow{\varkappa} \mathrm{Hom}_{S_i \otimes S_i}(X_i \otimes X_i, R_i \otimes_R R_i)$$
$$\xrightarrow{\lambda} \mathrm{Hom}_{S_i}(X_i, R_i \otimes_R R_i) \longrightarrow \mathrm{Hom}_{S_i}(X_i, R_i)$$

gives $\mathrm{Hom}_{S_i}(X_i, R_i)$ the structure of a differential graded algebra over R. The multiplication in this algebra is called the *cup product on the cochain level*. Note that by our construction it is naturally determined up to homotopy. It induces the unique cup product on the cohomology level. We will denote the cup product on both levels by $\cup$; confusion will not arise. Finally we will show that the morphism λ of Definition 1.2 preserves the cup product up to homotopy. In the diagram which we want to consider, we abbreviate $\mathrm{Hom}_{S_i}(X_i, R_i)$ by X_i^*, $\mathrm{Hom}_{S_i \otimes S_i}(X_i \otimes X_i, R_1 \otimes R_2)$ by $(X_i \otimes X_i)^*$. We use the notation and hypotheses of the introductory paragraphs of this section. The diagram on p. 103 commutes up to homotopy, since all individual parallelograms commute up to homotopy (or actually commute) either by the definitions of the maps involved, or by naturality, or by Lemma 1.3.

Assume now that all X_i^n, $i = 1, 2$, $n = 1, 2, \ldots$ are finitely generated free left S_i-modules. Also suppose that $R_1 = R_2 = R$. The natural map

$$\mathrm{Hom}_{S_i}(X_i, R) \otimes_R \mathrm{Hom}_{S_i}(X_i, R) \to \mathrm{Hom}_{S_i \otimes S_i}(X_i \otimes X_i, R)$$

is then an isomorphism since Hom is additive in the first argument.

The considerations concerning the change of rings in Lemma 1.3 also apply to the morphism of augmented rings

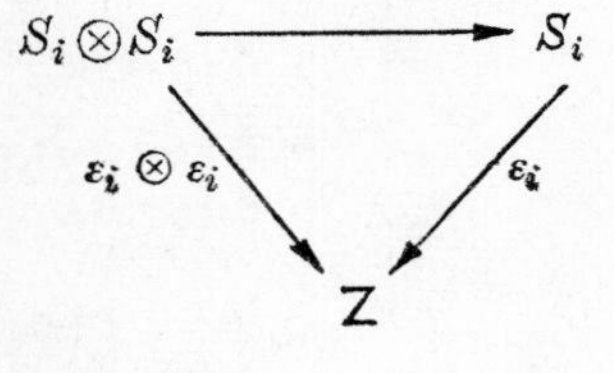

given by the multiplication in S_i. Thus, by Lemma 1.3 we obtain a morphism

$$\mathrm{Hom}_{S_i}(X_i, R) \to \mathrm{Hom}_{S_i \otimes S_i}(X_i \otimes X_i, R),$$

where we have identified $R \otimes_R R$ with R under the multiplication morphism. If we follow this morphism by the inverse of the isomorphism mentioned above, we have given $\mathrm{Hom}_{S_i}(X_i, R)$ the structure of a coalgebra and thus in fact of a Hopf algebra. Our considerations about the multiplication can be carried out for the comultiplication with the necessary modifications.

This is a good place to recall the Künneth theorem:

Lemma 1.5. *Let L and K be complexes over the commutative ring R. Let $B(K)$ denote the image and $Z(K)$ the kernel of the differential of K. Suppose that*

$$\mathrm{Tor}_1^R(B(K), B(L)) = 0 = \mathrm{Tor}_1^R(H(K), B(L)), \tag{1}$$

$$\mathrm{Tor}_1^R(B(K), Z(L)) = 0 = \mathrm{Tor}_1^R(H(K), Z(L)). \tag{2}$$

Then there is an exact sequence

$$0 \to H(K) \otimes_R H(L) \to H(K \otimes_R L) \to \mathrm{Tor}_1^R(H(K), H(L)) \to 0$$

where the first map is of degree 0 and the second of degree 1. If R is a principal ideal domain and K and L are free, then the sequence splits in the category of graded R-modules (but not naturally).

See Cartan-Eilenberg [11], p. 112.

We may apply this in the following fashion. Continuing our notation we assume that the complexes $K_i = L_i = \mathrm{Hom}_{S_i}(X_i, R)$, $i = 1, 2$, satisfy the conditions (1) and (2) of Lemma 1.5. If $\mathrm{Tor}_1^R(\mathrm{Ext}_{S_i}(\mathsf{Z}, R), \mathrm{Ext}_{S_i}(\mathsf{Z}, R)) = 0$ then $\mathrm{Ext}_{S_i}(\mathsf{Z}, R)$ has the structure of a Hopf algebra, where the comultiplication is defined by

$$\begin{aligned}\mathrm{Ext}_{S_i}(\mathsf{Z}, R) &= H(\mathrm{Hom}_{S_i}(X_i, R)) \to H(\mathrm{Hom}_{S_i}(X_i, R) \otimes_R \mathrm{Hom}_{S_i}(X, R))\\ &\to H(\mathrm{Hom}_{S_i}(X_i, R)) \otimes_R H(\mathrm{Hom}_{S_i}(X_i, R))\\ &= \mathrm{Ext}_{S_i}(\mathsf{Z}, R) \otimes_R \mathrm{Ext}_{S_i}(\mathsf{Z}, R).\end{aligned}$$

We have proved the following theorem:

Theorem 1.6. (a) *Let $S \xrightarrow{\varepsilon} \mathsf{Z}$ be an augmented Hopf algebra, R' an algebra over the commutative ring R and a left S-module such that $s \cdot (r\,r') = (s \cdot r)\,r'$, $s \in S, r \in R, r' \in R'$. Suppose that Z is a left S-algebra. Let $0 \leftarrow \mathsf{Z} \xleftarrow{\varepsilon} X$, $X^0 = S$ be a resolution of Z by projective left S-modules. Then $\mathrm{Hom}_S(X, R')$ is a differential graded algebra over R relative to a product which is functorial in S up to a homotopy. (In particular, if $\varphi\colon S \to S'$ is a morphism of augmented Hopf algebras and if $0 \leftarrow \mathsf{Z} \leftarrow X'$, $X'^0 = S'$, is a projective S'-resolution of Z, then there is a morphism h of differential graded R-modules*

$$\mathrm{Hom}_{S'}(X', R') \to \mathrm{Hom}_S(X, R')$$

such that $u \otimes_R v \to h(u \cup v)$ and $u \otimes_R v \to h(u) \cup h(v)$ are homotopic.) If X^n, $n = 0, 1, 2, \ldots$ is finitely generated free, then $\mathrm{Hom}_S(X, R')$ is a differential graded Hopf algebra over R, and the comultiplication is functorial up to a homotopy.

(b) *Let* $S_i \xrightarrow{\varepsilon_i} \mathsf{Z}$, $i = 1, 2$, *be augmented Hopf algebras, and let* R_i *be algebras over a commutative ring* R *and left* S_i*-modules with* $s \cdot (r\, r') = (s \cdot r)\, r'$. *Suppose that* Z *is a left* S_i *algebra for* $i = 1, 2$. *Let* $0 \leftarrow \mathsf{Z} \xleftarrow{\varepsilon_i} X_i$, $X_i^0 = S_i$, $S_3 = S_1 \otimes S_2$, $i = 1, 2, 3$, *be arbitrary projective resolutions of* Z *by left* S_i*-modules. Then there is a morphism of differential graded* R*-modules*

$$\varkappa\colon \operatorname{Hom}_{S_1}(X_1, R_1) \otimes_R \operatorname{Hom}_{S_2}(X_2, R_2) \to \operatorname{Hom}_{S_1 \otimes S_2}(X_3, R_1 \otimes_R R_2)$$

and this morphism is a morphism of differential graded algebras up to a homotopy (i. e. $u \otimes v \to \varkappa\,(u \cup v)$ *and* $u \otimes v \to \varkappa\,(u) \cup \varkappa\,(v)$ *are homotopic). If the* X_i^n, $i = 1, 2$, $n = 0, 1, \ldots$ *are finitely generated free then* $\varkappa$ *is a morphism of differential graded coalgebras over* R *up to a homotopy.*

(c) *Under the assumptions of* (a), $\operatorname{Ext}_S(\mathsf{Z}, R') = H(\operatorname{Hom}_S(X, R'))$ *is a graded algebra. Under the assumptions of* (b), *there are morphisms of graded* R*-algebras*

$$\operatorname{Ext}_{S_1}(\mathsf{Z}, R_1) \otimes_R \operatorname{Ext}_{S_2}(\mathsf{Z}, R_2) \xrightarrow{\alpha}$$

$$H(\operatorname{Hom}_{S_1}(X_1, R_1) \otimes_R \operatorname{Hom}_{S_2}(X_2, R_2)) \xrightarrow{H(\varkappa)} \operatorname{Ext}_{S_1 \otimes S_2}(\mathsf{Z}, R_1 \otimes_R R_2)$$

which are natural in S_1, S_2, R_1, R_2.

(d) *Suppose that under the assumptions of* (a) *there is a resolution* $0 \leftarrow \mathsf{Z} \leftarrow X$ *of finitely generated free left* S*-modules, and that* $R' = R$. *Further, suppose that the following conditions are satisfied*

$$\operatorname{Tor}_1^R(P, Q) = 0 \qquad (*)$$

for $\quad P = B\,(\operatorname{Hom}_S(X, R)),\quad \operatorname{Ext}_S(\mathsf{Z}, R),\quad Q = P,\quad Z\,(\operatorname{Hom}_S(X, R))$.

Then $\operatorname{Ext}_S(X, R)$ *is a Hopf algebra in natural way. If, in addition to the conditions of* (c), *we have the conditions of* (d) *above for* S_i, X_i, R_i, $i = 1, 2$, *in place of* S, X, R, *then the morphisms in* (c) *are morphisms of Hopf algebras.*

The question now arises whether one can find circumstances under which the important algebra morphism $H\,(\varkappa)$ of Theorem 1.6 (c) is in fact an isomorphism.

Let us now assume that $S_i \to \mathsf{Z}$, $i = 1, 2$, are augmented Hopf algebras (although a good deal of the following would go through for rings alone). Suppose that $T \to \mathsf{Z}$ is an augmented Hopf algebra and that $\varphi_i\colon S_i \to T$ are morphisms of augmented Hopf algebras. Observe that for any left S_i-module A_i there is a natural isomorphism of $T \otimes T$-modules

$$\chi\colon (T \otimes T) \otimes_{S_1 \otimes S_2} (A_1 \otimes A_2) \to (T \otimes_{S_1} A_1) \otimes (T \otimes_{S_2} A_2)$$

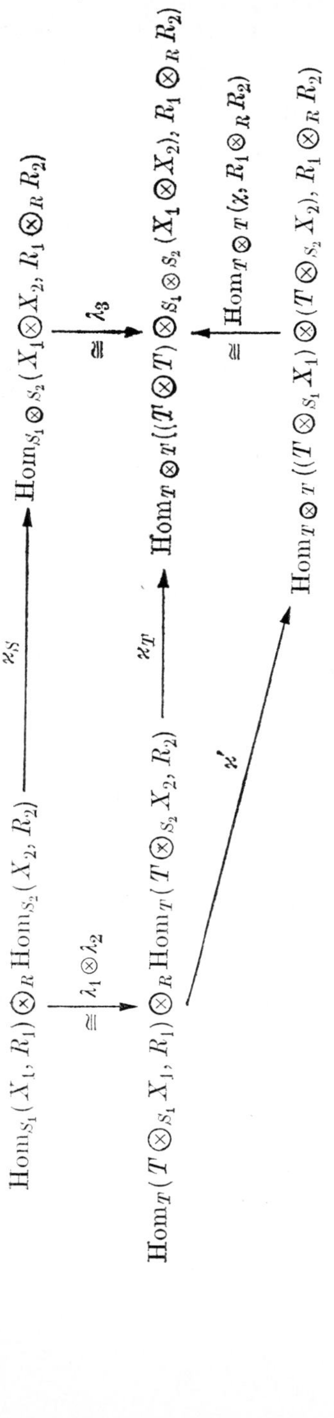

defined by $\chi\ ((t_1 \otimes t_2) \otimes (a_1 \otimes a_2)) = (t_1 \otimes a_1) \otimes (t_2 \otimes a_2)$. Further recall that for any left T-module R_i there is a natural isomorphism

$$\lambda_i \colon \mathrm{Hom}_{S_i}(A_i, R_i) \to \mathrm{Hom}_T(T \otimes_{S_i} A_i, R_i).$$

(In other words, the functor $T \otimes_{S_i} -$ from the category of S_i-modules to the category of T-modules is coadjoint to the forgetful functor from the category of T-modules to the category of S_i-modules.) Now we assume that R_i is an R-algebra as in Theorem 1.6 (b) and consider the diagram in the adjoining column, where $\varkappa'$ is defined by the diagram exactly as $\varkappa$ (see Def. 1.1). This diagram commutes by Lemma 1.3. Suppose now that in the resolutions X_i, every term is finitely generated and free over S_i. By the additivity of the tensor product, then, $T \otimes_{S_i} X_i^n$ is finitely generated and free over T for each n. By the additivity of Hom in the first argument, it follows easily that $\varkappa'$ is in fact an isomorphism. This then implies that $\varkappa_S$ is an isomorphism. Recall that up to homotopy we may replace $\mathrm{Hom}_{S_i}(X_i, R_i)$ by $\mathrm{Hom}_{S_i}(X_i', R_i)$ for any projective $S_i \otimes S_2$-resolution $0 \leftarrow \mathsf{Z} \leftarrow X_i'$. Thus we have the following Theorem:

Theorem II. *Let $S_i \to \mathsf{Z}$, $i = 1, 2$, and $T \to \mathsf{Z}$ be augmented Hopf algebras over Z with a suitable left T-module action and suppose that there are morphisms $S_i \to T$ of augmented Hopf algebras. Suppose further that R_i, $i = 1, 2$ are left T-modules and algebras over the commutative ring R such that $t \cdot (r\, r_i) = (t \cdot r)\, r_i$ for $r \in R$, $r_i \in R_i$, $t \in T$. Assume that Z has resolutions by finitely generated free left S_i-modules, $i = 1, 2$. Let $S_3 = S_1 \otimes S_2$ and let $0 \leftarrow \mathsf{Z} \leftarrow X_i$ be arbitrary projective S_i resolutions, $X_i^0 = S_i$, $i = 1, 2, 3$. Then the morphism*

$$\varkappa \colon \mathrm{Hom}_{S_1}(X_1, R_1) \otimes_R \mathrm{Hom}_{S_2}(X_2, R_2) \to \mathrm{Hom}_{S_1 \otimes S_2}(X_3, R_1 \otimes_R R_2)$$

of graded R-modules given in Definition 1.1 is a homotopy eqivalence and preserves cup products up to homotopy. If all X_i^n, $i = 1, 2$, $n = 0, 1, \ldots$ are finitely generated free and $X_3 = X_1 \otimes X_2$, then $\varkappa$ is an isomorphism. Further, there is an isomorphism of graded algebras

$$H(\varkappa): H(\mathrm{Hom}_{S_1}(X_1, R_1) \otimes_R \mathrm{Hom}_{S_2}(X_2, R_2)) \to \mathrm{Ext}_{S_1 \otimes S_2}(\mathbf{Z}, R_1 \otimes_R R_2)$$

which is natural in S_1, S_2, R_1, R_2, and there is a natural morphism of graded algebras

$$\mathrm{Ext}_{S_1}(\mathbf{Z}, R_1) \otimes_R \mathrm{Ext}_{S_1}(\mathbf{Z}, R_2) \to \mathrm{Ext}_{S_1 \otimes S_2}(\mathbf{Z}, R_1 \otimes_R R_2).$$

This theorem has a long series of very useful corollaries concerning the cohomology of groups.

Corollary 1.7. *Let $f_i: G_i \to G_3$, $i = 1, 2$, be morphisms of groups G_i into a group G_3. Let R_i, $i = 1, 2$, be left G_3-modules and thus left G_i-modules via f_i. Also assume that R_i, $i = 1, 2$, is an algebra over a commutative ring R such that $r(g \cdot r_i) = g \cdot (r\, r_i)$, $r \in R$, $r_i \in R_i$, $g \in \Gamma(G_i)$, where $\Gamma(G_i)$ is the integral group ring of G_i. Then there is an isomorphism of graded algebras*

$$H(\mathrm{Hom}_{\Gamma(G_1)}(X_1, R_1) \otimes_R \mathrm{Hom}_{\Gamma(G_2)}(X_2, R_2)) \to H(G_1 \times G_2, R_1 \otimes_R R_2),$$

where $0 \leftarrow \mathbf{Z} \leftarrow X_i$ is a finitely generated free $\Gamma(G_i)$-resolution of $\mathbf{Z}$. The isomorphism is natural in G_1, G_2, R_1 and R_2.

Proof. We apply Theorem II with $S_i = \Gamma(G_i)$, $i = 1, 2$, $T = \Gamma(G_3)$ and observe that $S_1 \otimes S_2 \cong \Gamma(G_1 \times G_2)$ naturally and that any group ring $\Gamma(G) \xrightarrow{\varepsilon} \mathbf{Z}$ with the natural augmentation defined by $\varepsilon(g) = 1$ for $g \in G$ is an augmented Hopf-algebra with the comultiplication defined by

$$\Gamma(G) \xrightarrow{\Gamma(\Delta)} \Gamma(G \times G) \longrightarrow \Gamma(G) \otimes \Gamma(G)$$

with the diagonal map $\Delta: G \to G \times G$.

Note that Corollary 1.7 applies for any pair of groups G_1 and G_2 with trivial action on R_1, resp. R_2: Simply let $G_3 = 1$.

Corollary 1.8. *Let everything be as in Corollary 1.7 and assume in addition that there are resolutions such that with the complexes $X_i^* = \mathrm{Hom}_{\Gamma(G_i)}(X_i, R_i)$, the following conditions are satisfied:*

$$\mathrm{Tor}_1^R(B(X_1^*), B(X_2^*)) = 0, \quad \mathrm{Tor}_1^R(H(G_1, R_1), B(X_2^*)) = 0,$$
$$\mathrm{Tor}_1^R(B(X_1^*), Z(X_2^*) = 0, \quad \mathrm{Tor}_1^R(H(G_1, R_1), Z(X_2^*)) = 0.$$

Then there is a natural exact sequence

$$0 \to H(G_1, R_1) \otimes_R H(G_2, R_2) \to H(G_1 \times G_2, R_1 \otimes_R R_2)$$
$$\to \operatorname{Tor}_1^R(H(G_1, R_1), H(G_2, R_2)) \to 0,$$

in which the first morphism is a morphism of graded algebras and the second is of degree 1. *The sequence splits non-naturally if R is a principal ideal domain and the X_i are free. If R is a field, then the last term is zero and the first morphism is an isomorphism.*

If R is a principal ideal domain and the resolutions X_i are free, then the hypotheses of the corollary are automatically satisfied.

Proof. This follows from Theorem II and the Künneth theorem (Lemma 1.5).

Corollary 1.9. *Let everything be as in Corollary* 1.7 *and assume in addition that* $\mathbf{Z}$ *has a projective resolution by left $\Gamma(G_1)$-modules such that*

$$H_1^* = \operatorname{Hom}_{\Gamma(G_1)}(X_1, R_1)$$

satisfies the following conditions:

(a) $B(X_1^*)$ *is projective as an R-module.*

(b) $H(G_1, R_1)$ *is a projective R-module.*

Then there is a natural isomorphism of graded algebras

$$H(G_1, R_1) \otimes_R H(G_2, R_2) \to H(G_1 \times G_2, R_1 \otimes_R R_2).$$

Proof. The hypotheses of Corollary 1.8 are satisfied and the last term in the Künneth sequence vanishes.

Corollary 1.10. *Let G_1 be a finite cyclic group, G_2 a group, and R_1, R_2 algebras over the commutative ring R such that R_1 is projective over R. If R_i is considered as a trivial G_i-left module, then there is a natural isomorphism of graded R-algebras*

$$H(G_1, R_1) \otimes_R H(G_2, R_2) \to H(G_1 \times G_2, R_1 \otimes_R R_2),$$

provided that $(\operatorname{ord} G_1) R_1 = 0$.

Proof. The hypotheses of Corollary 1.9 are satisfied with the familiar resolution of $\mathbf{Z}$ by $\Gamma(G_1)$-modules such as, e. g., given in Cartan-Eilenberg [11], p. 250ff.

Note that Corollary 1.10 of course allows some easy generalisations if one uses finite induction. The following is one such, where $H^+(G, R)$ denotes the ideal of $H(G, R)$ generated by elements of strictly positive degree.

Corollary 1.11. *Let G be a finite abelian group. Then there is a decomposition into a direct sum $G = G_1 \oplus \cdots \oplus G_n$ with cyclic groups G_i of order z_i with $z_i | z_{i+1}$*

and this decomposition is unique up to isomorphism. Let R_i, $i = 1, \ldots, n$ be commutative rings and R-algebras such that the characteristic of R_i divides z_i. Suppose that R_i is projective over R. Then

$$H(G_1, R_1) \otimes_R \cdots \otimes_R H(G_n, R_n) \cong H(G, R_1 \otimes_R \cdots \otimes_R R_n)$$

as graded rings. In particular

$$H(G_1, \mathbf{Z}/z\,\mathbf{Z}) \otimes \cdots \otimes H(G_n, \mathbf{Z}/z\,\mathbf{Z}) \cong H(G, \mathbf{Z}/z\,\mathbf{Z})$$

for any z which divides z_n. The exponent of $H^+(G, \mathbf{Z}/z\,\mathbf{Z})$ is z.

Proof by induction. Note that due to our knowledge of $H(G, R)$ for cyclic G (Cartan-Eilenberg [11], p. 250), the exponent of $H^+(G_i, \mathbf{Z}/z\,\mathbf{Z})$ is z.

If R is a commutative ring and M an R-module, we denote with $A(M) \subset R$ the annihilator ideal of M, i. e. $A(M) = \{r \in R: r \cdot M = \{0\}\}$. If $r \in R$ annihilates M or N, it also annihilates $M \otimes_R N$ and $\mathrm{Tor}_1^R(M, N)$. If

$$0 \to M_1 \to M_2 \to M_3 \to 0$$

is an exact sequence of R-modules, then $A(M_2) = A(M_1) \cap A(M_3)$.

Corollary 1.12. *Suppose that the hypotheses of Corollaries 1.7 and 1.8 are satisfied. Then*

$$A(H^+(G_1, R_1)) \cap A(H^+(G_2, R_2) \subset A(H^+(G_1 \times G_2, R_1 \otimes_R R_2)).$$

Proof. If we let $M = H^+(G_1, R_1)$, $N = H^+(G_2, R_2)$, then we have

$$\begin{aligned} A(M) \cap A(N) \subset A(M \otimes_R N) \cap A(\mathrm{Tor}_1^R(M, N)) \cap A(M \otimes_R R_2) \\ \cap A(\mathrm{Tor}_1^R(M, R_2)) \cap A(R_1 \otimes N) \cap A(\mathrm{Tor}_1^R(R_1, N)) \\ = A(H^+(G_1 \times G_2, R_1 \otimes_R R_2)), \end{aligned}$$

using Corollary 1.8 and the preceding remarks.

Corollary 1.13. *Let R be a principal ideal domain.*

(a) *Then the exponent of $H^+(G_1 \times G_2, R_1 \otimes R_2)$ divides the least common multiple of the exponents of $H^+(G_i, R_i)$, $i = 1, 2$.*

(b) *If G is a direct product of the groups G_i, $i = 1, \ldots, n$, then the exponent of $H^+(G, R)$ divides the least common multiple of the exponents of $H^+(G_i, R)$ for $i = 1, \ldots, n$.*

Proof. (a) is a consequence of Corollary 1.12 above. (b) follows by induction since for a principal ideal domain the Künneth theorem is applicable.

Corollary 1.14. *Let G be a finite abelian group and M an abelian group considered as a trivial G-module. Then the exponent of $H^+(G, M)$ is the greatest common divisor of the exponents of G and M.*

Proof. For cyclic G, $\exp H^+(G, \mathsf{Z}) = \exp G$ (Cartan-Eilenberg [11], p. 250ff.). Hence, by Corollary 1.13 (b), for any finite abelian group G, $\exp H^+(G, \mathsf{Z})$ divides $\exp G$. However, $H^2(G, \mathsf{Z}) \cong \hat{G} \cong G$ (MacLane [32], p. 117), and thus $\exp H^+(G, \mathsf{Z}) = \exp G$. Now for an arbitrary abelian group M, by the universal coefficient theorem we have a split exact sequence

$$0 \to H(G, \mathsf{Z}) \otimes M \to H(G, M) \to \operatorname{Tor}(H^+(G, \mathsf{Z}), M)) \to 0.$$

Thus the exponent of $H^+(G, M)$ is divided by the exponent of

$$H^1(G, M) = \operatorname{Hom}(G, M),$$

which is the greatest common divisor of $\exp G$ and $\exp M$, and divides the least common multiple of the exponents of $H^+(G, \mathsf{Z}) \otimes M$ and $\operatorname{Tor}(H^+(G, \mathsf{Z}), M)$. Since $-\otimes M$ and $\operatorname{Tor}(-, M)$ are additive and $H^i(G, \mathsf{Z})$ is finite for each i (see e.g. Evens [17]), the exponent of $H^+(G, M)$ divides the least common multiple of the exponents of $Z \otimes M$ and $\operatorname{Tor}(Z, M)$, where Z is a cyclic group of order $\exp H^+(G, \mathsf{Z})$. But $\operatorname{Tor}(Z, M)$ is naturally isomorphic to

$$\operatorname{Hom}(\operatorname{Hom}(Z, \mathsf{R}/\mathsf{Z}), M)$$

(see Cartan-Eilenberg [11], p. 137), hence (non-canonically) isomorphic to $\operatorname{Hom}(Z, M)$. The exponent of $Z \otimes M$ and $\operatorname{Hom}(Z, M)$ is the greatest common divisor of the order of Z and the exponent of M. Thus the assertion follows.

Parts of the discussion in the proof of Corollary 1.14 may be expanded slightly:

By the universal coefficient theorem, for any group G and any abelian group M, there is an exact sequence

$$0 \to H(G, \mathsf{Z}) \otimes M \to H(G, M) \to \operatorname{Tor}(H^+(G, \mathsf{Z}), M) \to 0.$$

For a finite abelian group A there is a natural isomorphism

$$\operatorname{Tor}(A, M) \to \operatorname{Hom}(\operatorname{Hom}(A, \mathsf{R}/\mathsf{Z}), M)$$

(see Cartan-Eilenberg [11], p. 137). For a finite group G, each $H^i(G, \mathsf{Z})$ is finite for $i > 0$ (see e.g. Evens [17]). Denoting $\operatorname{Hom}(-, \mathsf{R}/\mathsf{Z})$ with $\hat{}$, we thus obtain a split exact sequence

$$0 \to H(G, \mathsf{Z}) \otimes M \to H(G, M) \to \operatorname{Hom}(H^+(G, \mathsf{Z})\hat{}, M) \to 0.$$

We preserve exactness by tensoring with a second abelian group N:

$$\begin{aligned} 0 \to H(G, \mathsf{Z}) \otimes M \otimes N &\to H(G, M) \otimes N \\ &\to \operatorname{Hom}(H^+(G, \mathsf{Z})\hat{}, M) \otimes N \to 0. \end{aligned}$$

But likewise we have the exact sequence

$$0 \to H(G, \mathsf{Z}) \otimes M \otimes N \to H(G, M \otimes N)$$
$$\to \operatorname{Hom}(H^+(G, \mathsf{Z})\hat{}, M \otimes N) \to 0.$$

So far everything is natural. But now we observe, that under some non-natural isomorphism the last terms in these exact sequences are (non-naturally) isomorphic if M and N are finite: For if A is an abelian group we have the natural isomorphisms

$$\operatorname{Hom}(A, M) \cong \operatorname{Hom}(\hat{M}, \operatorname{Hom}(A, \mathsf{R}/\mathsf{Z}))$$
$$\cong \operatorname{Hom}(A \otimes \hat{M}, \mathsf{R}/\mathsf{Z}) \cong (A \otimes \hat{M})\hat{}.$$

So $\operatorname{Hom}(A, M) \otimes N \cong (A \otimes \hat{M})\hat{} \otimes N$ and

$$\operatorname{Hom}(A, M \otimes N) \cong (A \otimes (M \otimes N)\hat{})\hat{}$$

If everything in sight is finite, then under non-natural isomorphism we may omit all hats. Thus the non-natural isomorphy becomes apparent. In fact the last terms are, in a non-canonical fashion, isomorphic to

$$H^+(G, \mathsf{Z}) \otimes M \otimes N$$

under these circumstances. Thus

$$H^+(G, M) \otimes N \cong H^+(G, M \otimes N)$$

non-naturally, if M and N are finite abelian groups.

Corollary 1.15. *If G, M, and N are finite abelian groups and G acts trivially on M and N, then there is a non-canonical isomorphism of graded groups*

$$H(G, M) \otimes N \cong H(G, M \otimes N).$$

In particular, for any such group M,

$$H^+(G, M) \cong H^+(G, \mathsf{Z}(\exp G)) \otimes M,$$

and if $\exp G \mid z$, then

$$H(G, \mathsf{Z}/z\mathsf{Z}) \cong \mathsf{Z}/z\mathsf{Z} \oplus H(G, \mathsf{Z}(\exp G)).$$

Section 2

Special free resolutions for finite abelian groups

In this section, we will be concerned with abelian groups G, G' etc. However, since we will be talking about group algebras, it is convenient to consider multiplicatively written abelian groups M, M', etc. There are, of course, two completely equivalent ways of presentation of the group ring, the preference being determined only by the circumstances of the discussion. The first presentation of the group ring $\Gamma(M)$ of M is the one preferred by algebraists: $\Gamma(M)$ is considered to be the free abelian group generated by M, i. e. the set of elements $\sum \{a_m \cdot m \colon m \in M\}$ with multiplication defined by linear extension of the multiplication of the elements $m \in M$. In this case $M \subset \Gamma(M)$. The augmentation $\varepsilon' \colon \Gamma(M) \to \mathbf{Z}$ is defined by $\varepsilon'(m) = 1$ for each $m \in M$. The second presentation, preferred by analysts, is that the group ring $L(G)$ is to be considered as the additive group of all functions $a \colon G \to \mathbf{Z}$ with multiplication $a\,a'(g) = \sum \{a(h)\,a'(-h+g) \colon h \in G\}$, and augmentation ε' with $\varepsilon'(a) = \sum \{a(g) \colon g \in G\}$. In this case there is an injection of G into the multiplicative semi-group of $L(G)$ defined by $g \to e_g$ with $e_g(g) = 1$ and $e_g(h) = 0$ for $h \neq g$. For the present purposes we adopt the first presentation. A standard free resolution (I — 3.10) will then have the form $0 \longrightarrow F \xrightarrow{f} F \xrightarrow{\pi} M \to 0$ with basis elements e_i of F, $f = f_1 \oplus \cdots \oplus f_n$, where $f_i(e_i) = z_i\,e_i$ with natural numbers $z_i | z_{i+1}$. We will have $\pi = \pi_1 \cdots \pi_n$ and denote $\pi(e_i) = \pi_i(e_i)$ with u_i.

Definition 2.1. In the group ring $\Gamma(M)$, we denote with $\tilde{z}_i$ the element $u_i + u_i^2 + \cdots + u_i^{z_i}$; note $u_i^{z_i} = 1$. Observe that the augmentation ideal $\ker \varepsilon'$ is generated by the $u_i - 1$ and that $\tilde{z}_i(u_i - 1) = 0$.

Given a standard resolution $0 \longrightarrow F \xrightarrow{f} F \xrightarrow{\pi} M \longrightarrow 0$, we will denote with $\bar{F}$ the ground ring extension $\Gamma(M) \otimes F$, where as usual an unmarked tensor product sign refers to the tensor product over $\mathbf{Z}$. We will abbreviate the elements $1 \otimes x \in \bar{F}$ by $\bar{x}$. Thus $\bar{F}$ as a $\Gamma(M)$-module has a basis consisting of the elements $\bar{e}_i$. Let us abbreviate $\Gamma(M)$ by S. The S-module morphism $\bar{f} \colon \bar{F} \to \bar{F}$ will be defined by $\bar{f}(\bar{e}_i) = \tilde{z}_i \cdot \bar{e}_i = \tilde{z}_i \otimes e_i$. Then

$$E_2(\bar{f}) = P_S \bar{F} \otimes_S \wedge_S \bar{F}$$

is a well defined differential bi-graded algebra with differential, derivation and coderivation $d_{\bar{f}}$ of bi-degree $(2, -1)$ according to Section I — 2.

We will now define a second differential and derivative $\bar{\partial}_f$ by first defining a differential and coderivation $\bar{\partial}\colon \bigwedge_S \bar{F} \to \bigwedge_S \bar{F}$ by

$$\bar{\partial}(\bar{e}_s) = \sum \{(u_i - 1)\, \bar{e}_i \wedge \bar{e}_s \mid i = 1, \ldots, n\}, \quad s \in S(q).$$

Indeed, we have the following lemma:

Lemma 2.2. *Suppose that A is a finitely generated free S-module with a basis*

$$a_i,\ i = 1, \ldots, n,$$

and $\hat{A} = \mathrm{Hom}_S(A, S)$ its dual with the dual basis

$$f_i,\ i = 1, \ldots, n;$$

let us identify $(\bigwedge A)\hat{}$ with $\bigwedge \hat{A}$ under the natural isomorphism of I — 1.9, so that $f_s(a_t) = 0$ if $s \neq t$ and $= 1$ if $s = t$, where

$$f_s = f_{s(1)} \wedge \cdots \wedge f_{s(q)}, \quad a_t = a_{t(1)} \wedge \cdots \wedge a_{t(q)}.$$

Suppose that $v_i \in S$, $i = 1, \ldots, n$. Then $\partial\colon \bigwedge A \to \bigwedge A$ given by

$$\partial(a_s) = \textstyle\sum_{i=1}^{q} (-1)^{i-1}\, v_{s(i)} \cdot a(s_i)$$

is a differential and derivation of degree -1 by I — 2.2. Its dual $\bar{\partial}\colon \bigwedge \hat{A} \to \bigwedge \hat{A}$ is a differential and coderivation of degree $+1$ and is given by

$$\hat{\partial}(g) = \sum v_i \cdot f_i \wedge g \qquad (i = 1, \ldots, n), \quad g \in \bigwedge \hat{A}.$$

Proof. It is clear that the dual of a differential is a differential and that the dual of a derivation of degree -1 is a coderivation of degree $+1$. Let $s \in S(q)$ and $t \in S(q+1)$. Then

$$\begin{aligned}\hat{\partial}(f_s)(a_t) = f_s(\partial a_t) &= f_s\left(\textstyle\sum_{i=1}^{q} (-1)^{i-1}\, v_{t(i)} \cdot a(t_i)\right) \\ &= \textstyle\sum_{i=1}^{q} \{(-1)^{i-1}\, v_{t(i)} \colon s = t_i\}.\end{aligned}$$

On the other hand,

$$\begin{aligned}&\left(\textstyle\sum_{i=1}^{n} v_i \cdot f_i \wedge f_s\right)(a_t) \\ &= \textstyle\sum_{i=1}^{n} (-1)^{i-1}\, v_i f_{s(1)}(a_{t(1)}) \cdots f_{s(i-1)}(a_{t(i-1)})\, f_i(a_{t(i)})\, f_{s(i)}(a_{t(i+1)}) \cdots \\ &\qquad\qquad \cdots f_{s(q-1)}(a_{t(q)}) \\ &= \textstyle\sum_{i=1}^{q} \{(-1)^{i-1}\, v_{t(i)} \colon s = t_i\}.\end{aligned}$$

This finishes the proof.

Now we simply define $\partial_f\colon E_2(\bar{f}) \to E_2(\bar{f})$ by $\partial_f = 1_{P_S(\bar{F})} \otimes \partial$.

Lemma 2.3. $d_{\bar{f}}\, \partial_f + \partial_f\, d_{\bar{f}} = 0$.

Proof. Since ∂_f and $d_{\bar{f}}$ are both $P_S\bar{F}$-module endomorphisms, we need consider only elements of the form $1 \otimes \bar{e}_s$:

$$\begin{aligned} d_{\bar{f}}\,\partial_f(1 \otimes \bar{e}_s) &= d_{\bar{f}}\left(\textstyle\sum_{i=1}^{n} (u_i - 1) \cdot 1 \otimes (\bar{e}_i \wedge \bar{e}_s)\right) \\ &= \textstyle\sum_{i=1}^{n} ((u_i - 1)\, \tilde{z}_i \cdot \bar{e}_i \otimes \bar{e}_s \\ &\quad - \textstyle\sum_{j=1}^{q} (-1)^{j-1}(u_i - 1)\, \tilde{z}_{s(j)} \cdot \bar{e}_{s(j)} \otimes (\bar{e}_i \wedge \bar{e}(s_j)) \\ &= -\textstyle\sum_{i=1}^{n} \sum_{j=1}^{q} (u_i - 1)\, \tilde{z}_{s(j)} \cdot \bar{e}_{s(j)} \otimes (-1)^{j-1} \cdot \bar{e}_i \wedge e(s_j), \end{aligned}$$

where $s \in S(q)$. On the other hand we have

$$\begin{aligned} \partial_f\, d_{\bar{f}}\,(1 \otimes e_s) &= \partial_f\left(\textstyle\sum_{j=1}^{q} (-1)^{j-1}\, \tilde{z}_{s(j)} \cdot e_{s(j)} \otimes e\,(s_j)\right) \\ &= \textstyle\sum_{j=1}^{q} (-1)^{j-1}\, \tilde{z}_{s(j)} \cdot e_{s(j)} \otimes \sum_{i=1}^{n} (u_i - 1) \cdot e_i \wedge e\,(s_j) \end{aligned}$$

and this agrees with $-\, d_{\bar{f}}\,\partial_f\,(1 \otimes \bar{e}_s)$.

Lemma 2.4. *$D_f = d_{\bar{f}} + \partial_f$ is a differential and coderivation of $E_2(\bar{f})$.*

Proof. The sum of two coderivations is a coderivation, and the sum of two anticommuting differentials is a differential.

Definition 2.5. We will denote the differential bi-graded S-coalgebra $E_2(\bar{f})$ with coderivation D_f by $E(f)$. Note that $E(f)$ is a bi-graded S-algebra but that D_f is not a derivation. We let $\hat{E}(f) = \mathrm{Hom}_S(E(f), S)$ denote the dual S-Hopf algebra with differential and derivation $\hat{D}_f = \hat{d}_f + \hat{\partial}_f$.

Note that as a Hopf algebra $\hat{E}(f) = \hat{P}_S\,\bar{F} \otimes_S \wedge_S \hat{\bar{F}}$ by I — 1.22, and that this in turn may be expressed in the form

$$S \otimes \hat{P}F \otimes \wedge \hat{F} = S \otimes \hat{\mathbf{Z}}[X_1, \ldots, X_n] \otimes \Lambda(n).$$

The next goal is to show that $\hat{E}(f)$ is in fact exact. For this purpose we prove first that $\hat{E}$ is an exponential functor in the sense that

$$\hat{E}(f \oplus g) \cong \hat{E}(f) \otimes \hat{E}(g).$$

Lemma 2.6. *Let $0 \longrightarrow F_k \xrightarrow{f_k} F_k \longrightarrow G_k \longrightarrow 0$, $k = 1, 2$, two standard resolutions of the finite abelian groups G_k. Then $f = f_1 \oplus f_2$, $f\colon F \to F$, $F = F_1 \oplus F_2$ defines a standard resolution of $G_1 \oplus G_2$. Let $S_k = \Gamma(G_k)$ and*

$$S = S_1 \otimes S_2 \cong \Gamma(G_1 \oplus G_2)$$

be the group algebras. Then there is a natural isomorphism of differential graded S-algebras $E(f) \cong E(f_1) \otimes E(f_2)$ with $\otimes = \otimes_{\mathbf{Z}}$ and the S-module action on the tensor product given by

$$(s_1 \otimes s_2)\,(e_1 \otimes e_2) = s_1 \cdot e_1 \otimes s_2 \cdot e_2.$$

Proof. We have the following isomorphisms

$$\begin{aligned}\hat{E}(f) &= \mathrm{Hom}_S(P_S\bar{F} \otimes_S \wedge_S \bar{F}, S) \cong \mathrm{Hom}_S(S \otimes PF \otimes \wedge F, S)\\ &\cong \mathrm{Hom}\,(PF \otimes \wedge F,\ \mathrm{Hom}_S(S, S)) = \mathrm{Hom}(PF \otimes \wedge F, S)\\ &\cong S \otimes \mathrm{Hom}(PF \otimes \wedge F, \mathsf{Z})\\ &\cong S_1 \otimes S_2 \otimes \mathrm{Hom}\,((PF_1 \otimes \wedge F_1) \otimes (PF_2 \otimes \wedge F_2), \mathsf{Z})\\ &\cong (S_1 \otimes \mathrm{Hom}(PF_1 \otimes \wedge F_1, \mathsf{Z})) \otimes (S_2 \otimes \mathrm{Hom}(PF_2 \otimes \wedge F_2, \mathsf{Z})).\end{aligned}$$

By the same computation we have $\hat{E}(f_k) \cong S_k \otimes \mathrm{Hom}(PF_k \otimes \wedge F_k, \mathsf{Z})$. It follows that $\hat{E}(f) \cong \hat{E}(f_1) \otimes \hat{E}(f_2)$ as S-Hopf algebras. It remains to show that under this isomorphism, the differentials D_f and $D_{f_1} \otimes 1 + e \otimes D_{f_2}$ with $e(x) = (-1)^p x$ for homogeneous elements of total degree p correspond to each other.

We observe that there is a natural S-module morphism

$$\beta\colon \mathrm{Hom}_{S_1}(E(f_1), S_1) \otimes \mathrm{Hom}_{S_2}(E(f_2), S_2) \to \mathrm{Hom}_S(E(f_1) \otimes E(f_2), S)$$

given by

$$\beta(\varphi \otimes \psi)(a \otimes b) = \varphi(a) \otimes \psi(b) \in S_1 \otimes S_2 = S.$$

Since this is a restriction and corestriction of the natural isomorphism

$$\mathrm{Hom}\,(E(f_1), S_1) \otimes \mathrm{Hom}\,(E(f_2), S_2) \to \mathrm{Hom}\,(E(f_1) \otimes E(f_2), S)$$

(the $E(f_k)$ being free abelian groups), then β must be injective. On the other hand, the $S = S_1 \otimes S_2$-module $\mathrm{Hom}_S(E(f_1) \otimes E(f_2), S)$ is generated by the homomorphisms $\beta(f^{(1)}_{\sigma,s} \otimes f^{(2)}_{\sigma',s'})$ with

$$f^{(k)}_{\sigma,s}(\bar{e}^{(k)}_\tau \otimes_S \bar{e}^{(k)}_t) = 1$$

if $\sigma = \tau$ and $s = t$ and $= 0$ otherwise. We may conclude that β is surjective, hence an isomorphism.

In order to prove that the differential on $\hat{E}(f)$ corresponds to the differential $D_{f_1} \otimes 1 + e \otimes D_{f_2}$ on $\hat{E}(f_1) \otimes \hat{E}(f_2)$ it will suffice to show that the differential D_f on $E(f)$ corresponds to the differential $D_{f_1} \otimes 1 + e \otimes D_{f_2}$ on

$$E(f_1) \otimes E(f_2).$$

As far as the summand $d_{\bar{f}}$ is concerned, it can be shown that $d_{\bar{f}}$ on $E(f)$ corresponds to $d_{\bar{f}_1} \otimes 1 + e \otimes d_{\bar{f}_2}$ on $E(f_1) \otimes E(f_2)$ with a proof virtually identical to the proof of I — 2.21; the difference is only in the ring over which the tensor product is taken. We therefore concentrate on ∂_f. Here it is more convenient to show directly that $\hat{\partial}_f$ on $\hat{E}(f)$ corresponds to $\hat{\partial}_{f_1} \otimes 1 + e \otimes \hat{\partial}_{f_2}$

on $\hat{E}(f_1) \otimes \hat{E}(f_2)$. If we write

$$\hat{E}(f) \cong \hat{P}_S F \otimes_S \wedge_S \hat{\bar{F}} \cong \hat{P}_S \bar{F} \otimes_S (S \otimes \wedge \hat{F}) \cong (\hat{P}_S \bar{F} \otimes_S S) \otimes \wedge \hat{F}$$
$$= \hat{P}_S \bar{F} \otimes \wedge \hat{F}$$

then the differential $\hat{\partial}_f$ on $\hat{E}(f)$ corresponds to the derivation ∂ on $\hat{P}_S \bar{F} \otimes \wedge \hat{F}$ given by $\partial(\hat{P}\bar{F} \otimes 1) = 0$ and

$$\partial(1 \otimes \hat{e}_i) = (u_i - 1)(\hat{\bar{e}}_i \otimes 1)$$

with the generators $\hat{e}_i$ of $\hat{F}$ and $\hat{\bar{e}}_i$ of $\hat{\bar{F}}$. This is the derivation described in Lemma I — 2.2, and part (c) of that lemma applies now to show that under the isomorphism

$$(\hat{P}_{S_1} \hat{\bar{F}}^{(1)} \otimes \wedge \hat{F}^{(1)}) \otimes (\hat{P}_{S_2} \bar{F}^{(2)} \otimes \wedge \hat{F}^{(2)}) \to \hat{P}_S \bar{F} \otimes \wedge \hat{F}$$

the derivation $\partial^{(1)} \otimes 1 + e \otimes \partial^{(2)}$ corresponds to the derivation ∂. This finishes the proof of the Lemma.

Lemma 2.7. *Let $f\colon \mathsf{Z} \to \mathsf{Z}$ be any morphism which is not zero, and let*

$$f(1) = z.$$

Then the differential $\hat{D}_f$ of $\hat{E}(f)$ is exact.

Proof. The cokernel of f is the cyclic group G of order z with generator u. Let $\tilde{z} = u + u^2 + \cdots + u^z$ in $S = \Gamma(G)$. We observe $\wedge \mathsf{Z} = \mathsf{Z} \oplus \mathsf{Z}$ as groups so that $E(f)$ has just two rows and the differential D_f follows a zig-zag pattern between the elements of these two rows in such a fashion that the whole configuration may be displayed linearly into the complex

$$S \xleftarrow{u-1} S \xleftarrow{\tilde{z}} S \xleftarrow{u-1} S \xleftarrow{\tilde{z}} S \xleftarrow{u-1} S \cdots$$

It is well known (and easily checked) that this complex is exact. (See e. g. Cartan-Eilenberg [11], p. 251.)

Remark. In the complex appearing in the proof of the preceding lemma, the image of $S \xrightarrow{u-1} S$ is exactly the kernel of the augmentation $\varepsilon\colon S \longrightarrow \mathsf{Z}$ given by $\varepsilon(g) = 1$ for all $g \in G$. Hence $0 \longleftarrow \mathsf{Z} \xleftarrow{\varepsilon} E(f)$ is in fact a resolution of Z by free S-modules.

For arbitrary $f\colon F \to F$ we also consider the augmentation $\varepsilon\colon E(f) \to \mathsf{Z}$ which is zero on all components of positive degree and agrees with the natural augmentation $S \to \mathsf{Z}$ on the degree zero component. We decompose f into a direct sum $f = f_1 \oplus \cdots \oplus f_n$ of morphisms according to the canonical resolution of coker f. Then we have a commuting diagram

$$\begin{array}{ccccc} 0 & \longleftarrow & \mathbf{Z} & \xleftarrow{\varepsilon} & \hat{E}(f) \\ & & \uparrow \cong & & \uparrow \cong \\ 0 & \longleftarrow & \mathbf{Z} \otimes \cdots \otimes \mathbf{Z} & \xleftarrow{\varepsilon_1 \otimes \cdots \otimes \varepsilon_n} & \hat{E}(f_1) \otimes \cdots \otimes \hat{E}(f_n) \end{array}$$

Since $0 \leftarrow \mathbf{Z} \leftarrow \hat{E}(f_k)$ is a resolution of $\mathbf{Z}$ by free S_k-modules, then the top row is a resolution of $\mathbf{Z}$ by free $S = S_1 \otimes \cdots \otimes S_n$-modules.

Thus we have obtained the following crucial Lemma.

Lemma 2.8. *$0 \leftarrow \mathbf{Z} \leftarrow \hat{E}(f)$ is a resolution of $\mathbf{Z}$ by free $\Gamma(G)$-modules where $G = \operatorname{coker} f$.*

Now let R be an arbitrary S-module. Then $\operatorname{Hom}_S(\hat{E}(f), R)$ is a complex of R-modules whose homology is exactly the cohomology $H(G, R)$ of G with coefficients in the G-module R.

Now $\operatorname{Hom}_S(\hat{E}(f), R) = \operatorname{Hom}_S(\operatorname{Hom}_S(E(f), S), R)$, and in each homogeneous component, $E(f)$ is finitely generated free over S.

Lemma 2.9. *Let S be a commutative ring with identity and R an S-module. Let A be an S-module. Then there is a natural morphism of R-modules*

$$\varphi\colon R \otimes_S A \to \operatorname{Hom}_S(\operatorname{Hom}_S(A, S), R)$$

which sends $r \otimes_S a$ into $r \cdot a$, where $a\colon \operatorname{Hom}_S(A, S) \to R$ is the S-module morphism given by $a(f) = f(a)$. If A is finitely generated free, then φ is an isomorphism.

Proof. These assertions are straightforward.

This then implies that

$$\begin{aligned} \operatorname{Hom}_S(\hat{E}(f), R) &\cong R \otimes_S E(f) = R \otimes_S P_S \bar{F} \otimes_S \wedge_S \bar{F} \\ &\cong R \otimes_S (S \otimes PF \otimes \wedge F) = R \otimes PF \otimes \wedge F. \end{aligned}$$

The differential obtained from the differential on $\hat{E}(f)$ on $R \otimes RF \otimes \wedge F$ is given by $D = d + \partial$ with $d(R \otimes PF \otimes 1) = 0$ and

$$\begin{aligned} & d(r \otimes a \otimes e_s) \\ & = \sum \{(-1)^{i-1} \tilde{z}_{s(i)} \cdot r \otimes a\, e_{s(i)} \otimes e(s_i) \mid i = 1, \ldots, q\}, \; s \in S(q), \\ & \partial(r \otimes a \otimes x) \\ & = \sum \{(u_i - 1) \cdot r \otimes a \otimes (e_i \wedge x) \mid i = 1, \ldots, n\}, \end{aligned}$$

where the e_i are the generators of the free abelian group F.

If R is a ring, then $R \otimes PF \otimes \wedge F$ is a graded commutative R-algebra, d is a derivation and coderivation, and

$$R \otimes PF \otimes \wedge F \cong P_R\,(R \otimes F) \otimes_R \wedge_R(R \otimes F) = E_2(\varphi)$$

with derivation d_φ where $\varphi\colon R \otimes F \to R \otimes F$ is given by

$$\varphi(1 \otimes e_i) = \tilde{z}_i \otimes e_i.$$

If S acts trivially on R then $\partial = 0$ since $(u_i - 1) \cdot r = 0$ for all i and all $r \in R$, and $\tilde{z}_i \cdot r = z_i \cdot r$ for $r \in R$. We then have

$$\mathrm{Hom}_S\,(\hat{E}(f), R) \cong R \otimes E_2(f) = E_2(R \otimes f),$$

so that $H(G, R) \cong E_3(R \otimes f)$ follows, where $E_2(f) = PF \otimes \wedge F$ with derivation d_f and

$$E_2(R \otimes f) = P_R(R \otimes F) \otimes_R \wedge_R\,(R \otimes F)$$

with derivation $d_{R \otimes F}$. This ties in the cohomology theory of finite abelian groups with the theory of the spectral algebras which we developed in Chapter I, Sections 2 and 3.

The preceding discussion gives us a perfectly satisfactory isomorphism between the cohomology algebra of a finite abelian group G over a ring R and the spectral algebra $E_3(f)$ with $G = \operatorname{coker} f$. However, if we were to determine the functoriality of the isomorphism, then we would obtain the wrong variance. Firstly, f in the standard resolution $0 \longrightarrow F \xrightarrow{f} F \longrightarrow G \longrightarrow 0$ is indeed essentially uniquely determined by G, whereas a morphism

$$\varphi\colon G \to G'$$

of finite groups will determine a pair (α, β) of morphisms of free abelian groups in the commuting diagram

$$\begin{array}{ccccccccc} 0 & \longrightarrow & F & \xrightarrow{f} & F & \longrightarrow & G & \longrightarrow & 0 \\ & & \downarrow{\scriptstyle\alpha} & & \downarrow{\scriptstyle\beta} & & \downarrow{\scriptstyle\varphi} & & \\ 0 & \longrightarrow & F' & \xrightarrow{f'} & F' & \longrightarrow & G' & \longrightarrow & 0 \end{array}$$

only up to homotopy (see I — 3.9, I — 3.11). The pair (α, β) in any event determines a morphism of the spectral algebras

$$E_r(R \otimes F) \to E_r(R \otimes F'), \quad r = 2, 3,$$

whereas φ should determine an arrow in the other direction. If a finite group G is given, we therefore choose an essentially unique canonical resolution

$$0 \longrightarrow F_0 \xrightarrow{f_0} F_0 \longrightarrow G \longrightarrow 0$$

and define $F = \hat{F}_0 = \mathrm{Hom}(F_0, \mathsf{Z})$ and $f = \hat{f}_0 = \mathrm{Hom}(f_0, \mathsf{Z})$. Then, in a non-natural way we still have $G \cong \mathrm{coker}\, f$. If we now apply the entire preceding discussion to F and f, we obtain

$$R \otimes PF \otimes \wedge F = R \otimes P\hat{F}_0 \otimes \wedge \hat{F}_0$$

as a contra-variant functor of f. In case that R is a ring and G operates trivially, we obtain the isomorphic copies

$$\begin{aligned} &P_R(R \otimes F) \otimes_R \wedge_R(R \otimes F) \\ &= P_R(R \otimes \mathrm{Hom}(F_0, \mathsf{Z})) \otimes_R \wedge_R(R \otimes \mathrm{Hom}(F_0, \mathsf{Z})). \end{aligned}$$

But $R \otimes \mathrm{Hom}(F_0, \mathsf{Z})$ is naturally isomorphic to $\mathrm{Hom}(F_0, R)$, and under this isomorphism the endomorphism $R \otimes f$ becomes replaced by the endomorphism $\mathrm{Hom}(f_0, R)$. Thus $E_2(\varphi)$ is then isomorphic to $E_2\,(\mathrm{Hom}(f, R))$ and there is now a functorial isomorphism of graded R-algebras

$$H(\mathrm{coker}\, f, R) \cong E_3(\mathrm{Hom}(f, R))$$

of functors from the category of elementary morphisms f of free abelian groups into the category of graded R-algebras.

Let us collect all the relevant information in the following main theorem:

Theorem III (Fundamental Theorem of Cohomology of Finite Abelian Groups). *Let G be a finite abelian group and* $0 \longrightarrow F \xrightarrow{f} F \xrightarrow{\pi} G \longrightarrow 0$ *be exact with an elementary morphism f of free abelian groups* (I — 2.11). *Let S be the group ring of G,* let e_i, $i = 1, \ldots, n$, *be generators of F such that $u_i = \pi(e_i)$ are appropriate generators for the cyclic summands of G. Define $\tilde{z}_i$ in S to be $u_i + u_i^2 + \cdots + u_i^{z_i}$ where z_i is the order of the i-th direct summand in G. Denote* $\mathrm{Hom}(F, \mathsf{Z})$ *with $\hat{F}$ and* $\mathrm{Hom}(f, \mathsf{Z})$ *with $\hat{f}$. Let R be a G-module; then it is also an S-module. On $R \otimes P\hat{F} \otimes \wedge \hat{F}$ consider differentials d and ∂ defined as follows, where $\hat{e}_i \in \mathrm{Hom}(F, \mathsf{Z})$ are elements of a dual basis to the basis*

$$\{e_i \mid i = 1, \ldots, n\}$$

of F:

$$\begin{aligned} &d(r \otimes a \otimes 1) = 0, \\ &d(r \otimes a \otimes \hat{e}_s) = \sum \{(-1)^i\, \tilde{z}_i \cdot r \otimes a\, \hat{e}_{s(i)} \otimes \hat{e}(s_i) \mid i = 1, \ldots, q\} \end{aligned}$$

for $s \in S(\partial)$,

$$\partial(r \otimes a \otimes b) = \sum \{(u_i - 1) \cdot r \otimes a \otimes (\hat{e}_i \wedge b) \mid i = 1, \ldots, n\}$$

for all $r \in R$, $a \in PF$, $b \in \wedge F$. *(For the definition of* $\hat{e}_s$ *etc. see* I — 2.10.)

Then we have the following conclusions:

A) $d + \partial$ *is a differential of degree* -1 *on* $R \otimes P\hat{F} \otimes \wedge \hat{F}$ *relative to its total gradation, and its graded cohomology R-module is isomorphic to*

$$H(G, R) = H(\operatorname{coker} f, R).$$

The isomorphism is functorial in f.

B) *If* R *is a ring, then* d *is a derivation and if* d *is transported to the isomorphic copy* $P_R(R \otimes \hat{F}) \otimes_R \wedge_R(R \otimes \hat{F})$ *of* $R \otimes P\hat{F} \otimes \wedge \hat{F}$, *then it becomes the derivation and coderivation* d_φ *of* $E_2(\varphi) = P_R(R \otimes F) \otimes_R \wedge_R(R \otimes F)$ *given according to* I — 2.2 *by the R-module morphism* $\varphi\colon R \otimes \hat{F} \to R \otimes \hat{F}$ *defined by* $\varphi(1 \otimes \hat{e}_i) = \tilde{z}_i \otimes \hat{e}_i$.

C) *If the action of* G *on* R *is trivial then* $\partial = 0$, *and if* R *is a ring, then the bigraded differential module* $R \otimes P\hat{F} \otimes \wedge \hat{F}$ *relative to* $d + \partial$ *is a differential R-Hopf algebra isomorphic to*

$$R \otimes E_2(\hat{f}) \cong E_2(\operatorname{Hom}(f, R))$$

with $\hat{f}\colon \hat{F} \to \hat{F}$, *resp.* $\operatorname{Hom}(f, R)\colon \operatorname{Hom}(F, R) \to \operatorname{Hom}(F, R)$, *where*

$$E_2(\hat{f}) = P\hat{F} \otimes \wedge \hat{F}$$

with derivation and coderivation $d_{\hat{f}}$ *and* $E_2(\operatorname{Hom}(f, R))$ *is*

$$P_R(\operatorname{Hom}(F, R)) \otimes_R \wedge_R(\operatorname{Hom}(R, F))$$

with derivation and coderivation $d_{\operatorname{Hom}(f, R)}$ *(according to* I — 2.2). *In particular, the graded R-algebra* $H(G, R) = H(\operatorname{coker} f, R)$ *relative to the cup product is isomorphic to* $E_3(\operatorname{Hom}(f, R))$ *with its total gradation. The isomorphism is a natural isomorphism of cofunctors from the category of elementary morphisms* f *of free abelian groups of finite rank to the category of graded R-algebras.*

D) *If* $R = \mathsf{Z}$, *then* $H(G, \mathsf{Z}) \cong E_3(\hat{f})$ *as graded rings providing the action of* G *on* Z *is trivial.*

In particular all this applies to the coefficient rings Z and $\mathsf{Z}/z\mathsf{Z}$. Before we draw some conclusions concerning the coefficient group R/Z, we recall the following well known fact:

Lemma 2.10. *For a fixed ring* S, *for which* $\operatorname{Ext}^i_S(A, \mathsf{R}) = 0$ *for* $i > 0$, *there is an isomorphism of abelian groups* $\operatorname{Ext}^i_S(A, \mathsf{R}/\mathsf{Z}) \to \operatorname{Ext}^{i+1}_S(A, \mathsf{Z})$ *for* $i > 0$.

In particular, if S is the group ring of a finite group, we have

$$\mathrm{Ext}^i_S(\mathsf{Z}, \mathsf{R}) = H(G, \mathsf{R}) = 0$$

and obtain isomorphisms $H^i(G, \mathsf{R}/\mathsf{Z}) \to H^{i+1}(G, \mathsf{Z})$ for $i > 0$.

Proof. This follows directly from the long exact sequence for $\mathrm{Ext}_S(A, -)$ derived from the exact sequence $0 \to \mathsf{Z} \to \mathsf{R} \to \mathsf{R}/\mathsf{Z} \to 0$. As to the groups, compare e. g. MacLane [32], p. 117.

Now we have the following proposition.

Proposition 2.11. *Under the general hypotheses of Theorem* III, *we have the following isomorphisms, all of which are natural in f:*

(a) $H(G, \mathsf{R}/\mathsf{Z}) = H(\mathrm{coker}\, f, \mathsf{R}/\mathsf{Z}) \cong H(\mathsf{R}/\mathsf{Z} \otimes E_2(\mathrm{Hom}(f, \mathsf{Z})))$.

(b)
$$\begin{array}{ccc} H^i(G, \mathsf{R}/\mathsf{Z}) & \longrightarrow & H^{i+1}(G, \mathsf{Z}) \\ \downarrow\cong & & \downarrow\cong \\ H^i(\mathrm{coker}\, f, \mathsf{R}/\mathsf{Z}) & \cong & H^{i+1}(\mathrm{coker}\, f, \mathsf{Z}) \cong \Sigma\{E_3^{p,q}(\mathrm{Hom}(f, \mathsf{Z})), \\ & & p + q = i + 1\}. \end{array}$$

Proof. By Theorem III, $E_3(\mathrm{Hom}(f, \mathsf{Z})) \cong H(\mathrm{coker}\, f, \mathsf{Z})$. The assertions then follow readily from Lemma 2.10 above and I — 3.18, 3.17.

In the following discussion we utilize the next purely categorical lemma.

Lemma 2.12. *Let $\mathfrak{A}$, $\mathfrak{B}$, $\mathfrak{C}$ be categories and $S: \mathfrak{A} \to \mathfrak{B}$ and $T_1, T_2: \mathfrak{B} \to \mathfrak{C}$ be functors. Suppose that*

(i) *$S: \mathfrak{A}(A, A') \to \mathfrak{B}(SA, SA')$ is always surjective and S is surjective on objects.*

(ii) *There is a natural transformation $\eta: T_1 S \to T_2 S$.*

Then there is a natural transformation $\xi: T_1 \to T_2$ such that $\eta = \xi S$.

Proof. Let B be an object of $\mathfrak{B}$. Suppose that A, A' are objects in $\mathfrak{A}$ with $SA = SA' = B$. Then by (i) there is a morphism $f \in \mathfrak{A}$ with $Sf = 1_B$. By the naturality of η there is a commutative diagram

$$\begin{array}{ccc} T_1 SA & \xrightarrow{\eta_A} & T_2 SA \\ \downarrow 1_B = T_1 Sf & & \downarrow T_2 Sf = 1_B \\ T_1 SA' & \xrightarrow{\eta_{A'}} & T_2 SA' \end{array}$$

Thus, the morphism $\eta_A\colon T_1B \to T_2B$ is independent of the choice of A. We define it to be ξ_B. We have to show the naturality of ξ. Let $g\colon B \to B'$ be a morphism.

By (i) there is a morphism $f\colon A \to A'$ with $g = Sf$. By naturality of η we have $(T_2Sf)\,\eta_A = \eta_{A'}(T_1Sf)$. By definition this means $(T_2\,g)\,\xi_B = \xi_{B'}(T_1\,g)$. This shows the naturality of ξ.

We will apply this with the category of all elementary morphisms of finitely generated free abelian group as $\mathfrak{A}$, the category of finite abelian groups as $\mathfrak{B}$ and *coker* as S. By I — 3.9, this functor satisfies the conditions 2.12 (i). The functors T_i will be functors such as $P\,\mathrm{Ext}\,(-, R)$ and $H(-, R)$.

In the case of trivial action we now exploit our knowledge of the edge terms in the algebra $E_2(f)$.

Theorem 2.13. *Let G be a finite abelian group and A an abelian group. Then there is a pair of natural morphisms of abelian groups*

$$\mathrm{Hom}(\wedge G, A) \xrightarrow{\varrho} H(G, A) \xrightarrow{\varrho'} \mathrm{Hom}(\wedge G, A)$$

with $\varrho'\,\varrho = id$ and which are bijective in dimension 1. If $A = R$ is a ring ϱ and ϱ' are morphisms of graded algebras, and there is a pair of natural morphisms of graded algebras

$$P_R\,\mathrm{Ext}(G, R) \xrightarrow{\tau} H(G, R) \xrightarrow{\tau'} P_R\,\mathrm{Ext}(G, R)$$

with $\tau'\,\tau = id$. In this case, there is then a natural morphism

$$\omega\colon P_R\,\mathrm{Ext}(G, R) \otimes_R \mathrm{Hom}(\wedge G, R) \to H(G, R)$$

of graded algebras, where the domain algebra is given the total degree.

Proof. Let $0 \longrightarrow F \xrightarrow{f} F \xrightarrow{\pi} G \longrightarrow 0$ is a standard resolution of G (see I — 3.9, 3.10). Thus we have an isomorphism

$$H\,(A \otimes E_2(\mathrm{Hom}\,(f, \mathbf{Z}))) \to H\ (\mathrm{coker}\,f, A)$$

of graded abelian groups which is natural in f. The first result then follows from I — 3.14 and Lemma 2.12. If $A = R$ is a ring, then the morphism mentioned above preserves multiplication. There is, under these circumstances, also an injection of algebras

$$P\,\mathrm{Ext}(\mathrm{coker}\,f, R) \to E_3\,(\mathrm{Hom}\,(f, R))$$

which is natural in f (Proposition I — 3.14). The last algebra, as a graded algebra, is naturally isomorphic to $H(\mathrm{coker}\,f, R)$ by Theorem III. Thus there is a natural injection of algebras

$$P\,\mathrm{Ext}(\mathrm{coker}\,f,\,R) \to H(\mathrm{coker}\,f,\,R).$$

This implies that there is a natural injection

$$P\,\mathrm{Ext}(G,\,R) \to H(G,\,R)$$

for finite abelian groups G by Lemma 2.12. The last assertion concerning ω is a consequence of the preceding statements.

In concluding this section, we want to compare some aspects of the resolution used in the preceding with those of the bar resolution for a finite abelian group. These aside observations are not in any way necessary to the remainder of the theory. First we recall the definition of the bar resolution. Let G be a finite abelian multiplicatively written group, $S = \Gamma(G)$ its group ring. We define $X^0 = S$, $X^1 = S \otimes \mathsf{Z}^G, \ldots, X^n = S \otimes \mathsf{Z}^{G^n}$ so that we can form the graded free S-module $X^* = X^0 \oplus X^1 \oplus \cdots$. We let the augmentation $X^* \to \mathsf{Z}$ be defined by $\varepsilon(s) = \varepsilon'(s)$, where $\varepsilon' : S \to \mathsf{Z}$ is the augmentation of S and by $\varepsilon(x) = 0$ for $x \in X^i$, $i > 0$. Now we introduce a differentiation d on X^* of degree -1 by defining $d^i\colon X^i \to X^{i-1}$ in terms of the basis elements $[g_1|\cdots|g_i]$ of Z^{G^i} as follows: The basis element $[g_1|\cdots|g_i]$ is the function vanishing in all points of $G \times \cdots \times G = G^i$ except the point $(g_1, \ldots, g_i)$, where it takes the value 1; and

$$\begin{aligned} &d^i(1 \otimes [g_1|, \ldots, |g_i|) \\ &= g_1 \otimes [g_2|\cdots|g_i] \\ &\quad + \sum \{(-1)^j(1 \otimes [g_1|\cdots|g_j\,g_{j+1}|\cdots g_i]) : j = 1, \ldots, i-1\} \\ &\quad + (-1)^i[g_1|\cdots|g_{i-1}]. \end{aligned}$$

The comultiplication of augmented rings

$$\begin{array}{ccc} S & \longrightarrow & S \otimes S \\ \downarrow & & \downarrow \\ \mathsf{Z} & \longrightarrow & \mathsf{Z} \otimes \mathsf{Z} \end{array}$$

extends to a comultiplication of augmented differential graded S-modules

$$\begin{array}{ccc} X^* & \stackrel{\gamma}{\longrightarrow} & X^* \otimes X^* \\ \downarrow & & \downarrow \\ \mathsf{Z} & \longrightarrow & \mathsf{Z} \otimes \mathsf{Z} \end{array}$$

which is unique up to homotopy. Such a comultiplication defines the structure of a graded differemtial algebra on $\mathrm{Hom}_S(X^*, A)$, where A is any left S-module and R-algebra over a commutative ring R with identity satisfying

$$r \cdot (s \cdot a) = s \cdot (r \cdot a),$$

via

$$\mathrm{Hom}_S(X^*, A) \otimes_R \mathrm{Hom}_S(X^*, A) \to \mathrm{Hom}_{S \otimes S}\, X^* \otimes X^*, A \otimes_R A)$$
$$\xrightarrow{\mathrm{Hom}_S(\gamma, A \otimes_R A)} \mathrm{Hom}_S\,(X^*, A \otimes_R A) \longrightarrow \mathrm{Hom}_S\,(X^*, A).$$

(See also Definition 1.2.) This applies in particular to the case $A = R$.

If the action of G on A is trivial, then we have a natural isomorphism

$$\mathrm{Hom}_S(X^*, A) \cong \mathrm{Hom}\,(\mathbf{Z} \otimes_S X^*, A) \cong \mathrm{Hom}\,(Y^*, A),$$

where Y^* is the complex $\sum \{Z^{G^i} : i = 0, 1, \ldots\}$ with the differential d defined by

$$d^i [g_1 | \cdots | g_i]$$
$$= [g_2 | \cdots | g_i] + \sum \{(-1)^j [g_1 | \cdots | g_j g_{j+1} | \cdots | g_i] : j = 1, \ldots, j-1\}$$
$$+ (-1)^i [g_1 | \cdots | g_{i-1}].$$

It is clear that the complexes $\mathrm{Hom}_S(X^*, A)$, resp. $\mathrm{Hom}\,(Y^*, A)$, are naturally isomorphic to the complex $C(G, A) = \Sigma \{C^i(G, A) : i = 0, \ldots,\}$, where $C^i(G, A) = A^{G^i}$, and the differential is defined by

$$(df)\,(g_1, \ldots, g_i)$$
$$= g_1 f(g_2, \ldots, g_i) - f(g_1 g_2, g_3, \ldots, g_i) + - \cdots$$
$$+ (-1)^i f(g_1, \ldots, g_{i-1}),$$

resp.

$$= f(g_2, \ldots, g_i) - f(g_1 g_2, g_3, \ldots, g_i) + \cdots.$$

(Since the action of G on A is trivial, these two differentials are then the same.) The symmetric group S_q on the set $\{1, \ldots, q\}$ operates on the left on the module $X^q = S^{G^q}$ such that $\sigma^{-1}[g_1 | \cdots | g_q] = [g_{\sigma(1)} | \cdots | g_{\sigma(q)}]$. Let us define the antisymmetrisation endomorphism $a: X^n \to X^n$ by

$$a = \sum \{(\mathrm{sgn}\,\sigma)\,\sigma : \sigma \in S_q\}.$$

Lemma 2.14. $\sum \{\mathrm{sgn}\,\sigma [g_{\sigma(1)} | \cdots | g_{\sigma(i)} g_{\sigma(i+1)} | \cdots | g_q] : \sigma \in S_q\} = 0$ *in* X^{q-1}.

Proof. Let $\sigma \in S_q$ be the transposition $(i, i+1)$. Find a cross section $C \in S_q$ for the subgroup $\{1, (i, i+1)\}$ so that $C \cup (i, i+1)\, C = S_q$. In the sum, the pair of terms belonging to σ and $(i, i+1)\,\sigma$, for $\sigma \in C$, cancel out since they differ only by a sign.

With the aid of this lemma, it is easy to compute that

$$\begin{aligned} & da[g_1|\cdots|g_n] \\ & = \sum\{\operatorname{sgn}\sigma\, g_{\sigma(1)}[g_{\sigma(2)}|\cdots|g_{\sigma(q)}] + (-1)^q[g_{\sigma(1)}|\cdots|g_{\sigma(n-1)}]: \sigma\in S_q\} \\ & = \sum\{\operatorname{sgn}\sigma\,(g_{\sigma(1)}-1)\,[g_{\sigma(2)}|\cdots|g_{\sigma(q)}]: \sigma\in S_q\} \\ & = \sum\{(-1)^{i-1}\{\operatorname{sgn}\tau\,(g_{s(i)}-1)\,[g_{s_i(\tau(1))}|\cdots|g_{s_i(\tau(n-1))}]: \\ & \qquad \tau\in S_{q-1}, i = 1,\ldots,q\} \\ & = \textstyle\sum_1^n\{(-1)^{i-1}(g_i-1)\,a\,[g_i, g_2, \ldots, g_q]\}, \end{aligned}$$

where s is the identity function of $\{1, \ldots, n\}$ and where we use the notation of I—2.10. Now we formulate the following proposition:

Proposition 2.15. *Assume that* $f\colon F\to F$ *is an elementary morphism of finitely generated free abelian groups and* G *its cokernel in multiplicative form. Let* $X(f)$ *be the differential algebra* $S\otimes\wedge F$ *with* $d(1\otimes e_i) = ((u_i-1)\otimes 1)$. *Define the* S*-module morphism* $k\colon X(f)\to X^*$ *into the graded module of the bar resolution for* G *by* $k(1\otimes e_s) = a\,[u_{s(1)}|\cdots|u_{s(q)}]$ *for* $s\in S(q)$ *and by* $k(1\otimes 1) = 1\in S = X^0$. *Then* k *is a morphism of complexes over the identity map of* $\mathbf{Z}$. *Note that as such it is uniquely determined up to homotopy.*

Proof. We have

$$\begin{aligned} & k\,d\,(1\otimes e_s) \\ & = k\sum\{(-1)^{i-1}(u_{s(i)}-1)\otimes e(s_i): i = 1,\ldots,q\} \\ & = \sum\{(-1)^{i-1}(u_{s(i)}-1)\,a\,[u_{s_i(1)}|\cdots|u_{s_i(q-1)}]\} \\ & = da\,[u_{s(1)}|\cdots|u_{s(q)}] = dk\,(1\otimes e_s) \end{aligned}$$

by the preceding remarks. Further $\varepsilon_f(1\otimes 1) = 1 = \varepsilon^*(1)$.

Let us assume that A is a trivial left G-module. Then there is a morphism

$$\operatorname{Hom}_S(k, A)\colon \operatorname{Hom}_S(X^*, A)\to \operatorname{Hom}_S(X(f), A),$$

and since k is in fact a morphism of complexes over $\mathbf{Z}$, and since X^* is exact and $X(f)$ free, k as well as $\operatorname{Hom}_S(k, A)$ are unique up to homotopy. Because of trivial action, we then deduce a morphism

$$\operatorname{Hom}(\mathbf{Z}\otimes_S k, A)\colon \operatorname{Hom}(\mathbf{Z}\otimes X^*, A)\to \operatorname{Hom}(\mathbf{Z}\otimes_S X(f), A),$$

which may be written in the equivalent form $k^*\colon C(G, A)\to \operatorname{Hom}(\wedge F, A)$ since

$$\operatorname{Hom}(\mathbf{Z}\otimes_S X^*) \cong \operatorname{Hom}(Y^*, A) \cong C(G, A)$$

and

$$\mathbf{Z}\otimes_S X(f) \cong 1\otimes\wedge F \cong \wedge F.$$

It is now not difficult to observe that $(k^*f)(e_s) = af(u_{s(1)}, \ldots, u_{s(q)})$ for $s \in S(q)$, where as usual the e_i are the basis elements of F, where the $u_i = \pi(e_i)$ are their images in G, and where $e_s = e_{s(i)} \wedge \cdots \wedge e_{s(q)}$. It is, of course, understood that a is the antisymmetrisation operator on $C^q(G, A)$ defined by

$$(af)(g_1, \ldots, g_q) = \sum \{(\operatorname{sgn} \sigma) f(g_{\sigma(1)}, \ldots, g_{\sigma(q)}) : \sigma \in S_q\}.$$

Obviously k^* factors through $\operatorname{Hom}(\wedge G, A) \to \operatorname{Hom}(\wedge F, A)$. We denote with $k' : C(G, A) \to \operatorname{Hom}(\wedge G, A)$ the map so defined. We then have

$$(k'f)(u_{s(1)} \wedge \cdots \wedge u_{s(q)}) = (af)(u_{s(1)}, \ldots, u_{s(q)}).$$

Thus we have the following Corollary:

Corollary 2.16. *Let G be a finite abelian group and A a trivial left G-module. Then there is a surjective morphism*

$$k' : C(G, A) \to \operatorname{Hom}(\wedge G, A)$$

of graded abelian groups such that

$$(k'f)(u_{s(1)} \wedge \cdots \wedge u_{s(q)}) = (af)(u_{s(1)}, \ldots, u_{s(q)}).$$

The diagram

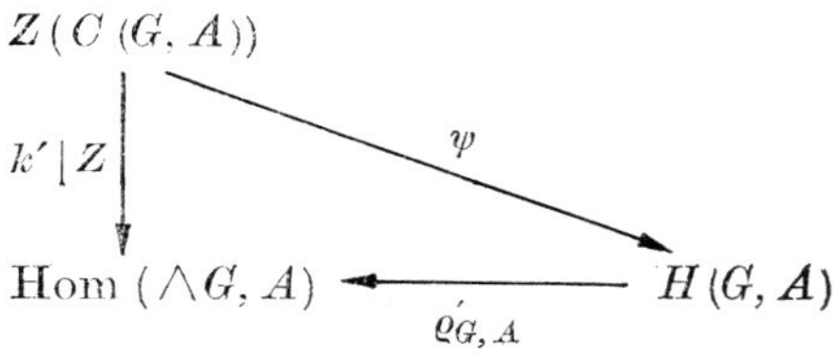

commutes, where ϱ' is the retraction given in Theorem 2.13.

Lemma 2.17. *If $f \in C^q(G, A)$ is multilinear, then*

(i) $df = 0$.

(ii) *For any $\sigma \in S_q$, one has $f - (\operatorname{sgn} \sigma)\sigma f \in B(C^q(G, A))$ provided $q > 1$ and f is multilinear.*

(iii) $2f \in B(C^q(G, A))$ *provided $q > 1$ and f is multilinear symmetric.*

(iv) *If f is multilinear, then $q!f - af \in B(C^q(G, A))$, $q > 1$.*

Proof. (i) is tedious but straightforward.

(ii) If a multilinear map f with $q > 1$ arguments is given and $\tau = (i, i+1)$ is a transposition of adjacent numbers, define

$$h \in C^{q-1}(G, A)$$

by

$$h(g_1, \ldots, g_{q-1}) = f(g_1, \ldots, g_i, g_i, \ldots, g_{q-1})$$

and compute $f + \tau f = dh$. Since every $\sigma \in S_q$ is a product of transpositions of adjacent numbers, the result follows by finite induction.

(iii) and (iv) follow directly from (ii).

We may identify Hom $(\wedge^q G, A)$ with the group of q-argument alternating multilinear functions in $C^q(G, A)$ and hence, by 2.17 (i), in $Z(C^q(G, A))$. If $\psi: Z(C(G, A)) \to H(G, A)$ again denotes the cohomology map, then one might conjecture at first glance that the restriction of ψ to Hom $(\wedge G, A)$ may be the natural morphism $\varrho_{G,A}$ of Theorem III. But this is not quite the case, for we have the following Corollary:

Corollary 2.18. *If* Hom $(\wedge^q G, A)$ *is identified with the group of alternating q-linear maps, and if* $f \in$ Hom $(\wedge^q G, A) \subset Z(C^q(G, A))$, *then* $q! = k'$ *and* $\psi f = q!\, \varrho_{G,A} f$ *for* $q > 1$.

Proof. The assertion follows directly from Corollary 2.16 and Lemma 2.17 (iv).

This shows that the natural transformation ϱ of Theorem III may not be retrieved from the bar resolution in the trivial fashion. The actual connection is given by the theory of divided powers developed by H. Cartan [10], 7—01 ff.

Section 3

About the cohomology of finite abelian groups in the case of trivial action

In this section, we exploit the results of Section 2 in Chapter I together with the findings of Section 2 in the present Chapter to investigate the structure of the cohomology ring of a finite group when the action on the coefficient ring is trivial.

For the sake of completeness we repeat the standard notation for the section in the first definition

Definition 3.1. Let G be a finite group and $0 \longrightarrow F \xrightarrow{f} F \xrightarrow{\pi} G \longrightarrow 0$ a standard resolution (I—3.10). We will write

$$F = F_1 \oplus \cdots \oplus F_n, \quad G = G_1 \oplus \cdots \oplus G_n, \quad f = f_1 \oplus \cdots \oplus f_n$$

and $\pi = \pi_1 \oplus \cdots \oplus \pi_n$ so that $f_i x = z_i x$ with $z_i | z_{i+1}$ for $i = 1, \ldots, n-1$. We may take $F = \mathsf{Z}^n$ and $F_i = \mathsf{Z}$. The natural number z_n is the *exponent* of G and z_1 is the *coexponent* of G.

Lemma 3.2. *Denote with* $\hat{G}$ *the character group* Hom $(G, \mathsf{R}/\mathsf{Z})$.

(a) *Let* z *be a natural number divisible by* z_n. *Then there are natural isomorphisms of functors*

$$\mathrm{Hom}\,(G, \mathsf{Z}(z)) \cong \hat{G} \cong \mathrm{Ext}\,(G, \mathsf{Z}(z)).$$

(b) *There is a natural isomorphism* $\hat{G} \cong \mathrm{Ext}\,(G, \mathsf{Z})$, *and under the assumption of* (a), *we have then a natural isomorphism*

$$\mathrm{Ext}\,(G, \mathsf{Z}(z)) \cong \mathrm{Ext}\,(G, \mathsf{Z}).$$

Proof. (a) Consider the exact sequence $0 \longrightarrow \mathsf{Z}(z) \xrightarrow{i} \mathsf{R}/\mathsf{Z} \xrightarrow{z} \mathsf{R}/\mathsf{Z} \longrightarrow 0$, and the long exact sequence derived from this sequence:

$$0 \to \mathrm{Hom}\,(G, \mathsf{Z}(z)) \to \hat{G} \to \hat{G} \to \mathrm{Ext}\,(G, \mathsf{Z}(z)) \to 0,$$

where we have $\mathrm{Ext}\,(G, \mathsf{R}/\mathsf{Z}) = 0$ because of the divisibility of R/Z. The morphism

$$\mathrm{Hom}\,(G, i)\colon \mathrm{Hom}\,(G, \mathsf{Z}(z)) \to \mathrm{Hom}\,(G, \mathsf{R}/\mathsf{Z})$$

is an isomorphism since every morphism $G \longrightarrow \mathsf{R}/\mathsf{Z}$ factors through $\mathsf{Z}(z) \xrightarrow{i} \mathsf{R}/\mathsf{Z}$ because G has exponent z and $z_n | z$. Hence the connecting morphism

$$\hat{G} \to \mathrm{Ext}\,(G, \mathsf{Z}(z))$$

is also an isomorphism.

(b) Recall that G is finite and consider the exact sequence

$$0 \to \mathsf{Z} \to \mathsf{R} \to \mathsf{R}/\mathsf{Z} \to 0$$

and a piece of the long exact sequence derived from it:

$$0 = \mathrm{Hom}\,(G, \mathsf{R}) \to \hat{G} \to \mathrm{Ext}\,(G, \mathsf{Z}) \to 0 = \mathrm{Ext}\,(G, \mathsf{R}/\mathsf{Z}).$$

Let R be a commutative ring with identity. We consider R as a trivial G left module. In Section 2, we have shown that there are isomorphisms

$$E_3(\mathrm{Hom}\,(f\, R)) \cong H(G, R)$$

of graded R-algebras and these isomorphisms are natural in f. First we discuss a variety of special results involving rings which are not integral domains but are otherwise quite special. In fact, we will first consider rings of the form $R = \mathsf{Z}/z\mathsf{Z}$.

Proposition 3.3. *Let R' be an integral domain of characteristic* 0 *and z a natural number dividing the coexponent z_1 of G. Set $R = R'/z\,R'$. Then the natural morphism of graded R-algebras*

$$\omega \colon P\,\mathrm{Ext}\,(G, R) \otimes_R \mathrm{Hom}\,(\wedge G, R) \to H(G, R)$$

of Theorem 2.13 *is an isomorphism. Moreover the natural morphism*

$$\varphi \colon \wedge\,\mathrm{Hom}\,(G, R) \to \mathrm{Hom}\,(\wedge G, R)$$

of I—1.6 *is an isomorphism* (I—1.14).

Proof. We have $\mathrm{Hom}\,(f, R) = 0$ under the present circumstances. But then

$$\begin{aligned} E_3(\mathrm{Hom}\,(f, R)) &\cong E_2(\mathrm{Hom}\,(f, R))) \\ &= P\,\mathrm{Hom}\,(F, R) \otimes \wedge\,\mathrm{Hom}\,(F, R). \end{aligned}$$

But the exact sequence of I—3.12 yields

$$\mathrm{Hom}\,(G, R) \cong \mathrm{Hom}\,(F, R) \cong \mathrm{Ext}\,(G, R)$$

and we have a commutative diagram

$$\begin{array}{ccccc} P\,\mathrm{Hom}\,(F, R) & \longrightarrow & E_2(\mathrm{Hom}\,(f, R)) & \longleftarrow & \wedge\,\mathrm{Hom}\,(F, R) \\ \downarrow{\cong} & & \downarrow{\cong} & & \downarrow{\cong} \\ P\,\mathrm{Ext}\,(G, R) & \longrightarrow & E_3(\mathrm{Hom}\,(f, R)) & \longleftarrow & \wedge\,\mathrm{Hom}\,(G, R) \\ \| & & \downarrow{\cong} & & \downarrow{\varphi} \\ P\,\mathrm{Ext}\,(G, R) & \xrightarrow{\ \tau\ } & H\,(G, R) & \xleftarrow{\ \varrho\ } & \mathrm{Hom}\,(\wedge G, R) \end{array}$$

φ is an isomorphism by I—1.14 and the top line is a coproduct diagram. Thus, so is the bottom line. But this is our assertion.

Definition 3.4. Let everything be as in Definition 3.1 and let z be any natural number. Define the elementary morphism $f' \colon F \to F$ (depending on z) by $f' = f'_1 \oplus \cdots \oplus f'_n$ with $f'_i(x) = (z_i, z)\,x$. Let $f' = \underline{f} \oplus \bar{f}$ be defined by $\underline{f} = f'_1 \oplus \cdots \oplus f'_{j-1}$, $\bar{f} = f'_j \oplus \cdots \oplus f'_n$ with $j = \min\{i \colon 1 \leq i \leq n,\ (z_i, z) = z\}$ and $G' = \operatorname{coker} f'$, $\underline{G} = \operatorname{coker} \underline{f}$, $\bar{G} = \operatorname{coker} \bar{f}$. Note that all these definitions depend on z. We have $G' = \underline{G} \oplus \bar{G}$ and $\bar{G} \cong (\mathsf{Z}/(z_n, z)\,\mathsf{Z})^{n-j+1}$. One might call G' the *z-constituent* of G.

Theorem 3.5. *Let R' be an integral domain of characteristic zero, z a natural number and $R = R'/z\,R'$. Then there is an isomorphism of graded algebras* *over R as follows:*

$$H(G, R) \cong H(G', R) \cong H(\underline{G}, R) \otimes P\bar{G}^\wedge \otimes \wedge \bar{G}^\wedge.$$

The isomorphisms are natural in f.

Proof. With the notation of Definition 1.1, we have

$$\operatorname{Hom}(f_i, R) = m_i \operatorname{Hom}(f_i', R)$$

with some unit m_i of R. Then, by Lemma I—2.31, we obtain

$$E_3(\operatorname{Hom}(f, R)) \cong E_3(\operatorname{Hom}(f', R)),$$

whence $H(G, R) \cong H(G', R)$ by Theorem III of Section 2. Further

$$\operatorname{Hom}(f', R) = \operatorname{Hom}(\underline{f}, R) \oplus \operatorname{Hom}(\bar{f}, R)$$

and $\operatorname{Hom}(\bar{f}, R) = 0$. Hence, by I—2.47, we arrive at

$$H(G', R) \cong E_3(\operatorname{Hom}(f', R)) \cong E_3(\operatorname{Hom}(\underline{f}, R) \otimes E_2(\operatorname{Hom}(\bar{f}, R))$$
$$\cong H(\underline{G}, R) \otimes_R P \operatorname{Ext}(\bar{G}, R) \otimes_R \wedge \operatorname{Hom}(\bar{G}, R)$$

by Proposition 3.3. Moreover

$$\operatorname{Ext}(\bar{G}, R) \cong \operatorname{Hom}(\bar{G}, R) \cong \operatorname{Hom}(\bar{G}, \mathsf{Z}/z\,\mathsf{Z} \otimes R')$$
$$\cong \operatorname{Hom}(\bar{G}, \mathsf{Z}/z\,\mathsf{Z}) \otimes R' \cong \bar{G}^\wedge \otimes R' \cong \bar{G}^\wedge \otimes R,$$

using Lemma 3.2. The assertion then follows.

Now we also specialize the group.

Theorem 3.6. *Let R' be an integral domain of characteristic 0, z a natural number, $R = R'/z R'$ and $G \cong \mathsf{Z}(z)^n$. Then there is a natural isomorphism*

$$H(G, R) \cong R \otimes P\hat{G} \otimes \wedge \hat{G} = R \otimes E_2(\varepsilon),$$

where ε is the identity on $\hat{G}$. The differential $R \otimes d_\varepsilon$ is exact and there is a natural isomorphism $H^+(G, R') \cong R \otimes \operatorname{im} d_\varepsilon$. If $H(G, R)$ and $R \otimes P\hat{G} \otimes \wedge \hat{G}$ are identified under the isomorphism above, then $R \otimes d_\varepsilon$ is the Bockstein differential on $H(G, R)$.

Proof. By Lemma 3.2 and Proposition 3.3, we have

$$H(G, R) \cong P \operatorname{Hom}(G, R) \otimes_R \wedge \operatorname{Hom}(G, R).$$

But again $\operatorname{Hom}(G, R)$ is naturally isomorphic to

$$\operatorname{Hom}(G, \mathsf{Z}(z)) \otimes R \cong \hat{G} \otimes R.$$

Further $P_R(R \otimes \hat{G}) = R \otimes P\hat{G}$ and $\wedge_R(R \otimes \hat{G}) \cong R \otimes \wedge \hat{G}$. Thus we obtain the natural isomorphism in question. As to the Bockstein, we apply I — 4.17 with $A = \operatorname{Hom}(F, R) \cong R^n$, $\varphi = z\,\varphi^*$, $\varphi^* = 1_A$. Then $\varphi^{*\prime} = 1_{A'}$, and with the necessary identifications, the assertion follows. The isomorphy $H^+(G, R') \cong R \otimes \operatorname{im} d_\nu$ is now just a reinterpretation of I—4.16.

The naturality in G follows from 2.12.

Corollary 3.7. *Under the hypotheses of Theorem 3.6, the Poincaré series of $H(G, R)$ is*

$$f(t) = 1 + r_1 t + r_2 t^2 + \cdots = \frac{1}{(1-t)^n}$$

and the Poincaré series of $H(G, R')$ is

$$g(t) = 1 + s_2 t^2 + \cdots = 1 + \frac{t}{1+t}\left(\frac{1}{(1-t^n)^n} - 1\right)$$

$$= \frac{1+t-t^n}{(1+t)(1-t^n)} \qquad (s_1 = 0),$$

where $H^i(G, R') = R'^{r_i}$, $H^i(G, R) = R'^{s_i}$, $i = 1, 2, \ldots$ Specifically

$$r_i = \binom{i+n-1}{n-1} = \binom{i+n-1}{i},$$

$$s_i = \sum_{k=0}^{i} (-1)^{i-k-1} \binom{k+n-1}{n-1}.$$

Remark. As a graded R-module, $R \oplus H^+(G, R)$ is isomorphic to the polynomial ring over R in n variables of degree 1.

Proof. The Poincaré series of $\wedge R^n$ is $(1+t)^n$ and that of PR^n is $\frac{1}{(1-t^2)^n}$. Thus

$$f(t) = \frac{(1+t)^n}{(1-t^2)^n} = \frac{1}{(1-t)^n} = \sum_{i=0}^{\infty} \binom{i+n-1}{n-1} t^i.$$

By Theorem 3.6 and Proposition I—4.15 we have a split exact sequence

$$0 \to H^i(G, R') \to (PR^n \otimes \wedge R^n)^i \to H^{i+1}(G, R') \to 0$$

for $i = 1, 2, \ldots$, so that $s_i + s_{i+1} = r_i$ for $i = 1, 2, \ldots, s_0 = s_1 = r_1 = 0$. Therefore we have

$$(g(t) - 1) + \frac{1}{t}(g(t) - 1) = f(t) - 1,$$

hence

$$g(t) = 1 + \frac{t}{1+t}(f(t) - 1),$$

which proves the assertion. The explicit forms of r_i and s_i follow by straightforward computation.

While the isomorphy $H(G, R) \cong R \otimes PG \otimes \wedge G$ of graded rings of Theorem 3.6 generally fails to be true, it may be interesting to observe that there is a non-canonical isomorphism of graded abelian groups of this sort for an arbitrary finite abelian group G. In fact we have the following

Corollary 3.8. *Let G be a finite abelian group and z a natural number which is divided by the exponent of G. Set $R = \mathsf{Z}/z\,\mathsf{Z}$. Then there is a non-natural isomorphism of graded abelian groups*

$$H(G, R) \cong R \otimes P\hat{G} \otimes \wedge \hat{G}.$$

Proof. (a) First we observe that if the assertion is true for $z = \exp G$, then it holds in general. For if the assertion is true with $\exp G$ in place of z, we have

$$H(G, \mathsf{Z}/z\,\mathsf{Z}) = \mathsf{Z}/z\,\mathsf{Z} \oplus H^+(G, \mathsf{Z}/z\,\mathsf{Z}) = \mathsf{Z}/z\,\mathsf{Z} \oplus H^+(G, \mathsf{Z}/(\exp G)\,\mathsf{Z})$$

by Lemma 2.17; but by our assumption

$$H^+(G, \mathsf{Z}/(\exp G)\,\mathsf{Z}) \cong (\mathsf{Z}/(\exp G)\,\mathsf{Z} \otimes P\hat{G} \otimes \wedge \hat{G})^+ = (P\hat{G} \otimes \wedge \hat{G})^+$$

so that indeed

$$H(G, \mathsf{Z}/z\,\mathsf{Z}) \cong \mathsf{Z}/z\,\mathsf{Z} \otimes P\hat{G} \otimes \wedge \hat{G}.$$

(b) We proceed by induction relative to the exponent of G. The basis of the induction is secure by Theorem 3.6 and (a). Now let G be arbitrary and write $G = K \oplus L$ with $L \cong \mathsf{Z}(e)^m$, $e = \exp G$ and $\exp K < e$. Then by Theorem 3.5, we have $H(G, \mathsf{Z}/e\,\mathsf{Z}) = H(K, \mathsf{Z}/e\,\mathsf{Z}) \otimes P\hat{L} \otimes \wedge \hat{L}$. By the induction hypothesis, we may write $H(K, \mathsf{Z}/e\,\mathsf{Z}) \cong \mathsf{Z}/e\,\mathsf{Z} \otimes P\hat{K} \otimes \wedge \hat{K}$. The exponentiality of P and $\wedge$ then implies

$$H(G, \mathsf{Z}/e\,\mathsf{Z}) \cong \mathsf{Z}/e\,\mathsf{Z} \otimes P\hat{G} \otimes \wedge \hat{G}.$$

By section (a) of the proof, if $R = \mathsf{Z}/z\,\mathsf{Z}$ with $e \mid z$, we have

$$H(G, R) \cong R \otimes PG \otimes \wedge G$$

so that induction is guaranteed.

We now turn to the discussion of the case of integral domains and principal ideal domains in particular as coefficient rings.

Proposition 3.9. *With the notation of Definition 3.1, define a sequence of groups $G = G^{(1)}, G^{(2)}, \ldots, G^{(n)}$ and a sequence of bi-graded differential rings $(E_2(i), d_i)$, $i = 1, \ldots, n$ as follows:*

(1) *Let $K_i = G_i \oplus \cdots \oplus G_n$, $i = 1, \ldots, n$ and define $G^{(i)} = K_i/z_{i-1}K_i$, $z_0 = 1$, $i = 1, \ldots, n$.*

(2) *Let* $z^{(i)} = z_i/z_{i-1}$, $i = 1, \ldots, n$, *and let*

$$\Phi_i\colon (\mathsf{Z}/z^{(i)}\mathsf{Z})^{n+1-i} \to (\mathsf{Z}/z^{(i)}\mathsf{Z})^{n+1-i}$$

be defined by

$$\Phi_i(\bar{a}_i, \ldots, \bar{a}_n) = (\bar{a}_i, z^{(i+1)}\bar{a}_{i+1}, \ldots, z^{(i+1)}\cdots z^{(n)}\bar{a}_n)$$

and put $(E_2(i), d_i) = E_2(\Phi_i)$.

Then there are exact sequences

$$0 \to \operatorname{im} d_i \to H(G^{(i)}, \mathsf{Z}) \to H(G^{(i+1)}, \mathsf{Z}) \to 0$$

with $i = 1, \ldots, n-1$, $H(G^{(1)}, \mathsf{Z}) = H(G, \mathsf{Z})$ *and*

$$0 \to \operatorname{im} d_n \to H(G^{(n)}, \mathsf{Z}) \to \mathsf{Z} \to 0.$$

The differentials d_i *also may be identified with the Bockstein differentials on* $H(G^{(i)}, \mathsf{Z}(z^{(i)}))$.

Remark. Note that $G^{(1)} = G$.

Proof. We apply Proposition I—2.35 of Section 2 in Chapter I. We let

$$A = \operatorname{Hom}(F, \mathsf{Z}) = \operatorname{Hom}(\mathsf{Z}^n, \mathsf{Z}) \cong \mathsf{Z}^n, \quad A^i = \mathsf{Z}, \quad i = 1, \ldots, n$$

and

$$\varphi = \operatorname{Hom}(f, \mathsf{Z}) = \varphi_1^1 \oplus \cdots \oplus \varphi_n^n$$

with

$$\varphi_i^i = \operatorname{Hom}(f_i, \mathsf{Z}), \quad \varphi_i^i(a) = z_i a = z^{(1)}\cdots z^{(i)}a.$$

We then define φ_i as in I—2.35. Then

$$\varphi_i\colon \mathsf{Z}^{n+1-i} \to \mathsf{Z}^{n+1-i}$$

with

$$\varphi_i = 1/z_{i-1}(\varphi^{(i)} \oplus \cdots \oplus \varphi^{(n)}),$$

i. e. with

$$\varphi_i(a_i, \ldots, a_n) = (z^{(i)}a_i, z^{(i)}z^{(i+1)}a_{i+1}, \ldots, z^{(i)}\cdots z^{(n)}a_n).$$

Thus, we have

$$\begin{array}{ccccccccc}
0 & \longrightarrow & \mathsf{Z}^{n+1-i} & \xrightarrow{\varphi^{(i)}\oplus\cdots\oplus\varphi^{(n)}} & \mathsf{Z}^{n+1-i} & \longrightarrow & K_i & \longrightarrow & 0\\
 & & \downarrow{\scriptstyle z_{i-1}} & & \| & & \downarrow & & \\
0 & \longrightarrow & \mathsf{Z}^{n+1-i} & \xrightarrow{\varphi_i} & \mathsf{Z}^{n+1-i} & \longrightarrow & G^{(i)} & \longrightarrow & 0
\end{array}$$

By Theorem III, we have $E_3(\varphi_i) \cong H(G^{(i)}, \mathsf{Z})$. Moreover, the diagram

$$\begin{array}{ccc} \mathsf{Z}^{n+1-i} & \xrightarrow{1(\mathsf{Z}) \oplus \varphi_{i+1}} & \mathsf{Z}^{n+1-i} \\ \downarrow & & \downarrow \\ (\mathsf{Z}/z^{(i)}\mathsf{Z})^{n+1-i} & \xrightarrow{\Phi_i} & (\mathsf{Z}/z^{(i)}\mathsf{Z})^{n+1-i} \end{array}$$

with $\varphi_{n+1} = 0$ commutes. The main assertion then follows from I—2.35. The remainder follows from I—4.17, since

$$H(G^{(i)}, \mathsf{Z}(z^{(i)})) \cong E_3(z^{(i)}\Phi_i) = E_2(z^{(i)}\Phi_i).$$

It might be observed that Z in Corollary 3.8 can be replaced by any principal ideal domain without major modifications.

Definition 3.10. We continue the notation of Definition 3.1. Let $A = \operatorname{Hom}(\mathsf{Z}^n, R)$ for some principal ideal domain R and let a_i, $i = 1, \ldots, n$, be the standard basis of A with $a_i(0, \ldots, 1, \ldots, 0) = 1$, if the 1 is in the i-th place, and $= 0$ otherwise. We define an exact sequence of bi-graded differential modules

$$0 \longrightarrow (E_2(\varphi), d_\varphi) \xrightarrow{\lambda} (E_2(\varphi), d') \longrightarrow E_2(\varphi)/\lambda E_2(\varphi) \longrightarrow 0$$

as in I—2.37, 38, 39; for completeness sake, we repeat:

$$d'(PA \otimes 1) = \{0\}, \quad d'(1 \otimes a_i) = a_i \otimes 1,$$

$$d'(1 \otimes a_r) = z_{r(2)} a_{r(1)} \otimes a(r_1) - z_{r(2)} a_{r(2)} \otimes a(r_2) + \cdots$$
$$+ (-1)^{n-1} z_{r(n)} a_{r(n)} \otimes a(r_n)$$
$$= (z_{r(2)} - z_{r(1)})\, a_{r(1)} \otimes a(r_1) + d_\varphi(1 \otimes a_r), \quad r \in S(q+1), \quad q > 1,$$

and

$$\lambda(a_\varrho \otimes a_r) = z_{r(1)} a_\varrho \otimes a_r.$$

We recall

$$E_2(\varphi)/\lambda E_2(\varphi) \cong PA \otimes \wedge^+ \operatorname{coker} \varphi$$
$$\cong P \operatorname{Hom}(\mathsf{Z}^n, R) \otimes \wedge^+ \operatorname{Ext}(G, R),$$

since $\operatorname{coker} \operatorname{Hom}(f, R) = \operatorname{Ext}(G, R)$.

Theorem 3.11. *With the assumptions and notation of Definition* 3.1, *there is a commutative diagram with exact rows and columns:*

$$\begin{array}{ccccc}
\ker d \cong \operatorname{im} d' & \rightarrowtail & E_2(\varphi) & \twoheadrightarrow & \operatorname{im} d = \operatorname{im} d \\
\downarrow & & \downarrow & & \downarrow \\
\ker d' & \rightarrowtail & E_2(\varphi) & \twoheadrightarrow & \operatorname{im} d' \cong \ker d \\
\downarrow & & \downarrow & & \downarrow \\
H(d') & \rightarrowtail & P\operatorname{Hom}(\mathsf{Z}^n, R) \otimes \wedge^+ \operatorname{Ext}(G, R) & \twoheadrightarrow & H(G, R)
\end{array}$$

and there is an exact sequence

$$0 \to E_2(\varphi) + \ker d' \to P\operatorname{Hom}(\mathsf{Z}^n, R) \otimes \wedge \operatorname{Hom}(\mathsf{Z}^n, R)$$
$$\to H(G, R) \to 0$$

in which the bi-degree of the last (non-zero) map is (2, —1). *Also*, $\ker d' = \ker d\lambda^{-1}$. *There is also an exact sequence*

$$0 \to H(d') \to PG^* \otimes \wedge^+ \hat{G} \to H(G, \mathsf{Z}) \to 0,$$

where $G^* = \operatorname{Hom}(F, G_n)$.

Proof. This follows directly from Theorem III in Section 2 and from I—2.42. The final statements follow from the fact that

$$\operatorname{Hom}(F, G_n) \approx \operatorname{Hom}(F, \mathsf{Z}(z_n)) \approx \mathsf{Z}(z_n)^n, \quad \operatorname{Ext}(G, \mathsf{Z}) \approx \hat{G},$$

and from the bottom exact sequence of the diagram.

Remark. Recall that for $R = \mathsf{Z}$ we may write $\hat{G}$ in place of $\operatorname{Ext}(G, R)$.

As far as the ring structure is concerned, the theorem which we are now about to discuss could very well be called the fundamental theorem for the cohomology of finite abelian groups over principal ideal domains.

Theorem IV. *Let R be a principal ideal domain of characteristic* 0 *considered as a trivial left G-module for a finite abelian group G. Let*

$$\tau_{G,A}\colon P\operatorname{Ext}(G, R) \to H(G, R)$$

be the natural injective morphism of graded algebras of Theorem 2.13. *Thus, in particular*, $H(G, R)$ *is an augmented* $P\operatorname{Ext}(G, R)$*-algebra under*

$$e \cdot h = \tau_{G,A}(e) \cup h.$$

There is a graded subgroup

$$M = M(G) = R \oplus M^2 \oplus \cdots \oplus M^{n+1}$$

(where n is the natural number defined in Definition 1) *of* $H(G, R)$ *which generates* $H(G, R)$ *as a ring and as a* $(P \operatorname{Ext}(G, R))$*-module and which is minimal relative to either one of these properties. In fact,* $H(G, R) = (P \operatorname{Ext}(G, R)) \cdot M(G)$. *If* $m, m' \in M$ *and* $m \cup m' \in M$, *then* $m \in R$ *or* $m' \in R$. *The group* $M(G)$ *may be obtained in any one of the following fashions:*

(1) *We may identify* $H(G, R)$ *with* $E_3(\operatorname{Hom}(f, R))$ *by Theorem* III. *By Proposition* I—2.18, *there is an isomorphism of R-modules*

$$\wedge^{q+1} \operatorname{Ext}(G, R) \to E_3^{2,q}(\operatorname{Hom}(f, R)), \quad q = 0, 1, \ldots$$

and

$$M^{q+1}(G) = E_3^{2,q}(\operatorname{Hom}(f, R)), \quad q = 0, 1, \ldots.$$

(2) *Let* $z \in R$ *be any element with* $z H^+(G, R) = 0$ *(which is certainly the case for* $R = \mathsf{Z}$ *and z divisible by the exponent of* G). *Let*

$$\Delta_z : H(G, R/z R) \to H(G, R)$$

be the morphism arising in the Bockstein formalism (I—4.15). *Then*

$$M^{q+1}(G) = R + \operatorname{im} \Delta_z^q \varrho_{G, R/zR}^q$$

with ϱ *as in Theorem* 2.13, *and* $\Delta_z^q \varrho_{G, R/zR}^q$ *is injective for* $q > 0$.

(3) *Let* Q *be the quotient field of* R, *and let* A *be the R-module* Q/R. *Then* $H^q(G, Q) = 0$ *for* $q > 0$ *and there is an isomorphism of R-modules*

$$\delta^q : H^q(G, A) \to H^{q+1}(G, R), \quad q = 1, 2, \ldots,$$

which then arises from the long exact sequence of the coefficient sequence

$$0 \to R \to Q \to A \to 0,$$

and

$$M^{q+1}(G) = R \oplus \operatorname{im} \delta^q \varrho_{G, A}^q$$

with ϱ *as in* 2.13, *and* $\delta^q \varrho_{G, A}^q$ *is injective for* $q > 0$.

Proof. (1) By Proposition I—2.38, $E_3(\operatorname{Hom}(f, R))$ is generated as a ring and as $E_3^{\mathrm{I}}(\operatorname{Hom}(f, R))$-module by the subgroup

$$E_3^{0,0} + \sum \{E_3^{2,q}(\operatorname{Hom}(f, R)) : q = 0, 1, \ldots\}, \quad E_3^{0,0} = R.$$

This subgroup is a minimal generating subgroup M' and satisfies the conditions that $m, m', m\,m' \in M'$ implies that at least one of m or m' is in $E_3^{0,0}$. By I—2.18 or the remark after I—2.42, there exists an isomorphism of R-modules

$$\wedge^{q+1} \operatorname{coker}(\operatorname{Hom}(f, R)) \to E_3^{2,q}(\operatorname{Hom}(f, R)).$$

But coker $\operatorname{Hom}(f, R) = \operatorname{Ext}(G, R)$ by I—3.12. This establishes (1).

In order to prove (2) and (3) we reduce the situation to case (1).

(2) The isomorphism $E_3(\mathrm{Hom}(f, -)) \to H(G, -)$ of Theorem III in Section 2 is functorial on the category of commutative rings. It is therefore compatible with the Bockstein formalism of Section 4 in Chapter 3. There is an isomorphism of exact sequences

$$\begin{array}{ccccccccc} 0 & \longrightarrow & H(G, R) & \longrightarrow & H(G, R/zR) & \xrightarrow{\Delta_z} & H^+(G, R) & \longrightarrow & 0 \\ & & \downarrow & & \downarrow & & \downarrow & & \\ 0 & \longrightarrow & E_3(\mathrm{Hom}(f, R)) & \longrightarrow & E_3(\mathrm{Hom}(f, R/zR)) & \xrightarrow{\Delta'_z} & E_3^+(\mathrm{Hom}(f, R)) & \longrightarrow & 0 \end{array}$$

Since $E_3^{\mathrm{II}}(\mathrm{Hom}(f, R)) = 0$ (I—2.18), Δ'_z induces an injection

$$E_3^{\mathrm{II}}(\mathrm{Hom}(f, R/aR)) \to E_3^+(\mathrm{Hom}(f, R)).$$

We let a_i be the standard basis elements of $\mathrm{Hom}(F, R)$ and a'_i the corresponding basis elements of $\mathrm{Hom}(F, R/zR)$ so that with the quotient map $\zeta: R \to (R/zR)$ we have $\mathrm{Hom}(F, \zeta)(a_i) = a'_i$. The elements a_s, a'_s of $\mathrm{Hom}(\wedge^q F, R)$, resp. $\mathrm{Hom}(\wedge^q F, R/zR)$, are defined correspondingly, for $s \in S(q)$, $q = 1, 2, \ldots$ The element

$$b_{\sigma s} = \mathrm{Hom}(F, \zeta)(a_\sigma \otimes (z/z_{s(1)})\, a_s)$$

is obviously in $\ker d_{\mathrm{Hom}(f, R/zR)}$. On the other hand the element

$$d_{\mathrm{Hom}(f, R)}(a_\sigma \otimes (z/z_{s(1)})\, a_s)$$

is divisible by z and

$$(1/z)\, d_{\mathrm{Hom}(f, R)}(a_\sigma \otimes (z/z_{s(1)})\, a_s) = d_{\mathrm{Hom}(f, R)}\, \lambda^{-1}(a_\sigma \otimes a_s);$$

and the latter elements generate $\oplus \{\ker d^{2p, q}_{\mathrm{Hom}(f, R)} : q = 0, 1, \ldots\}$ if σ ranges through all $\Sigma(p)$ and s through all $S(q)$. On the other hand, by definition of the Bockstein map Δ', we have

$$\Delta' b_{\sigma s} = \psi^{2p, q}\, d_{\mathrm{Hom}(f, R)}\, \lambda^{-1}(a_\sigma \otimes a_s).$$

Since

$$R \oplus \bigoplus \{E_3^{2, q}(\mathrm{Hom}(f, R)) : q = 0, 1, \ldots\}$$

is a minimal generating submodule of the ring $E_3(\mathrm{Hom}(f, R))$, the assertion now follows in view of the commuting diagram

$$\begin{array}{ccccc} \mathrm{Hom}(\wedge G, R/zR) & \xrightarrow{\varrho_{G, R/zR}} & H(G, R/zR) & \xrightarrow{\Delta} & H(G, R) \\ \downarrow \cong & & \downarrow \cong & & \downarrow \\ E_3^{\mathrm{II}}(\mathrm{Hom}(f, R/zR)) & \longrightarrow & E_3(\mathrm{Hom}(f, R/zR)) & \xrightarrow{\Delta'} & E_3(\mathrm{Hom}(f, R)) \end{array}$$

(3) There is an exact sequence of complexes

$$0 \to E_2(\mathrm{Hom}(f, R)) \to E_2(\mathrm{Hom}(f, Q)) \to E_2(\mathrm{Hom}(f, A)) \to 0.$$

The middle complex is acyclic by I—2.24. Hence there is a morphism

$$E_3^+(\mathrm{Hom}(f, A)) \to E_3^+(\mathrm{Hom}(f, R))$$

of bidegree $(2, -1)$ as a consequence of the long exact sequence on the cohomology level derived from the short exact sequence above, and it is an isomorphism at each degree of $E_3^+(\mathrm{Hom}(f, A))$. Thus, in particular,

$$E_3^{0,q+1}(\mathrm{Hom}(f, A))$$

maps isomorphically onto $E_3^{2,q}(\mathrm{Hom}(f, R)) = M'^{q+1}(G)$. If via Theorem III we pass to $H(G, -)$, in view of what was said under (1), we get exactly the assertion.

Note that Theorem IV *in particular gives the structure of $M(G)$ as a graded module; as such it is isomorphic (with a degree shift) to* $\wedge$ Ext(G, R). *Note also that $M(G)$ is finite. That $H(G, \mathsf{Z})$ is finitely generated as a $PH^2(G, \mathsf{Z})$-module was observed by Evens in the Erratum to* [17], *where indeed he observed that a finite group is abelian if and only if this is the case.*

The theorem applies in particular to the case $R = \mathsf{Z}$, $Q = \mathsf{Q}$ and $A = \mathsf{Q}/\mathsf{Z}$. In this case one observes that $H(G, \mathsf{Q}/\mathsf{Z}) \cong H(G, \mathsf{R}/\mathsf{Z})$, and $\mathrm{Ext}(G, \mathsf{Z}) \cong \hat{G}$ (Lemma 3.2). Further information centering around Theorem IV will be given in Propositions 3.19, 3.20, but first we will draw a few quick conclusions from the theorem:

Definition 3.12. If $S = S^0 \oplus S^1 \oplus \cdots$ is a graded R-algebra, we call the smallest natural number m such that $S^0 \oplus \cdots \oplus S^m$ generates S as an R-algebra the *generating degree* of S.

Corollary 3.13. *The generating degree of $H(G, R)$ under the hypotheses of Theorem* IV *is $n + 1$, where n is the unique natural number associated with G in Definition* 1.

Corollary 3.14. *If Π is any (not necessarily abelian) finite group and if $H(\Pi, \mathsf{Z})$ has generating degree $\leqq 3$, then Π is cyclic of the direct sum of two cyclic groups.*

Proof. Evens [17] has shown that Π is abelian under the present assumptions. By Corollary 3.12, $n = 1$ or $n = 2$.

Corollary 3.15. *As a ring, $H(G, \mathsf{Z})$ has a minimal generating set of 2^n elements.*

Proof. By Theorem IV, $H(G, \mathbf{Z})$ has a minimal generating subgroup isomorphic to $\wedge \operatorname{Ext}(G, \mathbf{Z})$, which has a basis of 2^n elements, and none of these elements is in a subring generated by the others.

We next turn to the case that R is a field. Since in Theorem IV we have already observed that for fields of characteristic 0, we have $H(G, R) = R$, only the prime characteristic case remains.

Theorem V. *Let R be a commutative field with prime field K and G a finite abelian group. Then the natural morphisms*

$$\varphi\colon \wedge_R \operatorname{Hom}(G, R) \to \operatorname{Hom}(\wedge G, R) \qquad \textit{of } \mathrm{I}\text{—}1.6$$

and

$$\omega\colon P_R \operatorname{Ext}(G, R) \otimes_R \operatorname{Hom}(\wedge G, R) \to H(G, R) \qquad \textit{of Theorem } 2.13$$

are isomorphisms. Moreover, there are natural isomorphisms of graded algebras

$$\begin{aligned} R \otimes P\hat{G} \otimes \wedge \hat{G} &\cong P_R(R \otimes \hat{G}) \otimes_R \wedge_R (R \otimes \hat{G}) \\ &\cong R \otimes P \operatorname{Tor}(\hat{G}, K) \otimes \wedge \operatorname{Tor}(\hat{G}, K) \\ &\cong R \otimes P(G \otimes K)^\wedge \otimes \wedge (G \otimes K)^\wedge \cong H(G, R). \end{aligned}$$

Remark. Parts of this theorem overlap with results of H. Cartan [10], 9—08 (Thm. 2).

Proof. If the characteristic of R is zero or is relatively prime to the order of G, then $H^i(G, R)$, $i > 0$, $\operatorname{Hom}(G, R)$, $\operatorname{Ext}(G, R)$, $\operatorname{Tor}(G, K)$, $G \otimes K$ are all zero (see corollary 1.14). Hence there is nothing to prove. We may therefore assume for the remainder of the proof that $\operatorname{char} R = p$ and $K = GF(p)$. By the universal coefficient theorem, $H(G, R) \cong H(G, K) \otimes R$ for any group G. In dimension 1 (in the present case of trivial action of G) we obtain

$$\operatorname{Hom}(G, R) \cong \operatorname{Hom}(G, K) \otimes R.$$

From this it also follows that

$$\operatorname{Ext}(G, R) \cong \operatorname{Ext}(G, K) \otimes R,$$

e. g. via the exact sequence of Lemma I—3.12. That φ is an isomorphism as asserted is clear by I—1.14. For $R = GF(p)$ the isomorphy of ω follows from Proposition 3.3. Since

$$R \otimes P \operatorname{Ext}(G, K) \otimes \wedge \operatorname{Hom}(G, K)$$

is naturally isomorphic to

$$P_R(\operatorname{Ext}(G, K) \otimes R) \otimes_R \wedge_R (\operatorname{Hom}(G, K) \otimes R),$$

the bijectivity of ω then follows from the previous remarks. In order to prove the final assertion, we take a standard resolution $0 \to F \to F \to G \to 0$ of G.

From I—3.12, we obtain isomorphisms

$$\mathrm{Hom}\,(G, GF\,(p)) \to \mathrm{Hom}\,(F, GF\,(p))$$

and

$$\mathrm{Hom}\,(F, GF\,(p)) \to \mathrm{Ext}\,(F, GF\,(p)),$$

both of which are natural in f. Thus there is a natural isomorphism

$$\mathrm{Hom}\,(\mathrm{coker}\, f, GF\,(p)) \to \mathrm{Ext}\,(\mathrm{coker}\, f, GF\,(p))$$

and by Lemma 2.12, there is then a natural isomorphism

$$\mathrm{Hom}\,(G, GF\,(p)) \cong \mathrm{Ext}\,(G, GF\,(p)).$$

But now there are natural isomorphisms

$$\begin{aligned}\mathrm{Hom}\,(G, GF\,(p)) &\cong \mathrm{Hom}\,(G, \mathrm{Hom}\,(GF\,(p),\ GF\,(p)))\\ &\cong \mathrm{Hom}\,(G \otimes GF\,(p), GF\,(p)) = (G \otimes GF\,(p))^{\wedge}\end{aligned}$$

(the last isomorphy by Lemma 3.2). There is a natural exact sequence

$$0 \longrightarrow \mathrm{Tor}\,(G, GF\,(p)) \longrightarrow G \xrightarrow{\ p\ } G \to G \otimes GF\,(p) \to 0\,.$$

The dual sequence compared with the same sequence for $\hat{G} \xrightarrow{\ p\ } \hat{G}$ shows that there is a natural isomorphism

$$(G \otimes GF\,(p))^{\wedge} \cong \mathrm{Tor}\,(\hat{G}, GF\,(p))\,.$$

This then clearly establishes the claim for characteristic p in view of the isomorphisms

$$\mathrm{Ext}\,(G, R) \cong R \otimes \mathrm{Ext}\,(G, GF\,(p))$$

and from

$$\mathrm{Hom}\,(G, R) \cong R \otimes \mathrm{Hom}\,(G, GF\,(p))$$

used before. Finally, there is a natural isomorphism

$$(GF\,(p) \otimes G)^{\wedge} = GF\,(p) \otimes \hat{G}$$

since there is a commutative functor diagram of left adjoint functors

$$\begin{array}{ccc}\mathfrak{A} & \xrightarrow{\ GF\,(p) \otimes -\ } & \mathfrak{A}_{GF(p)}\\ \downarrow{\scriptstyle\wedge} & & \downarrow{\scriptstyle\wedge}\\ \mathfrak{A}^{*} & \xrightarrow{\ GF\,(p) \otimes -\ } & \mathfrak{A}^{*}_{GF(p)}\end{array}$$

where $\mathfrak{A}$ is the category of finite abelian groups and $\mathfrak{A}_{GF(p)}$ is the category of finite dimensional $GF\,(p)$-vector spaces.

Remark. It should be pointed out that in the case of $R = \mathbf{Z}/p^2\mathbf{Z}$, the natural morphism of graded algebras $\varrho_{G, \mathbf{Z}/p^2\mathbf{Z}}$ of Theorem 2.13 is neither injective nor surjective, as the following example shows:

Example. Let $R = \mathbf{Z}(p^{2+k})$, $k > 1$, $G = \mathbf{Z}(p) \oplus \mathbf{Z}(p^2)$. With the notation of Definition 1, we have $z_1 = p$, $z_2 = p^2$ (in R) and $w_1 = p^{k+1}$, $w_2 = p^k$. We refer to the example following I—2.17. The computations described there apply to this example. They show that

$$\ker \omega \mid \mathrm{Ext}(G, R) \otimes \mathrm{Hom}(G, R) \cong R/p^k R$$
$$\cong \operatorname{coker} \omega \mid \mathrm{Ext}(G, R) \otimes \mathrm{Hom}(G, R).$$

It is also useful to recall the following fact.

Proposition 3.16. *If R is a principal ideal domain and G_1, G_2 are finite abelian groups of relatively prime order, then*

$$H(G_1 \times G_2, R) \cong H(G_1, R) \otimes H(G_2, R)$$
$$\cong R \oplus H^+(G_1, R) \oplus H^+(G_2, R).$$

In particular, for each finite abelian group G, $H(G, R)$ is the sum of R and all $H^+(S, R)$, where S ranges through the set of Sylow groups of G. The $H^+(S, R)$ are the Sylow subgroups of $H^+(G, R)$.

Proof. Since $\mathrm{Tor}(H^+(G_1, R), H^+(G_2, R))$ is (non-canonically) isomorphic to $H^+(G_1, R) \otimes H^+(G_2, R)$, we obtain

$$\mathrm{Tor}(H(G_1, R), H(G_2, R)) = 0$$

from 1.14. Hence Corollary 1.8 proves the assertion in view of the fact that the tensor product of two finitely generated torsion modules is zero if the sum of their annihilators is the whole ring. The remainder is clear.

By Proposition 3.16, the computation of $H(G, \mathbf{Z})$ is reduced to the case of p-groups G.

The Bockstein formalism has already appeared in this section with Theorem 3.6. But there is still more that can be said, even in a more general setting.

Proposition 3.17. *Let R be a commutative ring with identity whose additive group is torsion free and let a be an element such that $a\,x = 0$ if $x = 0$. (In $\mathbf{Z}$ this holds for any $a \neq 0$.) For every commutative group G, we have the diagrams with exact rows and columns. The morphism d is the standard connecting morphism in the* Hom-Ext *sequence.*

$$
\begin{array}{ccccccccccccc}
 & & & & & & & & 0 & & & & \\
 & & & & & & & & \downarrow & & & & \\
 & & & & & & & & aH^2(G, R) & & & & \\
 & & & & & & & & \downarrow & & & & \\
 & & & & 0 & \longrightarrow & \mathrm{Hom}(G, R/aR) & \xrightarrow{D^1} & H^2(G, R) & \xrightarrow{a} & H^2(G, R) & \longrightarrow & H^2(G, R/aR) \xrightarrow{d^2} \\
 & & & & & & \| & & \downarrow & & \downarrow & & \downarrow \\
0 \longrightarrow \mathrm{Hom}(G, R/aR) & \longrightarrow & \mathrm{Hom}(G, R/a^2 R) & & & \longrightarrow & \mathrm{Hom}(G, R/aR) & \xrightarrow{d^1} & H^2(G, R/aR) & \longrightarrow & H^2(G, R/a^2R) & \longrightarrow & H^2(G, R/aR) \xrightarrow{d^2} \\
 & & & & & & & & \downarrow & & & & \\
 & & & & & & & & \ker D^2 & & & & \\
 & & & & & & & & \downarrow & & & & \\
 & & & & & & & & 0 & & & &
\end{array}
$$

$$
\begin{array}{ccccccccccccc}
 & 0 & & 0 & & 0 & & 0 & & 0 & & 0 & \\
 & \downarrow & & \downarrow & & \downarrow & & \downarrow & & \downarrow & & \downarrow & \\
0 \longrightarrow & \mathrm{Hom}(G, R/aR) & \longrightarrow & \mathrm{Hom}(G, R/a^2R) & \longrightarrow & \mathrm{Hom}(G, R/aR) & \xrightarrow{\delta} & \mathrm{Ext}(G, R/aR) & \longrightarrow & \mathrm{Ext}(G, R/a^2R) & \longrightarrow & \mathrm{Ext}(G, R/aR) & \longrightarrow 0 \\
 & \downarrow & & \downarrow & & \downarrow & & \downarrow & & \downarrow & & \downarrow & \\
0 \longrightarrow & H^1(G, R/aR) & \longrightarrow & H^1(G, R/a^2R) & \longrightarrow & H^1(G, R/aR) & \xrightarrow{d^1} & H^2(G, R/aR) & \longrightarrow & H^2(G, R/a^2R) & \longrightarrow & H^2(G, R/aR) & \xrightarrow{d^2} \\
 & \downarrow & & \downarrow & & \downarrow & & & & & & & \\
 & 0 & & 0 & & 0 & & & & & & &
\end{array}
$$

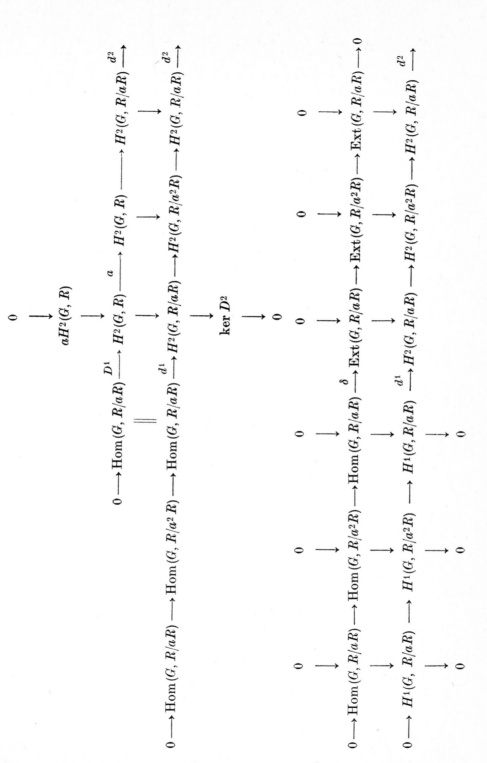

Proof. We apply I—4.15 and observe the natural morphisms

$$\mathrm{Hom}(G, M) \to H^1(G, M)$$

and

$$\mathrm{Ext}(G, M) \to H^2(G, M).$$

The commuting of the second diagram follows from the uniqueness of connecting morphisms.

However, it may be instructive, to check the commuting of the square involving δ and d^1 directly; we will do this in the proof of the following Proposition, which is slightly more general in some respects:

Proposition 3.18. *Let*

$$0 \longrightarrow A \xrightarrow{\varphi} B \xrightarrow{\psi} C \to 0$$

be an exact sequence of abelian groups. Then there is a commutative diagram of natural morphisms (see the adjoining column) *in which all vertical maps are injections and the ones so marked are isomorphisms. The morphisms d^1 and δ are the connecting morphisms.*

$$\begin{array}{ccccccccccccccc}
0 & \to & \mathrm{Hom}(G, A) & \to & \mathrm{Hom}(G, B) & \to & \mathrm{Hom}(G, C) & \xrightarrow{\delta} & \mathrm{Ext}(G, A) & \to & \mathrm{Ext}(G, A) & \to & \mathrm{Ext}(G, C) & \to & 0 \\
 & & \downarrow\cong & & \downarrow\cong & & \downarrow\cong & & \downarrow & & \downarrow & & \downarrow & & \\
0 & \to & H^1(G, A) & \to & H^1(G, B) & \to & H^1(G, C) & \xrightarrow{d^1} & H^2(G, A) & \to & H^2(G, B) & \to & H^2(G, C) & \to & 0
\end{array}$$

Proof. Although the commuting of the square involving the connecting morphisms follows from the uniqueness and functorial properties of the connecting morphisms, we will nevertheless give a direct proof. Let

$$f \in \mathrm{Hom}(G, C)$$

and let $c: G \to B$ be a cochain with $\psi c = f$. Then the function

$$(g, h) \to c(g) - c(g + h) + c(h)$$

is the image of a unique 2-cocycle $F: G \times G \to A$ under the map induced by φ, i. e. we have

$$\varphi F(g, h) = c(g) - c(g + h) + c(h).$$

Since φ is injective, obviously $F \in Z^+$. Thus F defines an element

$$F' = 0 \to A \to X \to G \to 0 \text{ in } \mathrm{Ext}(G, A).$$

This element maps onto $d^1 f$ under the injection

$$\mathrm{Ext}(G, A) \to H^2(G, A).$$

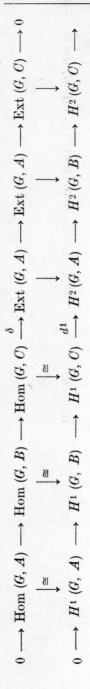

The group X may be represented as the set $A \times G$ with the multiplication

$$(a, g)\,(a', g') = (a + a' + F(g, g'),\; g + g').$$

If one defines a map $\varrho\colon X \to B$ by $\varrho(a, g) = \varphi f(g)$, then this map is a morphism, and the diagram

$$\begin{array}{ccccccccc}
0 & \longrightarrow & A & \longrightarrow & X & \longrightarrow & G & \longrightarrow & 0 \\
 & & \| & & \downarrow{\scriptstyle\varrho} & & \downarrow{\scriptstyle f} & & \\
0 & \longrightarrow & A & \xrightarrow{\varphi} & B & \xrightarrow{\psi} & C & \longrightarrow & 0
\end{array}$$

commutes. Hence $F' = \delta f$ as we wanted to show.

Proposition 3.19. *Let R be a commutative ring with identity whose additive group is torsion free. Let $z \in R$ be such that $z\,a = 0$ only if $a = 0$. Let G be a finite abelian group, and denote with*

$$\delta\colon \mathrm{Hom}(G, R/zR) \to \mathrm{Ext}(G, R/zR)$$

the connecting morphism derived from the short exact sequence

$$0 \to R/zR \to Rz^2R \to R/zR \to 0.$$

Consider the bi-graded algebra

$$E_2(\delta) = (P_R\,\mathrm{Ext}(G, R/zR) \otimes_R \wedge_R \mathrm{Hom}(G, R/zR), d_\delta)$$

according to Section 2 of Chapter I. *On the bi-graded algebra*

$$E_3(\mathrm{Hom}(f, R/zR)) \cong H(G, R/zR)$$

(*see Definition* 3.1) *we consider the Bockstein differential according to* I — 4.16. *Then there is a natural morphism of bi-graded differential algebras*

$$\begin{array}{ccc}
m\colon P_R\,\mathrm{Ext}(G, R/zR) \otimes_R \wedge_R \mathrm{Hom}(G, R/zR) & \longrightarrow & E_3(\mathrm{Hom}(f, R/zR)) \\
\| & & \| \\
(E_2(\delta), d_\delta) & & (H(G, R/zR), \text{Bockstein}).
\end{array}$$

In case $\mathbb{Z} = R$, $z = p$, *a prime,* m *is the isomorphism of Theorem* V, *and in the general case, there is a commutative diagram*

$$\begin{array}{ccc}
P_R\,\mathrm{Ext}(G, R/zR) \otimes_R \wedge_R \mathrm{Hom}(G, R/zR) & & \\
\downarrow{\scriptstyle 1 \otimes \varphi} & \searrow{\scriptstyle m} & \\
P_R\,\mathrm{Ext}(G, R/zR) \otimes_R \mathrm{Hom}(\wedge G, R/zR) & \xrightarrow{\ \omega\ } & H(G, R/zR)
\end{array}$$

Proof. The existence of m follows immediately from the fact that the isomorphism

$$\mathrm{Hom}(G, R/zR) \to H^1(G, R/zR)$$

extends to a morphism of graded algebras

$$\wedge \mathrm{Hom}(G, R/zR) \to H(G, R/zR),$$

that

$$\mathrm{Ext}(G, R/zR) \to H^2(G, R/zR)$$

extends to a morphism of graded algebras

$$P\,\mathrm{Ext}(G, R/zR) \to H(G, R/zR),$$

and that the tensor product of graded commutative algebras is a coproduct. Since d_δ and the Bockstein morphism β on $H(G, R/zR)$ are derivations, in order to show $m\,d_\delta = \beta\,m$ it suffices to establish this equality on a generating set of dom m. But

$$\mathrm{Ext}(G, R/zR) \otimes 1 \oplus 1 \otimes \mathrm{Hom}(G, R/zR)$$

is such a generating set and the coincidence of the maps $m\,d_\delta$ and $\beta\,m$ on this set has been established in Proposition 3.17. (We have argued by identifying $E_3(\mathrm{Hom}(f, R/zR))$ and $H(G, R/zR)$. By Theorem III of Section 2 above, this is legitimate.) The last two assertions follow from Theorem III and Theorem V.

Proposition 3.20. *Let R be a principal ideal domain of characteristic 0 and $z \in R$ an element such that $zH^+(G, R) = 0$. Let $\pi\colon R \to R/zR$ denote the quotient morphism. Then*

$$\mathrm{Ext}(G, \pi)\colon \mathrm{Ext}(G, R) \to \mathrm{Ext}(G, R/zR)$$

is an isomorphism and

$$P\,\mathrm{Ext}(G, \pi)\colon P_R\,\mathrm{Ext}(G, R) \to P_{R/zR}\,\mathrm{Ext}(G, R/zR)$$

has kernel zR. We consider $H(G, R/zR)$ as a $P\,\mathrm{Ext}(G, R)$-module under

$$e \cdot h = H(G, \pi)\,(\tau_{G,R}\,e) \cup h.$$

Then

$$H(G, \pi)\colon H(G, R) \to H(G, R/zR)$$

is a morphism of $P\,\mathrm{Ext}(G, R)$-modules with kernel zR. If β is the Bockstein derivation of $H(G, R/zR)$, we define the degree one endomorphism of R/zR-modules d of $H(G, R/zR)$ by $d^i(h^i) = (-1)^i\,\beta(h^i)$, $h^i \in H^i(G, R/zR)$. Then d is an endomorphism of $P\,\mathrm{Ext}(G, R)$-modules and $\mathrm{im}\,d \cong H^+(G, R)$ as

$P\,\mathrm{Ext}(G, R)$-modules.

If $b \in \operatorname{Hom}(\wedge^i G, R/zR)$, $i > 0$, *is a basis element corresponding to a basis element* $a'_\sigma \in E_3^{0,i}(\operatorname{Hom}(f, R/zR))$ *under the isomorphisms of Theorems* III *and* 3.11, *and if* $h = \varrho^i_{G,R/zR}(b)$, $m = \Delta^i_z h \in M(G) \subset H(G, R)$, *using the notation of Theorem* V, *then the commutative diagram*

$$\begin{array}{ccc} P\operatorname{Ext}(G, R/zR) \otimes \operatorname{Hom}(\wedge G, R/zR) & \xrightarrow{\ \omega\ } & H(G, R/zR) \\ \uparrow{\scriptstyle P\operatorname{Ext}(G,\pi)\otimes(\Delta_z \varrho_{G,R/zR})^{-1}} & & \downarrow{\scriptstyle H^+(G,\pi)^{-1} d} \\ P\operatorname{Ext}(G, R) \otimes M^+(G) & \xrightarrow{\ e \otimes m \longrightarrow e \cdot m\ } & H^+(G, R) \end{array}$$

defines injective module maps

$$P\operatorname{Ext}(G, R/zR) \otimes b \to H^+(G, R/zR),$$
$$P\operatorname{Ext}(G, R/zR) \otimes m \to H^+(G, R).$$

Proof. From the hypothesis about z and the exact Hom-Ext sequence, it follows readily that $\operatorname{Ext}(G, \pi)$ is an isomorphism. The injectivity of $H^i(G, \pi)$ for $i > 0$ was shown in I — 4.15, and the definition of the module actions and the fact that $H(G, \pi)$ is a ring morphism make $H(G, \pi)$ a morphism of $P\operatorname{Ext}(G, R)$-modules. β vanishes on im τ since Δ'_z in the proof of Theorem IV vanishes on the horizontal edge term of $E_3(\operatorname{Hom}(f, R/zR))$. Now let

$$e^i \in P^i\operatorname{Ext}(G, R), \quad h^j \in H^j(G, R/zR).$$

Then

$$\begin{aligned} &d^{i+j}(e^i \cdot h^j) \\ &= (-1)^{i+1}(\beta^i(H^i(G, \pi)(e^i) \cup h^j + (-1)^j H^i(G, \pi)(e^i) \cup \beta^j h^j)) \\ &= H^i(G, \pi)(e^i) \cup d^j h^j = e^i \cdot d^j h^j. \end{aligned}$$

The equality im d = im $H^+(G, \pi)$ is clear from I — 4.15.

By I — 2.16,

$$e \otimes h \to e \cdot h\colon\ P\operatorname{Ext}(G, R/zR) \otimes h \to H(G, R/zR)$$

is injective in view of Theorem III. Now consider the commutative diagram

$$\begin{array}{ccc} P\operatorname{Ext}(G, R) \otimes m & \xrightarrow{\ e \otimes m \longrightarrow e \cdot m\ } & H^+(G, R) \\ \downarrow{\scriptstyle P\operatorname{Ext}(G,\pi)} & & \downarrow{\scriptstyle H^+(G,\pi)} \\ P\operatorname{Ext}(G, R/zR) \otimes h & \xrightarrow{\ e \otimes h \longrightarrow e \cdot h\ } & H^+(G, R/zR) \end{array}$$

The bottom horizontal and the right vertical maps are injective and the left vertical map is injective since $z\,m = 0$, $z\,h = 0$. Thus the top horizontal map is injective.

Corollary 3.21. *Under the hypotheses of Proposition* 3.20, *the natural* $P\,\mathrm{Ext}(G, R)$*-module morphism*

$$\varphi\colon P\,\mathrm{Ext}(G, R) \otimes \mathrm{Hom}(\wedge G, R/zR) \to H(G, R)$$

defined by $\varphi(e \otimes b) = e \cdot (\Delta_z\, \varrho_{G, R/zR}(b))$ *is surjective and injective on the submodules* $P\,\mathrm{Ext}(G, R/zR) \otimes b$ *for all elements* b *of a suitable basis of*

$$\mathrm{Hom}(\wedge G, R/zR).$$

If Q *is the quotient field of* R *and* A *the* R*-module* Q/R, *then a similar statement holds for the map*

$$P\,\mathrm{Ext}(G, R) \otimes \mathrm{Hom}(\wedge G, A) \to H(G, A).$$

Note in particular, that for $R = \mathbf{Z}$, we have

$$\mathrm{Ext}(G, R/zR) \cong \hat{G}$$

and

$$\mathrm{Hom}(\wedge G, R/zR) \cong \mathrm{Hom}(\wedge G, A) \cong (\wedge G)\hat{}\,.$$

Section 4

Appendix to Section 3: The low dimensions

In this section, we list the low dimensional cohomology groups and briefly examine the significance of the first two in terms of central extensions.

We close the Section with a list of low dimensional cohomology groups:

Proposition 4.1. *Let* G *be a finite abelian group,* R *a weakly principal ideal ring. Then we have the following table of cohomology modules in low dimensions:*

i	$H^i(G, R)$
0	R
1	$\mathrm{Hom}(G, R)$
2	$\mathrm{Ext}(G, R) \oplus \mathrm{Hom}(\wedge^2 G, R)$
3	$E_3^{2,1}(\varphi) \oplus \mathrm{Hom}(\wedge^3 G, R)$
4	$P^2\,\mathrm{Ext}(G, R) \oplus E_3^{2,2}(\varphi) \oplus \mathrm{Hom}(\wedge^4 G, R)$
5	$E_3^{4,1}(\varphi) \oplus E_3^{2,3}(\varphi) \oplus \mathrm{Hom}(\wedge^5 G, R)$

where the $E_3^{2,i}(\varphi)$, $i = 1, 2, 3$ *and* $E_3^{4,1}(\varphi)$ *are explicitly computed in* I — 2.17, *and where* $\varphi =$ Hom(f, R) *with the notation of Definition* 3.1.

If R is a principal ideal domain and $z_i \neq 0$ in R for all i, one has the following list:

i	$H^i(G, R)$	$H^i(G, \mathsf{Z})$
0	R	Z
1	0	0
2	$\mathrm{Ext}(G, R)$	$\hat{G}$
3	$\wedge^2 \mathrm{Ext}(G, R)$	$\wedge^2 \hat{G}$
4	$P^2 \mathrm{Ext}(G, R) \oplus \wedge^3 \mathrm{Ext}(G, R)$	$P^2 \hat{G} \oplus \wedge^3 \hat{G}$
5	$E_3^{4,1}(\varphi) \oplus \wedge^4 \mathrm{Ext}(G, R)$	$E_3^{4,1}(\varphi) \oplus \wedge^4 \hat{G}$

where

$$E_3^{4,1}(\varphi) \approx \oplus \{(R/z_i R)^{\binom{n-i+1}{2}} : i = 1, \ldots, n-1\}.$$

Moreover, the isomorphisms which give the terms in the table are natural.

Proof. All these have been given in I — 2.17 and I — 2.18 in terms of $E_3(\varphi) \cong H(G, R)$. The specific form of the edge terms has been discussed in Proposition I — 3.14, Theorem II — 2.13 and Lemma II — 3.2.

It is an interesting side remark that our theory provides natural isomorphisms of graded abelian groups

$$\mathrm{Hom}(\wedge G, \mathsf{R}/\mathsf{Z}) \cong \wedge \mathrm{Hom}(G, \mathsf{R}/\mathsf{Z}).$$

We formulate this as a proposition:

Proposition 4.2. *Let G be a finite abelian group. Then there is a natural isomorphism of graded groups* $(\wedge G)\hat{} \cong \wedge \hat{G}$.

Proof. Let $0 \longrightarrow F \xrightarrow{f} F \xrightarrow{\pi} G \longrightarrow 0$ be a standard resolution for G. By I — 3.15, there is a natural isomorphism

$$\mathrm{Hom}(\wedge \operatorname{coker} f, \mathsf{R}/\mathsf{Z}) \cong H^{\mathrm{II}}(\mathsf{R}/\mathsf{Z} \otimes E_2(\mathrm{Hom}(f, \mathsf{Z})))$$

where H^{II} denotes the second edge term of the bigraded cohomology module of $\mathsf{R}/\mathsf{Z} \otimes E_2(\mathrm{Hom}(f, \mathsf{Z}))$. There is a natural isomorphism of bi-degree $(2, -1)$

$$H(\mathsf{R}/\mathsf{Z} \otimes E_2(\mathrm{Hom}(f, \mathsf{Z}))) \to E_3(\mathrm{Hom}(f, \mathsf{Z}))$$

except for bi-degree $(0, 0)$ arising from the coefficient sequence

$$0 \to \mathsf{Z} \to \mathsf{R} \to \mathsf{R}/\mathsf{Z} \to 0$$

in view of the fact that

$$H(\mathsf{R} \otimes E_2(\mathrm{Hom}(f, \mathsf{Z}))) \cong H(G, \mathsf{R}) = \mathsf{R}.$$

Thus we have a natural isomorphism from $\mathrm{Hom}(\wedge^{q+1} \operatorname{coker} f, \mathsf{R}/\mathsf{Z})$ onto $E_3^{2,q}(\mathrm{Hom}(f, \mathsf{Z}))$ for $q = 0, 1, \ldots$

On the other hand, by I — 2.18, there is a natural isomorphism

$$\wedge^{q+1} \operatorname{coker} \mathrm{Hom}(f, \mathsf{Z}) \cong E_3^{2,q}(\mathrm{Hom}(f, \mathsf{Z})).$$

But there is a natural isomorphism

$$\operatorname{coker} \mathrm{Hom}(f, \mathsf{Z}) \cong \mathrm{Ext}(\operatorname{coker} f, \mathsf{Z}).$$

Now we invoke Lemma 2.12 and obtain a natural isomorphism

$$\mathrm{Hom}(\wedge^{q+1} G, \mathsf{R}/\mathsf{Z}) \to \wedge^{q+1} \mathrm{Ext}(G, \mathsf{Z}) \quad \text{for} \quad q = 0, 1, \ldots$$

Since $\mathrm{Ext}(G, \mathsf{Z}) \cong \mathrm{Hom}(G, \mathsf{R}/\mathsf{Z})$ naturally by Lemma 3.2, we have the assertion.

Some remarks are in place about the cohomology of a finite abelian group with coefficients in R/Z.

Proposition 4.3. *Let G be a finite abelian group and let $f: F \to F$ be an elementary morphism of free abelian groups with* $\operatorname{coker} f = G$. *Then there are isomorphism*

$$\begin{aligned} \mathrm{Hom}(E_3(\mathrm{Hom}(f, \mathsf{Z})), \mathsf{Z}/\mathsf{R}) &\cong H^i(\mathsf{R}/\mathsf{Z} \otimes E_2(\mathrm{Hom}(f, \mathsf{Z}))) \\ &\cong H^i(G, \mathsf{R}/\mathsf{Z}) \cong H^{i+1}(G, \mathsf{Z}),\ i = 1, 2, \ldots \end{aligned}$$

In particular we have, with $E_3^{4,1}(\varphi)$ as in Proposition 4.1:

i	$H^i(G, \mathsf{R}/\mathsf{Z})$
0	R/Z
1	$\hat{G} = \mathrm{Hom}(G, \mathsf{R}/\mathsf{Z})$
2	$\hat{G} \wedge \hat{G} \cong (G \wedge G)\hat{}$
3	$P^2\hat{G} \oplus \wedge^3 \hat{G}$
4	$E_3^{4,1}(\varphi) \oplus \wedge^4 \hat{G}$

Proof. We refer to I — 3.17, 3.18 and II — 2.11 for the existence of the isomorphisms and to Proposition 4.1 above for the explicit form of some of the groups.

The isomorphism expressing $H^2(G, \mathsf{R}/\mathsf{Z})$ in the last proposition in terms of $\hat{G}$ is of interest in the theory of Schur's multiplicator. In fact, for every finite group (abelian or not), there exists a central non-commutative extension

$$0 \to H^2(G, \mathsf{R}/\mathsf{Z})\hat{} \to \bar{G} \to G \to 0$$

such that every projective representation $G \to Gl(n, \mathsf{C})/\text{center } Gl(n, \mathsf{C})$ lifts uniquely to a linear representation $\tilde{G} \to Gl(n, \mathsf{C})$ so as to make the diagram

$$\begin{array}{ccc} \tilde{G} & \longrightarrow & Gl(n, \mathsf{C}) \\ \downarrow & & \downarrow \\ G & \longrightarrow & Gl(n, \mathsf{C})/\text{center } Gl(n, \mathsf{C}) \end{array}$$

commutative. The image of $H^2(G, \mathsf{R}/\mathsf{Z})\hat{}$ in $\tilde{G}$ is contained in the commutator group of $\tilde{G}$. In particular, if G is abelian, this image is exactly the commutator group. (See [22, 27].) Thus for a finite abelian G, the multiplicator $H^2(G, \mathsf{R}/\mathsf{Z})$ is isomorphic to $\hat{G} \wedge \hat{G} \cong (G \wedge G)\hat{}$ and the representation group $\tilde{G}$ is a central extension of $G \wedge G$ by G.

If K is any algebraically closed field and $K^{\times}$ its multiplicative group, then $H^2(G, K^{\times}) \cong \hat{G} \wedge \hat{G} \cong (G \wedge G)\hat{}$ for a finite abelian group [27], pp. 631, 652.

In the context of the catalogue of cohomology groups in Propositions 4.1 and 4.3, one should mention that if one does not insist on natural isomorphy, a complete list of the cohomology groups $H^i(G, \mathsf{Z})$ (and hence of the groups $H^i(G, \mathsf{R}/\mathsf{Z})$ can be obtained with the results of Section 1 and 3).

By 3.8 we have a non-natural isomorphism of graded groups

$$\mathsf{Z} \oplus H^+(G, \mathsf{Z}(n)) = P\hat{G} \oplus \wedge \hat{G}$$

for any n divided by $\exp G$. By the universal coefficient theorem (or the Bockstein formalism), there is a non-naturally split exact sequence

$$0 \to H(G, \mathsf{Z}) \otimes \mathsf{Z}(n) \to H(G, \mathsf{Z}(n)) \to \operatorname{Tor}(H(G, \mathsf{Z}), \mathsf{Z}(n)) \to 0$$

which, in view of the non-natural isomorphism

$$\operatorname{Tor}(H^i(G, \mathsf{Z}), \mathsf{Z}(n)) \cong H^i(G, \mathsf{Z}) \otimes \mathsf{Z}(n) \quad \text{for} \quad i > 0$$

and the isomorphies $H^i(G, \mathsf{Z}) \otimes \mathsf{Z}(n) \cong H^i(G, \mathsf{Z})$ for $i > 0$, amounts to a non-natural isomorphism

$$H^i(G, \mathsf{Z}(n)) = H^i(G, \mathsf{Z}) \oplus H^{i+1}(G, \mathsf{Z}), \quad i = 1, 2, \ldots$$

Consider the discrete category of all abelian groups of the form

$$\mathsf{Z}^m \oplus \mathsf{Z}(n_1) \oplus \cdots \oplus \mathsf{Z}(n_k).$$

(A discrete category has no morphisms except the identity morphisms.) This category is isomorphic to the free abelian semigroup generated by Z, $\mathsf{Z}(n)$, $n = 2, 3, \ldots$, and forms a semiring relative to the tensor product, with identity Z. We embed this semiring into the ring R obtained by considering the free semigroup as subsemigroup of the free abelian group generated by Z,

$\mathsf{Z}(n)$, $n = 2, 3, \ldots$ and by extending the multiplication. This ring has the identity Z, and satisfies the relations $\mathsf{Z}(n) \otimes \mathsf{Z}(m) = \mathsf{Z}((n, m))$ with the greatest common divisor (n, m) of m and n. The ring R is the semi-direct sum of the subring $R_0 \cong \mathsf{Z}$ generated by the identity Z and the ideal I generated by the $\mathsf{Z}(n)$. The ideal I is the direct sum of the ideals I_p, p a prime, where I_p is generated by $\mathsf{Z}(p)$, $\mathsf{Z}(p^2)$, ... It is therefore sufficient to discuss the rings $R(p) = R_0 \oplus I_p$. Each such ring is isomorphic to the semigroup $\Gamma(S)$ over the integers of the semigroup $S = \{0, 1, 2, \ldots\}$ with multiplication

$$(m, n) \to \min(m, n).$$

Now we consider the ring of formal power series $\Gamma(S)\{X\}$ in one variable over $\Gamma(S)$. We define two functions

$$e, p\colon \Gamma(S) \to \Gamma(S)\{X\}$$

by

$$e(x) = \sum e_i(x) X^i, \quad p(x) = \sum p_i(x) X^{2i}$$

with elements $e_i(x), p_i(x) \in \Gamma(S)$ for $x \in \Gamma(S)$ defined as follows: For

$$s(1), \ldots, s(n) \in S \subset \Gamma(S), \quad s(1) \leqq \cdots \leqq s(n),$$

we have

$$\begin{Bmatrix} e_i\,(s(1) + \cdots + s(n)) \\ p_i\,(s(1) + \cdots + s(n)) \end{Bmatrix} = \sum f(1),$$

where the summation is extended over all

$$\begin{Bmatrix} \text{strictly decreasing} \\ \text{non-decreasing} \end{Bmatrix} \text{ functions } f\colon \{1, \ldots, i\} \to \{s(1), \ldots, s(n)\}.$$

Thus, if under the isomorphism $R(p) \to \Gamma(S)$, the element $g \in \Gamma(S)$ represents the group $G \in R(p)$, then $e_i(g)$ represents $\wedge^i G$ and $p_i(g)$ represents $P^i G$. The i-th coefficient in the power series $p(g)\,e(g)$ represents the i-th homogeneous component of $PG \otimes \wedge G$. Now we let $h_i \in \Gamma(S)$ represent the groups $H^i(G, \mathsf{Z})$, $i = 1, 2, \ldots$ Then $h_1 = 0$ and the elements h_i are completely determined by the equation

$$\sum (h_i + h_{i+1}) X^i = p(g)\,e(g)$$

where we set $h_0 = 1$. If we let $h = \sum h_i X^i$, this means $Xh + h - 1 = Xp(g)\,e(g)$ or, equivalently

$$h = (1 + Xp(g)\,e(g))\,(1 + X)^{-1}$$

which determines the h_i completely.

It is fairly instructive to check out the low dimensions in the cohomology of a finite abelian group for their concrete significance in terms of central extensions. Let G be a finite abelian group and let A be a left G-module.

For a concrete interpretation of $H^1(G, A)$ and $H^2(G, A)$, one uses the bar resolution as we described it in a paragraph preceding 2.14. In fact we will adopt the notation of Section 2, notably the conventions of Definition 2.1 and of our discussion of the bar resolution after Theorem 2.13. We will only consider the case of trivial action, which corresponds to the situation of central extensions. Under these circumstances, $H^1(G, A)$ may be identified with $\mathrm{Hom}\,(G, A)$. Further it is convenient to identify the group of all alternating bilinear maps $G \times G \to A$ with $\mathrm{Hom}(\wedge^2 G, R)$; in other words we do not distinguish between $f(g_1, g_2)$ and $f(g_1 \wedge g_2)$ for an alternating bilinear function f. The group $Z^2(C(G, A))$ of 2-cocycles will shortly be denoted with Z. Then $\mathrm{Hom}\,(\wedge^2 G, A) \subset Z$. The group of 2-coboundaries

$$(g_1, g_2) \to h(g_1) - h(g_1 + g_2) + h(g_2)$$

is denoted with B.

As in the general case, the permutation group S_2 on two elements acts on $C^2(G, A)$. We denote the involution in S_2 by σ, so that $(\sigma f)(g_1, g_2) = f(g_2, g_1)$. The antisymmetrisation operator $a = \sum \{(\mathrm{sgn}\,\sigma)\,\sigma : \sigma \in S_2\}$ in our situation simply yields

$$(a f)(g_1, g_2) = f(g_1, g_2) - f(g_2, g_1).$$

The symmetrisation operator $s = \sum \{\sigma : \sigma \in S_2\}$ gives

$$(s f)(g_1, g_2) = f(g_1, g_2) + f(g_2, g_1).$$

Note the $2f = sf + af$.

Lemma 4.4. *If $f \in Z(C^2(G, A))$, then $a f$ is bilinear.*

Proof. f is a factor set and defines a central extension $R \times G$ of R by G via the multiplication

$$(r, g)(r', g') = (r + r' + f(g, g'), g g').$$

The commutator

$$[(r, g), (r', g')] = (r, g)^{-1}(r', g')^{-1}(r, g)(r', g')$$

equals

$$(f(g, g') - f(g', g), 1) = ((a f)(g, g'), 1).$$

The commutator in a nilpotent group of class 2 depends only on the elements modulo the center and is bilinear. This shows that $a f$ is bilinear.

Recall that in II—2.17 it was established that for any bilinear $f \in Z$, we have $2\,\psi f = 0$ if f is symmetric, where ψf denotes the cohomology class of f. Since one has $2f = sf + af$, we have $2\,Z \subset Z_+ + \mathrm{Hom}\,(\wedge^2 G, R)$, where Z_+ denotes the fixed point set of σ, i. e. the group of all symmetric cocycles. It is clear that $B \subset Z_+$ and that $Z_+/B \cong \mathrm{Ext}\,(G, R)$, since the symmetric cocycles obviously define abelian extensions. We assume for the remainder of our discussion, that $A = R$ is a commutative ring with identity and is uniquely divisible by 2. If R is finite this is tantamount to saying that R has no elements of order 2. Then

$$Z_+ \cap \mathrm{Hom}\,(\wedge^2 G, R) = \{0\} \quad \text{and} \quad f = (1/2)\, sf + (1/2)\, af,$$

whence

$$Z = Z_+ \oplus \mathrm{Hom}\,(\wedge^2 G, R),$$

and

$$H^2(G, R) = Z/B \cong \mathrm{Ext}\,(G, R) \oplus \mathrm{Hom}\,(\wedge^2 G, R).$$

The cup product situation is as follows: On the cochain level, there is a cup product

$$\vee : \mathrm{Hom}\,(G, R) \times \mathrm{Hom}\,(G, R) \to C^2(G; R)$$

defined by

$$f \vee f'(g, g') = f(g)\, f'(g').$$

Clearly $f \vee f'$ is bilinear, so $s(f \vee f')$ is a boundary. Consequently

$$\psi \vee : \mathrm{Hom}\,(G, R) \times \mathrm{Hom}\,(G, R) \to H^2(G, R)$$

is alternating bilinear, and it therefore factors through $\wedge^2\,\mathrm{Hom}\,(G, R)$. In fact, if

$$\varphi : \wedge^2\,\mathrm{Hom}\,(G, R) \to \mathrm{Hom}\,(\wedge^2 G, R)$$

is the natural transformation of I—1.6 and

$$i : \mathrm{Hom}\,(\wedge^2 G, R) \to H^2\,(G, R)$$

is the inclusion map specified earlier, then we have

$$\varphi\,(f \wedge f')\,(g \wedge g') = f(g)\, f'(g') - f(g')\, f'(g)$$

and the relation $\psi \vee = i\, \varphi\, \wedge$.

Chapter III

The cohomology of classifying spaces of compact groups

This Chapter is of a more expository character and serves the purpose of making our later discussions of the cohomology of classifying spaces of compact abelian groups a little more self-contained. For the more experienced reader it will suffice to peruse this Chapter in order to be familiarized with our notation and definitions. The first Section discusses Milnor's construction of the universal and the classifying space of a compact group, and our definition of the functor $h(-, R)$. The second Section briefly discusses the fact that for finite groups there is a natural isomorphism between $h(-, R)$ and $H(-, R)$, where H is the algebraic cohomology with trivial action discussed in Chapter II. The reader may read Chapter VI (the appendix by E. Nummela) in conjunction with the current material; there he will find an alternate functorial construction of a classifying space.

Section 1

The functor *h*

Throughout this section, G will denote a compact group. Cohomology H on spaces will be Čech cohomology. The coefficient module will be held fixed throughout our discussion and so will be suppressed from the notation. Later, when we refer to the cohomology algebra, this coefficient group will be in fact a commutative ring with identity.

Definition 1.1. Let X, Y be a compact G-space. The *join* $X * Y$ is the quotient space of the space $X \times [0, 1] \times Y$ modulo the relation which collapses all sets $\{x\} \times \{0\} \times Y$, $x \in Y$ and all sets $X \times \{1\} \times \{y\}$, $y \in Y$ to points. The obvious action of G on $X \times [0, 1] \times Y$ defined by $g \cdot (x, r, y) = (g \cdot x, r, g \cdot y)$ induces an action of G on $X * Y$. By induction we define a spectrum of compact G-spaces $X^{(1)} \to X^{(2)} \to X^{(3)} \to \cdots$ by $X^{(1)} = X$, $X^{(n)} = X^{(n-1)} * X$ for $n > 1$, where the maps are the obvious inclusion maps $X^{(n-1)} \to X^{(n-1)} * X$. With X^∞ we denote the direct limit of the spectrum.

Proposition 1.2. *For compact spaces X, Y there is a commutative diagram of exact sequences:*

$$\begin{array}{ccccccccc}
0 & \longrightarrow & HX \otimes HY & \longrightarrow & H(X \times Y) & \longrightarrow & H(X * Y) & \longrightarrow & 0 \\
 & & \downarrow & & \| & & \downarrow & & \\
0 & \longrightarrow & HX \otimes HY & \longrightarrow & H(X \times Y) & \longrightarrow & HX * HY & \longrightarrow & 0,
\end{array}$$

$$HX * HY = \mathrm{Tor}\,(HX, HY).$$

Proof. We define A, resp. B, to be the images of $X \times [0, 1/2] \times Y$, resp. $X \times [1/2, 1] \times Y$ in $X * Y$. Now we consider the commutative diagram

$$\begin{array}{ccccc}
X \times Y & \longrightarrow & X \times \{1/2\} \times Y & \longrightarrow & A \cap B \\
\| & & \downarrow & & \downarrow{\scriptstyle i_A} \\
X \times Y & \longrightarrow & X \times [0, 1/2] \times Y & \longrightarrow & A \\
\downarrow{\scriptstyle p_X} & & \downarrow & & \downarrow{\scriptstyle \varphi_A} \\
X & \longrightarrow & X \times [0, 1/2] \times \{y_0\} & \longrightarrow & A
\end{array}$$

where i_A is an inclusion map and p_X a projection map, and φ_A is the map induced by p_X on A. Denote the map across the top with j and the one across the bottom with k. The map

$$H\varphi_A \colon HA \to HA$$

is an isomorphism as is readily seen from the definition of A and the homotopy axiom. Thus we have a commutative diagram

$$\begin{array}{ccc}
HA & \xrightarrow{H(i)_A} & H(A \cap B) \\
\downarrow{\scriptstyle H(k)\, H(\varphi_A)^{-1}} & & \downarrow{\scriptstyle H(j)} \\
HX & \xrightarrow{H(k)} & H(X \times Y)
\end{array}$$

in which the vertical maps are isomorphisms. A similar diagram holds for B in place of A. These two diagrams define canonically a commutative diagram

$$\begin{array}{ccccc} & & HA \oplus HB & \longrightarrow & H(A \cap B) \\ & & \downarrow & & \downarrow \\ 0 & \longrightarrow & HX \otimes HY & \longrightarrow & H(X \times Y) \end{array}$$

in which the bottom row is exact. The top horizontal map is defined by the coproduct property as $H(i_A)+H(i_B)$, the left vertical map is the isomorphism

$$H(k_A)H(\varphi_A)^{-1} \oplus H(k_B)H(\varphi_B)^{-1}$$

followed by the natural included $HX \oplus HY \to HX \otimes HY$, the right vertical map is the isomorphism $H(j)$ and the bottom map is the injection appearing in the Künneth sequence

$$0 \to HX \otimes HY \to H(X \times Y) \to HX * HY \to 0.$$

Now the top map can be embedded into the Mayer-Vietoris exact sequence

$$\xrightarrow{D} H(A \cup B) \longrightarrow HA \oplus HB \longrightarrow H(A \cap B) \xrightarrow{D}$$

with a degree 1 morphism D. By the last commutative diagram,

$$HA \oplus HB \to H(A \cap B)$$

is monic, whence

$$H(A \cup B) = H(X * Y) \to HA \oplus HB$$

is zero. In view of the isomorphisms

$$HA \to HX \quad \text{and} \quad HB \to HY,$$

this proves our assertion.

Corollary 1.3. *If, under the conditions of Proposition 1.2, we have* $H^i(X) = 0$ *for* $i = 1, \ldots, n$, *then* $H^i(X * Y) = 0$ *for* $i = 1, \ldots, n+1$.

Corollary 1.4. *For any compact space,* $H^i(X^{(n)}) = 0$ *for* $i = 1, \ldots, n-1$.

Definition 1.5. Any spectrum of compact G-spaces and injections

$$E(G) = E^1(G) \xrightarrow{\varepsilon^1} E^2(G) \xrightarrow{\varepsilon^2} \cdots \longrightarrow E^\infty(G)$$

such that $H^i(E^n(G)) = 0$ for $0 < i < n$ and such that G acts freely on $E^n(G)$ is called a *spectrum of universal spaces for* G. The space $E^n(G)$ is called a *universal space up to* n. Note that we give $E^\infty(G)$ the direct limit topology and that

$E^\infty(G)$ is not necessarily compact. The sequence derived from such a sequence by passing to the orbit spaces

$$B(G) = B^1(G) \xrightarrow{\beta^1} B^2(G) \xrightarrow{\beta^2} \cdots \longrightarrow B^\infty(G)$$

is called a *spectrum of classifying spaces for G*. The space $B^n(G)$ is called a *classifying space up to n*.

Lemma 1.6. *For every compact group G, there exists always a spectrum of universal spaces and a spectrum of classifying spaces.*

Proof. This follows from Lemma 1.3 by taking $X = G$.

Definition 1.7. The spectrum constructed according to Definition 1.1 with $X = G$ will be called the *Milnor spectrum of universal spaces for G*, resp. the *Milnor spectrum of classifying spaces for G*.

Other spectra have been constructed by Dold and Lashov [12], Milgram, Steenrod [43], and others. (See also McCord [34] and Steenrod and Rothenberg [40].) The Dold-Lashof construction is described in Chapter IV.

Proposition 1.8. *For the Milnor spectrum of universal spaces the morphisms*

$$H^i(B^{m+1}(G)) \xrightarrow{H^i(\beta^m)} H^i(B^m(G))$$

are isomorphisms for $i < m$.

Proof. For the Milnor spectra we have $E^{n+1}(G) = E^n(G) * G$, whence $B^{n+1}(G) = \bar{A} \cup \bar{B}$, where $\bar{A}$ is the image of $(G^{(n)} \times [0, 1/2] \times G)/G$ and $\bar{B}$ the image of $(G^{(n)} \times [1/2, 1] \times G)/G$, and where G acts on $G^{(n)} \times [0, 1] \times G$ under $(x, y, z) \cdot g = (x \cdot g, y, z \cdot g)$. Clearly, $\bar{B}$ is just the cone over $(G^{(n)} \times G)/G$, whereas $\bar{A}$ is the mapping cylinder for the map $(G^{(n)} \times G)/G \to B^n$ defined by passing to the quotients from the projection $G^{(n)} \times G \to G^{(n)}$. However, $(G^{(n)} \times G)/G$ is naturally and equivariantly homeomorphic to $G^{(n)}$ and is, therefore, acyclic in dimensions $1, \ldots, n-1$ by Corollary 1.4. The Mayer-Vietoris sequence

$$\to H(\bar{A} \cap \bar{B}) \to H(\bar{A} \cup \bar{B}) \to H\bar{A} \oplus H\bar{B} \to H(\bar{A} \cap \bar{B}) \to$$

therefore provides us with a sequence

$$0 \to H^i(B^{n-1}) \to H^i \bar{A} \oplus H^i \bar{B} \to 0 \quad \text{for} \quad i = 1, \ldots, n-1.$$

But $H\bar{B}$, being the cohomology of a cone, is zero in positive dimensions. On the other hand, $H\bar{A}$ is naturally isomorphic to HB^n in such a fashion that the isomorphism $H^i B^{n+1} \to H^i \bar{A}$ yields an isomorphism $H^i \beta^n : HB^{n+1} \to HB^n$.

This leads us to the following definition.

Definition 1.9. For any compact group G and any commutative ring with identity, we let

$$h(G, R) = \lim H(B^n(G, R)).$$

Thus $h(G, R)$ is the graded R-algebra whose i-th homogeneous component $h^i(G, R)$ is canonically isomorphic to $H^i(B^n(G), R)$ for all $n > i$. In other words, the graded algebra $h(G, R)$ is isomorphic to the graded algebra $H(B^n(G), R)$ up to dimension $n - 1$. From the definitions, it is clear that there is a natural morphism of graded algebras

$$\lambda_{G,R}: H(B^\infty(G), R) \to h(G, R)$$

for Čech cohomology on $B^\infty(G)$.

It is useful to observe the following fact:

Remark. Let $E_i(G)$, $B_i(G)$, $i = 1, 2$, be two spectra of universal and classifying spaces for G. Then there is a natural isomorphism

$$\varphi_{E_1^n, E_2^n}: H(B_1^n(G)) \to H(B_2^n(G))$$

of graded algebras up to dimension $n - 1$. Moreover, $\varphi_{E_2^n, E_3^n} \circ \varphi_{E_1^n, E_2^n} = \varphi_{E_1^n, E_3^n}$. (Borel [3], p. 166.)

We will show that under favorable circumstances (such as prevail if the Milnor spectrum is used), λ is a natural isomorphism; thus h is actually obtained as some space cohomology.

Let $X_0 \subset \cdots \subset X$ be an ascending sequence of compact subspaces of a space X such that X is the colimit in the category of spaces (i. e. that $U \subset X$ is open if and only if $U \cap X_i$ is open for all i).

The *telescope* over this sequence is the locally compact space

$$T(\{X_i\}) = [0, 1] \times X_0 \cup [1, 2] \times X_1 \cup \cdots \subset [0, \infty[\times X.$$

Let

$$T_0 = \bigcup \{[n, n+1] \times X_n : n = 0, 2, 4, \ldots\},$$

$$T_1 = \bigcup \{[n, n+1] \times X_n : n = 1, 3, 5, \ldots\}.$$

Then, not counting natural isomorphisms,

$$H(T_1) = \prod \{H(X_n) : n = 2m + i, m = 0, 1, 2, \ldots\}, \quad i = 0, 1 \ldots$$

where H denotes Čech cohomology with an arbitrary but fixed coefficient group. In a similar vein, we may identify $H(T_0 \cap T_1)$ with

$$\prod \{H(X_n) : n = 0, 1, \ldots\}$$

since $T_0 \cap T_1 = \{1\} \times X_0 \cup \{2\} \times X_1 \cup \cdots$.

Let $i_n\colon X_n \to X_{n+1}$ be the inclusion map and set $H(i_n) = i_n^*$. Then the inclusion $T_0 \cap T_1 \to T_0$ induces the morphism

$$s_0\colon H(T_0) \to H(T_0 \cap T_1)$$

defined by

$$s_0(h_0, h_2, h_4, \ldots) = (h_0, i_1^* h_2, h_2, i_3^* h_4, \ldots),$$

and the inclusion $T_0 \cap T_1 \to T_1$ induces the morphism

$$s_1\colon H(T_1) \to H(T_0 \cap T_1)$$

defined by

$$s_1(h_1, h_3, \ldots) = (i_0^* h_1, h_1, i_2^* h_3, h_3, \ldots).$$

Thus the map

$$\psi\colon H(T_0) \oplus H(T_1) \to H(T_0 \cap T_1)$$

defined by

$$\psi(h, h') = s_0(h) - s_1(h')$$

is given by

$$\psi((h_0, h_2, \ldots), (h_1, h_3, \ldots))$$
$$= (h_0 - i_1^* h_1, -(h_1 - i_1^* h_2), h_2 - i_2^* h_3, -(h_3 - i_3^* h_4), \ldots)$$

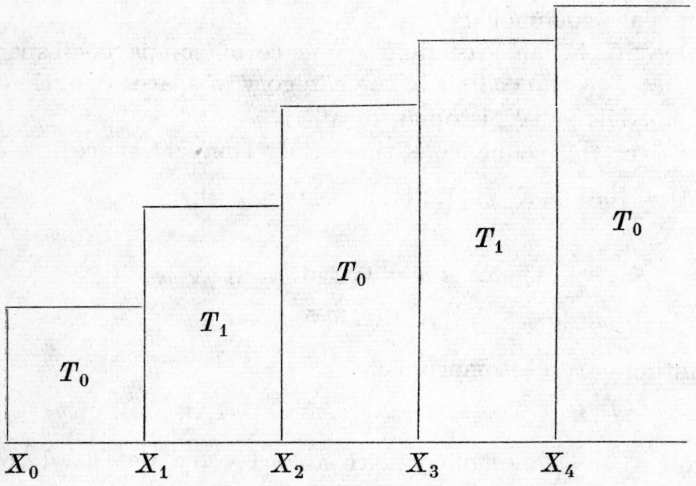

The kernel of ψ is exactly the limit of the inverse system

$$H(X_0) \xleftarrow{i_0^*} H(X_1) \xleftarrow{i_1^*} \cdots,$$

i. e. $\ker \psi = \lim_n H(X_n)$. The cokernel is called the *derived limit functor*, and we write

$$L'_n H(X_n) = \operatorname{coker} \psi.$$

Thus we have an exact sequence

$$0 \to \lim_n H(X_n) \to H(T_0) \oplus H(T_1) \xrightarrow{\psi} H(T_0 \cap T_1) \to L'_n H(X_n) \to 0.$$

On the other hand, there is the exact Mayer-Vietoris sequence

$$\longrightarrow H^{n-1}(T_0 \cap T_1) \xrightarrow{\Delta^{n-1}} H^n(T) \xrightarrow{\varphi^n} H^n(T_0) \oplus H^n(T_1) \xrightarrow{\psi^n} H^n(T_0 \cap T_1) \xrightarrow{\Delta^n}.$$

Hence $\operatorname{coker} \psi^{n-1} = \operatorname{im} \Delta^{n-1} = \ker \varphi^n$, and $\ker \psi^n = \operatorname{im} \varphi^n$. Thus we have an exact sequence

$$0 \to L'_n H^{i-1}(X_n) \to H^i(T) \to \lim_n H^i(X_n) \to 0,$$

and this sequence is naturally defined.

Now suppose that $i^n_k \colon H^n(X_{k+1}) \to H^n(X_k)$ is surjective for $k \geqq k_0$. Then equations

$$a_k = h_k - i^n_k h_{k+1}, \quad k = k_0, k_0 + 1, \ldots$$

may be successively solved for h_k, if h_{k_0} is given. The finitely many equations

$$a_k = h_k - i^n_k h_{k+1}, \quad k = k_0 - 1, k_0 - 2, \ldots$$

can then also be solved for h_k. Thus ψ^n is surjective. Hence we have the following:

Proposition 1.10a. *Let $B^n(G)$ be a spectrum of compact classifying spaces of G. Let $T(\{B^n(G)\})$ be the (locally compact) telescope of the spectrum. Then $h(G, R)$ is naturally isomorphic to $H(T(\{B^n(G)\}), R)$, where H denotes Čech cohomology with closed supports.*

Moreover, the telescope $T(\{E^n(G)\})$ is a locally compact acyclic space on which G acts freely.

Thus $h(G, R)$ can be obtained as the space cohomology of some suitably chosen locally compact space. However, one would like to obtain $h(G, R)$ as the space cohomology of $B^\infty(G)$ itself.

To prove this, we recall that a map $i \colon X \to Y$ of topological spaces is called a *cofibration* if any homotopy $f \colon X \times [0, 1] \to Z$ for which there is a map $F_0 \colon Y \to Z$ with $f(x, 0) = F_0(i(x))$ for all $x \in X$ extends to a homotopy $F \colon Y \times [0, 1] \to Z$ with $F(i(x), s) = f(x, s)$ for all $x \in X$, $s \in [0, 1]$. Note that every cofibration is injective.

We establish the following Lemma:

Lemma. *Let X and Y be compact spaces. Let $i\colon X \to X * Y$ be the injection defined by $i(x) = \langle x, 0, b\rangle$ (where $\langle x, r, y\rangle$ denotes the class of*

$$(x, r, y) \in X \times [0, 1] \times Y \text{ in } X * Y,$$

and where b is an arbitrary base point in Y; note that the choice of b is irrelevant for the definition of i). Then i is a cofibration.

Proof. Let $F_0\colon X * Y \to Z$ and $f\colon i(X) \times [0, 1] \to Z$ be given so that

$$f(\langle x, 0, b\rangle, 0) = F_0(\langle x, 0, y\rangle).$$

Define $F\colon X * Y \times [0, 1] \to Z$ by

$$F(\langle x, r, y\rangle, s) = \begin{cases} F_0\left(\left\langle x, \dfrac{2r - s}{2 - s}, y\right\rangle\right) & \text{for } 0 \leqq s \leqq 2r \leqq 2, \\ f\left(\langle x, 0, b\rangle, \dfrac{s - 2r}{1 - r}\right) & \text{for } 0 \leqq 2r \leqq s \leqq 1. \end{cases}$$

Then F is the desired extension of f and F_0.

Lemma. *If both X and Y are G-spaces and $X * Y$ is made into a G-space in the obvious fashion such that with the product action on $X \times [0, 1] \times Y$ the quotient map onto $X * Y$ is equivariant, then the natural inclusion map of X/G into $(X * Y)/G$ is a cofibration.*

Proof. This follows straightforwardly as in the preceding lemma.

Next we observe that an inclusion $i\colon X \to Y$ is a cofibration iff

$$([0, 1] \times X) \cup \{1\} \times Y$$

is a retract of $[0, 1] \times Y$. Now let us assume that in some category we have a commuting diagram

$$\begin{array}{ccccccccc} T^0 & \xrightarrow{t_0} & T^1 & \xrightarrow{t_1} & T^2 & \longrightarrow & T^3 & \longrightarrow \cdots \longrightarrow & P \\ \downarrow & & \downarrow & & \downarrow & & \downarrow & & \\ T^0 & \xleftarrow{p_0} & T^1 & \xleftarrow{p_1} & T^2 & \longleftarrow & T^3 & \longleftarrow \cdots & \end{array}$$

such that all vertical maps are identity maps and P is the colimit of the top line. Then by the definition of the colimit, there is a morphism $P \xrightarrow{p^i} T^i$ commuting with the diagram. If $t^i\colon T^i \to P$ denotes the obvious morphism, then t^i is the right inverse of p^i, and is therefore a coretraction, and p^i is a retraction. We apply this to the following situation:

Let $X_0 \to X_1 \to \cdots \to X$ be a diagram of cofibrations of compact spaces with the colimit X. We have the inclusion map

$$t\colon T(\{X_i\}) \to [0, \infty[\times X.$$

We define $T^0 = T(\{X_i\})$, $T^1 = [0, 2] \times X_1 \cup T^0$, $T^2 = [0, 3] \times X_2 \cup T^0$, ... and let $t_i\colon T^i \to T^{i+1}$ be the inclusion maps. Since the t_i are cofibrations, there exist retractions $p_i\colon T^{i+1} \to T^i$. The space $P = [0, \infty[\times X$ is the colimit of the system $T^0 \to T^1 \to \cdots$ (we leave the verification of this assertion as an exercise). Thus, by the preceding, there is a retraction $p^0\colon P \to T(\{X_i\})$ which is a left inverse of the inclusion map t^0. In fact we will show that p^0 is a weak deformation retraction; specifically, we will show that $t^0 p^0$ is homotopic to the identity of P, which will establish the assertion in view of the previous observation that p^0 is a retraction.

It is easy to see that each p_i is in fact a weak homotopy retraction; in particular $t_i p_i$ is homotopic to the identity of T^{i+1}. We utilize this in defining a function $q_i\colon T^i \times [0, \infty] \to T^i$ with the properties $q_i(x, s) = x$ for $s \geqq i + 1$ and $t_i p_i(x) = q_i(x, s)$ for $s \leqq i$. Now we let $h_i\colon P \times [0, \infty] \to P$ be defined by

$$h_i(x, s) = t^i q_i(p^i(x), s),$$

i. e. $h_i(\cdot, s) = t^i q_i(\cdot, s)\, p^i$. Note that $h_i(\cdot, s) = t^i p^i$ for $s \geqq i + 1$. The function $p\colon P \times [0, \infty] \to P$ may therefore be well defined as the 'infinite' product $p(\cdot, s) = h_0(\cdot, s)\, h_1(\cdot, s) \cdots$; thus $p(x, s) = h_0^s h_1^s \cdots h_i^s t^i p^i$ for $i + 1 \leqq s$ with $h_k^s(x) = h_k(x, s)$. It is now clear that $p(x, 0) = t^0 p^0(x)$ and $p(x, \infty) = x$. Moreover, the function p is clearly continuous when restricted to any one of the spaces $t^i(T) \times [0, \infty]$; but $P \times [0, \infty]$ is the colimit of the subspaces $t^i(T) \times [0, \infty]$. Hence p is continuous and thus p^0 is a homotopy retraction.

The inclusion map $t\colon T(\{X_i\}) \to X \times [0, \infty[$ therefore induces an isomorphism $H(i)\colon H(X \times [0, \infty[) \to H(T)$. The projection $X \times [0, \infty[\to X$ induces an isomorphism in cohomology. Hence the projection $\pi\colon T(\{X_i\}) \to X$ induces an isomorphism $H(\pi)$ in cohomology. We have therefore proved the following

Lemma. *Let* $X_0 \xrightarrow{i_0} X_1 \xrightarrow{i_1} \cdots \longrightarrow X$ *be a diagram of cofibrations of compact spaces* X_i *(compactness is not essential!) such that* X *is the colimit of the* X_i. *Then for any coefficient group we have* $H(T(\{X_i\})) \cong H(X)$ *under a natural isomorphism when Čech cohomology with closed supports is used.*

We have the following proposition.

Proposition 1.10b. *If* $\{B^i(G)\}$ *is a compact spectrum of classifying spaces for the compact group* G, *then* $h(G, R) \cong H(B^\infty(G), R)$ *under a natural isomorphism with Čech cohomology* H, *provided that* $B^\infty(G)$ *is the colimit of the* $B^i(G)$

and that all injections $B^i(G) \to B^{i+1}(G)$ are cofibrations. The latter is the case in particular for the Milnor spectrum. The isomorphism is the composition of the isomorphism of 1.10a *and $H(\pi)^{-1}$, where $\pi\colon T(\{B^n(G)\}) \to B^\infty(G)$ is the projection. Its inverse equals the map $\lambda_{G,R}$ of p.* 159.

Proposition 1.11. *The functor $h(-, R)$ transforms projective limits into direct limits.*

Proof. Let $\{G_1, \varphi_{ij}\}$ be an inverse system of compact groups. Then

$$\{E^n(G_i), E^n(\varphi_{ij})\} \quad \text{and} \quad \{B^n(G_i), B^n(\varphi_{ij})\}$$

are inverse systems of compact spaces for $n = 1, 2, \ldots$ If $G = \varprojlim G_i$, then $E^n(G) = \varprojlim E^n(G_i)$ and $B^n(G) = \varprojlim B^n(G_i)$, and hence

$$H(B^n(G), R) = \varinjlim H(B^n(G_i), R)) \quad \text{for} \quad n = 1, 2, \ldots$$

Thus in view of the natural isomorphism $h^i(G, A) \cong H^i(B^n(G), R)$ for $i < n$ we obtain $h(G, R) = \varinjlim h(G, R)$.

Corollary 1.12. *For any compact group G, there is a directed (under inclusion) system $\mathfrak{N}$ of normal subgroups with G/N a Lie group for each*

$$N \in \mathfrak{N}, \ \cap \mathfrak{N} = \{1\},$$

and $h(G, R) = \varinjlim \{h(G/N, R)\colon N \in \mathfrak{N}\}$.

A discussion of the functorial aspects of this important proposition will follow in Chapter IV.

Proposition 1.13. *For two compact groups G_1 and G_2 and a principal ideal domain R there is an exact sequence*

$$0 \to h(G_1, R) \otimes h(G_2, R) \to h(G_1 \times G_2, R) \to \mathrm{Tor}^R(h(G_1, R), h(G_2, R)) \to 0.$$

This sequence splits non-naturally.

Proof. The spectrum $G_1^{(n)} \times G_2^{(n)}$, with the group $G_1 \times G_2$ acting componentwise is a spectrum of universal spaces, since by the Künneth theorem for compact spaces the n-th term is acyclic in dimensions $1, \ldots, n-1$. The associated spectrum $(G_1^{(n)} \times G_2^{(n)})/G$ of classifying spaces is naturally homeomorphic to the spectrum $B^n(G_1) \times B^n(G_2)$. By the Künneth theorem for compact spaces, we have a split exact sepuence

$$0 \to H(B^n(G_1), R) \otimes H(B^n(G_2), R) \to H(B^n(G_1) \times B^n(G_2), R)$$
$$\to \mathrm{Tor}^R(H(B^n(G_1), R), H(B^n(G_2), R) \to 0.$$

But then because of Proposition 1.10b and the definition of h, the assertion follows.

Corollary 1.14. *If $h(G_i, R)$ has no torsion for $i = 1$ or $i = 2$, there is a natural isomorphism*

$$h(G_1, R) \otimes h(G_2, R) \to h(G_1 \times G_2, R).$$

Corollary 1.15. *Let G be a compact connected Lie group such that $H(G, \mathbf{Z})$ (resp. $H(G, GF(p))$) is an exterior algebra of a free abelian group (resp. a $GF(p)$-vector space) generated by elements of odd degree. Then for any principal ideal domain R and for any compact group G', there is a natural isomorphism*

$$h(G', R) \otimes h(G, R) \to h(G' \times G, R)$$

(resp. a natural isomorphism

$$h(G', GF(p)) \otimes h(G, GF(p)) \to h(G' \times G, GF(p))).$$

This assertion maintains if G is a projective limit of Lie groups satisfying the conditions spelled out for G above.

Proof. By a result of Borel's, $h(G, \mathbf{Z})$ (resp. $h(G, GF(p))$) is a polynomial algebra (Borel [3], p. 171). By the Universal Coefficient Theorem, $h(G, R)$ is then torsion free over R, and Corollary 1.14 applies. (In the case of $GF(p)$, it applies directly.) If G is a projective limit, then we have to appeal to Proposition 1.11, and the fact that a direct limit of torsion free R-modules is torsion free.

Proposition 1.16. *Let $G = G_1 \times G_2$, where G_1 and G_2 are compact groups, and let A be an abelian group. Let $\pi\colon G \to G_1$ be the natural projection, and suppose that $h(G_2, A) = 0$. Then $h(\pi)\colon h(G_1, A) \to h(G, A)$ is an isomorphism. If A is a commutative ring with identity, then $h(\pi)$ is a ring isomorphism relative to the cup product.*

Proof. As previously observed, there is a classifying space for dimension n of G of the form $B^n_{G_1} \times B^n_{G_2}$. Apply the Vietoris-Begle theorem and observe that $h(\pi)$ commutes with cup products.

Proposition 1.17. *Let G be a compact group. Let R be a commutative ring with identity and $a \in R$ an element satisfying $a\,x = 0$ if and only if $x = 0$. Then there is a derivation and differential $d_a\colon h(G, R/aR) \to h(G, R/aR)$ such that all assertions of Proposition I — 4.15 are satisfied with h in place of H.*

Proof. The assertions of Proposition I — 4.15 hold for $H(B^n(G), R)$ in place of $H(G, R)$, $H(B^n(G), R/aR)$ in place of $H(G, R/aR)$, $H(B^n(G), R/a^2R)$ in place of $H(G, R/a^2R)$, etc. and suitable maps d^n_a, D^n_a in place of d_a and D_a respectively. The morphism

$$H(\varrho^m, R/aR)\colon H(B^{m+1}(G), R/aR) \to H(B^m(G), R/aR)$$

of Proposition 1.8 is a morphism of graded differential algebras, i. e.

$$d_a^m \, H(\beta^{m+1},\, R/aR) = H(\beta^{m+1},\, R/aR)\, d_a^{m+1}.$$

Thus $d_a = \lim d_a^n$ exists. In fact, $d_a x$ for a given element x of dimension n may be computed by evaluating $d_a^{n+2} x$, where x' is the image of x in

$$H(B^{n+2}(G),\, R/aR)$$

(Proposition 1.8). Clearly d_a is a differential and a derivation. Since the taking of the limit in Definition 1.9 is exact in view of Proposition 1.8, the remaining assertions of Proposition I — 4.15 with h in place of H are obtained by passing to the limits. — In view of Proposition 1.10 b, the limit arguments are not absolutely necessary for the proof.

Definition 1.18. The derivation and differential d_a defined in Proposition 1.17 will be called the *Bockstein differential* on $h(G, R/aR)$.

Proposition 1.19. *Let R be a ring and $a \in R$ satisfying the conditions given in Proposition 1.17. Consider the functor $h(-, R/aR)$ as a functor taking values in the category of differential graded R-algebras. Then $h(-, R/aR)$ transforms projective limits into direct limits.*

Proof. This follows from Proposition 1.11 and the definition of the Bockstein differential.

Lemma 1.20. *Let the conditions of Proposition 1.19 be satisfied. Suppose that there is a natural transformation of functors $n\colon F \to h^2(G, -)$ from the category of R-modules into itself such that*

$$FR \xrightarrow{Fa} FR \longrightarrow F(R/aR) \longrightarrow 0$$

is exact. Then $d_a^2\, n_{R/aR} = 0$. (Note: d^2 stands for d in degree two and not for $d \cdot d$.)

Proof. Apply the definition of d_a in Proposition 1.17 above and Lemma I — 4.18.

Section 2

The functor h for finite groups

For finite groups the functor h can be described differently. For the following definition we adopt the notation of MacLane [32], p. 233.

Definition 2.1. Let N denote the set of non-negative integers. For each $p \in \mathsf{N}$, let $[p] = \{0, 1, \ldots, p\}$ with the natural order. Let $\mathfrak{M}$ denote the category of all $[p]$ as objects with weakly monotonic maps as morphisms.

For any category $\mathfrak{C}$, a *simplicial object in* $\mathfrak{C}$ is a contravariant functor $A: \mathfrak{M} \to \mathfrak{C}$. For notational convenience, we write A_p for $A([p])$, and μ^* for $A(\mu)$ where $\mu \in \mathfrak{M}([p], [q])$ provided there is no danger of confusion. A *simplicial mapping* f of simplicial objects A and B in $\mathfrak{C}$ is, thus, a natural transformation of contravariant functors; i.e., for each $p \in \mathsf{N}$ there exists $f_p: A_p \to B_p$ in $\mathfrak{C}$ such that the diagrams

$$\begin{array}{ccc} A_p & \xrightarrow{f_p} & B_p \\ \downarrow{\scriptstyle \mu^*} & & \downarrow{\scriptstyle \mu^*} \\ A_q & \xrightarrow{f_q} & B_q \end{array}$$

are commutative for all $\mu \in \mathfrak{M}([q], [p])$.

Remarks. Let $\mathfrak{T}$ denote the category of topological spaces and Set $\mathfrak{M}^*$ the category of simplicial sets. Let $|\ |$: Set $\mathfrak{M}^* \to \mathfrak{T}$ be the geometric realization functor and $S: \mathfrak{T} \to$ Set $\mathfrak{M}^*$ the usual simplicial set functor (see e. g. K. Lamotke, *Semisimpliziale algebraische Topologie*, Springer-Verlag, Berlin-Heidelberg-New York 1968, pp. 34, 6). Then $|-|$ is left adjoint to S (see K. Lamotke, p. 46). Moreover, for a simplicial set A, the unit (front adjunction)

$$\eta_A: A \to S(|A|)$$

induces isomorphisms in cohomology (J. Milnor, The geometric realization of a semi-simplicial complex, Ann. Math. **65** (1957), 357—362).

Let $\Delta(n)$ denote the *n-model*; i. e., the simplicial set defined by setting $\Delta(n)_p = \mathfrak{M}([p], [n])$ and setting $\mu^*(\alpha) = \alpha\, \mu$ for all

$$\mu \in \mathfrak{M}([p], [q]), \quad \alpha \in \Delta(n)\, q.$$

Then $|\Delta(n)|$ is the standard n-dimensional simplex in $n+1$ dimensional Euclidean space (see K. Lamotke, pp. 3, 8, 37).

We state the following proposition without proof.

Proposition 2.2. (1) *If a category* $\mathfrak{C}$ *has products (resp., coproducts, equalizers, coequalizers), then the category of simplicial objects in* $\mathfrak{C}$ *has products (resp., coproducts, equalizers, coequalizers).*

(2) *The category* Set $\mathfrak{M}^*$ *thus has products, which are not in general preserved by the functor* $|\ |$. *However, if A and B are simplicial sets, then*

$$|A \times B| \cong |A| \times |B|$$

if either $|A|$ *or* $|B|$ *is locally compact.*

(3) *All colimits which exist in the category* Set $\mathfrak{M}^*$ *are preserved by the functor* $|\ |$, *since it is a coadjoint.*

Definition 2.3. Let G be a discrete group. By abuse of notation we denote also by G the simplicial group defined by

$$G_p = G \quad \text{for all} \quad p \in \mathsf{N},$$
$$\mu^* = \mathrm{id}_G \quad \text{for all} \quad \mu \text{ in } \mathfrak{M}.$$

The mapping $G \times G \to G$ of simplicial sets determined by the group multiplication on each level is a simplicial mapping such that $|G| = G$ both topologically and algebraically.

We say that G *acts* on a simplicial set A if there is a simplicial mapping $G \times A \to A$ such that the mapping at each level $G_p \times A_p \to A_p$ is a group action in the usual sense. We define the *quotient* simplicial set A/G to be the functor which associates with $[p]$ the set A_p/G_p and with each morphism $\mu\colon [p] \to [q]$ the induced function $A_q/G_q \to A_p/G_p$.

Lemma 2.4. *Let A be a simplicial set, and G a group of automorphisms of A. Let $\mathsf{Z}A$ be the simplicial Z-module generated freely over Z by A. If S is the group ring of G over Z, then $\mathsf{Z}A$ is a simplicial left S-module in the obvious fashion. This module is free, and if A_p is finite for all p, then $K(A)$, the simplicial complex over Z generated by A, is a finitely generated free S-complex.*

Proof. The only assertion which is perhaps not obvious is the freeness of $\mathsf{Z}A$ over S. But this claim follows immediately from the following

Lemma. *Let M be a set on which the group G of bijections of M operates on the left. Let F be the free abelian group over M on which the group ring S of G over the integers operates in the obvious fashion. Then the S-module F is free.*

Proof of the lemma. Since freeness is a property of finitely generated submodules, we may assume that M is a finite union of orbits

$$G \cdot m_i, \quad i = 1, \ldots, k.$$

Then F is a coproduct of the cyclic subgroups $\mathsf{Z} \cdot (g \cdot m_i)$, $i = 1, \ldots, k$, $g \in G$. The coproducts $\bigoplus \{\mathsf{Z} \cdot (g \cdot m_i)\colon g \in G\}$ are monogenic S-submodules, each isomorphic to S under $s \mapsto s \cdot m_i$. But F is the coproduct of k of these submodules and is, therefore, free over S.

Lemma 2.5. *Under the conditions of Lemma 2.4, let $\pi\colon A \to A/G$ be the natural transformation defined by taking the orbit map $\pi_p\colon A_p \to A_p/G_p$ for each p.*

This natural transformation defines a natural transformation

$$\mathsf{Z}\pi\colon \mathsf{Z}A \to \mathsf{Z}(A/G).$$

If R is a commutative ring, S is the group ring of G over the integers, and R is given the trivial structure of an S-left module, then there is a natural isomorphism of R-complexes

$$\mathrm{Hom}_S(K(A), R) \to \mathrm{Hom}(K(A/G), R).$$

Proof. Since $\mathsf{Z}A$ is freely generated over Z by A, there is a natural isomorphism $\mathrm{Set}(A, R) \to \mathrm{Hom}(\mathsf{Z}A, R)$, where $\mathrm{Set}(A, R)$ denotes the set of natural transformations in the category of functors from $\mathfrak{M}$ into sets (R being identified with the constant functor with value R) and where $\mathrm{Hom}(\mathsf{Z}A, R)$ is defined similarly for group valued functors. If $\mathrm{Set}_G(A, R)$ denotes the subset of transformations $A \to R$ which are constant on the orbits of G, then this isomorphism obviously induces an isomorphism

$$\mathrm{Set}_G(A, R) \to \mathrm{Hom}_S(\mathsf{Z}A, R),$$

since R is a trivial S-module. Now $\mathrm{Set}_G(A, R)$ and $\mathrm{Set}(A/G, R)$ are naturally isomorphic. But since there is a natural isomorphism between $\mathrm{Set}(A/G, R)$ and $\mathrm{Hom}(\mathsf{Z}(A/G), R)$, the assertion follows.

For finite groups the functor h is in fact the algebraic cohomology of G for trivial G-modules, as we will indicate now for the sake of completeness.

Proposition 2.6. *If G is a finite group, then there is a natural isomorphism $H(G, R) \to h(G, R)$, where H is defined as follows: Take any free resolution*

$$0 \leftarrow \mathsf{Z} \leftarrow X^0 \leftarrow X^1 \leftarrow$$

of G-modules. Then

$$H(G, R) = H(\mathrm{Hom}_G(X, R)),$$

where R is a trivial G-module.

Proof. Define simplicial sets $A^n(G)$ as follows:
$A^1(G) = G$ (the simplicial group of Definition 2.3); for $n > 1$,

$$A^n(G) = A^{n-1}(G) * G,$$

where $*$ denotes the simplicial analog of the join defined in Definition 1.1 above. That is, $\Delta(1)$ replaces $[0, 1]$, and 0 and 1 are ambiguously used to denote the functions in $\mathfrak{M}$ with codomain 0 and 1, respectively. The prescribed identifications, i. e. the collapsing of sets $\{x\} \times \{0\} \times G_p$ and

$$A_p^{n-1}(G) \times \{1\} \times \{g\}$$

to points, takes place on each level. Since all simplicial sets appearing in this definition are finite, their geometric realizations are compact. Hence the functor $|\ |$ preserves the products which appear. The functor $|\ |$, being a coadjoint, also preserves the quotient operation involved in the join. Moreover, the canonical action of G on each $A^n(G)$ is simplicial and induces the expected action on the geometric realization level. That is, $|A^n(G)| \cong E^n(G)$ and

$$|A^n(G)/G| \cong B^n(G)$$

as G-spaces for all $n \geqq 1$.

Let $K(A^n(G))$ and $K(A^n(G)/G)$ denote the corresponding simplicial complexes over Z. Let S denote the group ring $\Gamma(G)$.

We define a complex $K(i)$ of S-modules as follows: Let

$$K^0(i) \leftarrow K^1(i) \leftarrow \cdots \leftarrow K^{i-1}(i)$$

be the acyclic complex

$$K^0(A^i(G)) \leftarrow \cdots \leftarrow K^{i-1}(A^i(G))$$

and extend that complex in any fashion to an exact S-free complex

$$K^{i-1}(A^i(G)) \leftarrow K^i(i) \leftarrow .$$

There is a morphism of complexes $K(i) \to K(A^i(G))$ which is an isomorphism up to dimension i. Finally let $K^0(i) \xrightarrow{\varepsilon_i} \mathsf{Z}$ be the augmentation which maps all the generators of $K^0(i)$ onto 1. Then

$$0 \longleftarrow \mathsf{Z} \xleftarrow{\varepsilon_i} K(i)$$

is a free resolution of S-modules. Hence there are chain equivalences

$$f_i\colon X \to K(i), \quad f_{ij}\colon K(j) \to K(i)$$

for $i < j$ which make the diagrams

$$\begin{array}{ccccc} 0 \longleftarrow & \mathsf{Z} & \xleftarrow{\varepsilon} & X \\ & \Vert & & \downarrow f_i \\ 0 \longleftarrow & \mathsf{Z} & \xleftarrow{\varepsilon_i} & K(i) \end{array} \qquad \begin{array}{ccccc} 0 \longleftarrow & \mathsf{Z} & \xleftarrow{\varepsilon_j} & K(j) \\ & \Vert & & \downarrow f_{ij} \\ 0 \longleftarrow & \mathsf{Z} & \xleftarrow{\varepsilon_i} & K(i) \end{array}$$

commute and which are unique up to chain homotopy. Moreover, we have that f_i and $f_{ij} f_i$ are chain homotopic for all $i \leqq j$, and $f_{ki} f_{ij}$ and f_{kj} are chain homotopic for $k \leqq i \leqq j$. Hence there are cochain equivalences

$$\operatorname{Hom}_S(f_i, R)\colon \operatorname{Hom}_S(K(i), R) \to \operatorname{Hom}_S(X, R),$$

$$\operatorname{Hom}_S(f_{ij}, R)\colon \operatorname{Hom}_S(K(i), R) \to \operatorname{Hom}_S(K(j), R), \quad i \leqq j.$$

They yield morphisms of graded abelian groups

$$f_i^*: H(\mathrm{Hom}_S(K(i), R)) \to H(G, R),$$

$$f_{ij}^*: H(\mathrm{Hom}_S(K(i), R)) \to H(\mathrm{Hom}_S(K(j), R)), \quad i \leqq j,$$

such that $f_i^* = f_j^* f_{ij}^*$ and $f_{ik}^* = f_{jk}^* f_{ij}^*$. Thus

$$H(G, R) = \varinjlim H(\mathrm{Hom}_S(K(n), R)).$$

But $H^m(\mathrm{Hom}_S(K(i), R))$ is isomorphic to $H^m(\mathrm{Hom}_S(K(j), R))$ for $m < i \leqq j$ under f_{ij}^*, and these groups are in turn isomorphic to $H^m(\mathrm{Hom}_S(K(A^i(G)), R))$. Thus $H^m(G, R)$ is in fact isomorphic to any of the groups

$$H^m(\mathrm{Hom}_S(K(A^i(G)), R)).$$

But by Lemma 2.5 we know that

$$H(\mathrm{Hom}_S(K(A^i(G)), R))$$

is naturally isomorphic to $H(\mathrm{Hom}(K(A^i(G))/G, R))$, and by the result of Milnor's mentioned in the Remarks following Definition 2.1, the latter is naturally isomorphic to $H(B^i(G), R)$ for all i, since $|A^i(G)/G| = B^i(G)$ and Čech cohomology coincides with singular cohomology on finite polyhedra. By the definition of h, the assertion now follows.

Corollary 2.7. *Let R be an arbitrary abelian group and let G be a totally disconnected compact (i. e. profinite) group. Then*

$$h(G, R) = \varinjlim H(G/N, R),$$

where the limit is taken over the direct system arising from the inverse system of all G/N, where N runs through all normal subgroups with finite index.

Proof. This assertion follows immediately from Proposition 2.4 and Corollary 1.12.

It should be pointed out that in general $h(G, R)$ will not be the cohomology of a compact space, since by Conner's lemma, every cohomology ring of a compact space is locally nilpotent, whereas even for a cyclic group G of two elements $h(G, \mathbf{Z})$ is a polynomial ring.

Chapter IV

Kan extensions of functors on dense categories

In deriving cohomology algebras of arbitrary compact groups from the cohomology algebras of Lie groups where they are reasonably well understood one has to pass through limit operations. However, these limit operations need to be controlled to such an extent that one can see how morphisms are lifted to the limits. In other words one has to pass to the limit functorially. In order to do this efficiently, we develop a theory which connects our situation with the more general concept of Kan extensions of functors. Indeed we derive a quite general existence and uniqueness theorem for Kan extensions—one which covers much more ground than we actually need for our current applications. Among other things we will be able to show as an application that the Čech cohomology of a compact group space is the Kan extension of any of the standard cohomologies on manifolds applied to Lie group spaces, and that the cohomology of a classifying space of a compact totally disconnected group is the Kan extension of the algebraic cohomology of a finite group.

The essential use we are making of our categorical results is that certain functors are sometimes determined by their action on very small subcategories of their domains. In Section 1 we show that functors that are continuous in a suitable sense are determined on subcategories which are dense in a suitable sense. In Section 2 we show in essence that additive functors are determined on very small generating subcategories of their domain. The details are somewhat involved and will be explained and illustrated by various examples.

Section 1

Dense categories and continuous functors

We start by defining a particular 2-category which is useful in the applications. Let $\mathfrak{D}$ be a full subcategory of the category of small categories; thus the objects of $\mathfrak{D}$ are certain small categories, the morphisms of $\mathfrak{D}$ are the functors between these. A typical example which we shall use in the applications is the category $\mathfrak{D}$ of directed sets and order preserving maps; indeed any partially ordered set is a category in which each hom set has at most one morphism, and the order preserving maps are then exactly the functors. We shall speak of $\mathfrak{D}$ as a *category of small categories*, meaning always the full subcategory of the category of all small categories generated by the objects of $\mathfrak{D}$.

Let $\mathfrak{A}$ be an arbitrary category. We proceed to describe a new category $\mathfrak{A}^{\mathfrak{D}}$ by giving its objects, its morphisms, and the composition of its morphisms. Objects of $\mathfrak{A}^{\mathfrak{D}}$ are functors $F\colon X \to \mathfrak{A}$, where X is an object of $\mathfrak{D}$. The morphisms $F \to G$ of $\mathfrak{A}^{\mathfrak{D}}$ are pairs $(\varkappa, f)$, where f is a functor $f\colon \operatorname{dom} G \to \operatorname{dom} F$ (where $\operatorname{dom} F$ denotes the domain category of F; note that $\operatorname{dom} G$ comes first!) and where $\varkappa$ is a natural transformation $\varkappa\colon Ff \to G$ of functors $\operatorname{dom} G \to \mathfrak{A}$. Finally let $(\varkappa, f)\colon F \to G$ and $(\lambda, g)\colon G \to H$ be two morphisms; then its composition is given by

$$(\lambda, g)\,(\varkappa, f) = (\lambda(\varkappa\, g), f\, g).$$

Notice that if $\mathfrak{D}'$ is a (full) subcategory of $\mathfrak{D}$, then $\mathfrak{A}^{\mathfrak{D}'}$ is a full subcategory of $\mathfrak{A}^{\mathfrak{D}}$.

Remark. If one reads the composition of $\mathfrak{D}$ in its opposite category (i. e. $f\,g = g \circ f$ with the composition $\circ$ of $\mathfrak{D}^{op}$) and if one further writes $g \circ \varkappa = \varkappa\, g$ with functors g of $\mathfrak{D}$ and natural transformations $\varkappa$, then the rule of composition takes the more familiar form

$$(\lambda, g)\,(\varkappa, f) = (\lambda(g \circ \varkappa),\; g \circ f)$$

occurring in this form in semidirect products of algebraic structures. However, in our applications the composition suggested above is more convenient.

It is straightforward to check, that $\mathfrak{A}^{\mathfrak{D}}$ is indeed a category, in fact a category with small hom sets. Note that if $\mathfrak{D}$ has only one object then $\mathfrak{A}^{\mathfrak{D}}$ is the conventional functor category of all functors $\mathfrak{D} \to \mathfrak{A}$ as objects and all natural transformations as morphisms.

We say that $\mathfrak{A}$ is $\mathfrak{D}$-*complete* if any $F \in \mathfrak{A}^{\mathfrak{D}}$ has a limit in $\mathfrak{A}$. If $\mathfrak{D}$ is the category of all small categories, then $\mathfrak{A}$ is *complete*. Let now $\mathfrak{A}$ be a $\mathfrak{D}$-complete category. We define a functor LIM: $\mathfrak{A}^{\mathfrak{D}} \to \mathfrak{A}$ as follows: If F is an object of $\mathfrak{D}$, then LIM F is the vertex of the limit cone $\lambda\colon \lim F \to F$ of F, where $\lim F$ it the constant functor whose value then is LIM F. If $(\varkappa, f)\colon F \to G$ is a morphism in $\mathfrak{A}^{\mathfrak{D}}$, then by the universal property of the limit $\varrho\colon \lim G \to G$ there is a unique constant natural transformation $\varkappa'\colon (\lim F) f \to \lim G$ such that the diagram

$$\begin{array}{ccc} (\lim F)\, f & \xrightarrow{\varkappa'} & \lim G \\ \downarrow{\scriptstyle \lambda f} & & \downarrow{\scriptstyle \varrho} \\ Ff & \longrightarrow & G \end{array}$$

commutes.

The vertex of the cone $\lambda f\colon (\lim F) f \to Ff$ is also LIM F, the vertex of the cone $\varrho\colon \lim G \to G$ is LIM G; thus the constant natural transformation $\varkappa'$ is given by a unique morphism LIM $F \to$ LIM G which we call LIM $(\varkappa, f)$. Thus, in terms of diagrams in $\mathfrak{A}$, we can write a commutative diagram

$$\begin{array}{ccc} \mathrm{LIM}\, F & \xrightarrow{\mathrm{LIM}\,(\varkappa, f)} & \mathrm{LIM}\, G \\ \downarrow{\scriptstyle \lambda f} & & \downarrow{\scriptstyle \varrho} \\ Ff & \longrightarrow & G \end{array}$$

where here the vertical arrows indicate morphisms LIM $F \to Ff(x)$, $x \in \operatorname{dom} G$ and Ff represents $Ff(\operatorname{dom} G)$.

Lemma 1.1. LIM *is a functor.*

Proof. We have to show that

$$\mathrm{LIM}\,(\lambda, g)\, \mathrm{LIM}\,(\varkappa, f) = \mathrm{LIM}\,(\lambda(\varkappa\, g), f\, g).$$

By the definition of LIM we have commutative diagrams

$$\begin{array}{ccc} \mathrm{LIM}\, F & \xrightarrow{\mathrm{LIM}\,(\varkappa, f)} & \mathrm{LIM}\, G \\ \downarrow{\scriptstyle \lambda f} & & \downarrow{\scriptstyle \varrho} \\ Ff & \xrightarrow{\varkappa} & G \end{array} \tag{1}$$

$$\begin{array}{ccc} \mathrm{LIM}\, G & \xrightarrow{\mathrm{LIM}\,(\lambda,\, g)} & \mathrm{LIM}\, H \\ \downarrow{\scriptstyle \varrho g} & & \downarrow{\scriptstyle \eta} \\ Gg & \xrightarrow{\lambda} & H \end{array} \tag{2}$$

From diagram (1) we derive the commuting diagram

$$\begin{array}{ccc} \mathrm{LIM}\, F & \xrightarrow{\mathrm{LIM}\,(\varkappa,\, f)} & \mathrm{LIM}\, G \\ \downarrow{\scriptstyle \lambda f g} & & \downarrow{\scriptstyle \varrho g} \\ F f g & \xrightarrow{\varkappa g} & Gg \end{array} \tag{3}$$

The composition of (2) and (3), and the uniqueness of the fill-in morphism in the limit situation yields the desired result.

Lemma 1.2. *Let* $\mathfrak{D}$ *be a category of small categories and let* $S\colon \mathfrak{A} \to \mathfrak{B}$ *be a functor. Then there is a functor* $S^{\mathfrak{D}}\colon \mathfrak{A}^{\mathfrak{D}} \to \mathfrak{B}^{\mathfrak{D}}$ *given by* $S^{\mathfrak{D}} F = SF$ *for an object* $F\colon X \to \mathfrak{A}$ *of* $\mathfrak{A}^{\mathfrak{D}}$, *and by* $S^{\mathfrak{D}}(\varkappa, f) = (S\varkappa, f)$ *for a morphism* $(\varkappa, f)\colon F \to G$ *in* $\mathfrak{A}^{\mathfrak{D}}$.

The proof is straightforward.

Let now $\mathfrak{D}$ be a small category and let $S\colon \mathfrak{A} \to \mathfrak{B}$ be a functor of $\mathfrak{D}$-complete categories. Then there is a natural transformation

$$\sigma^S\colon S\,\mathrm{LIM}_{\mathfrak{A}} \to \mathrm{LIM}_{\mathfrak{B}}\, S^{\mathfrak{D}}$$

of functors $\mathfrak{A}^{\mathfrak{D}} \to \mathfrak{B}$ such that for any object $F\colon X \to \mathfrak{A}$ of $\mathfrak{A}^{\mathfrak{D}}$ the diagram

$$\begin{array}{ccc} S \lim F & \xrightarrow{\sigma^S_F} & \lim SF \\ & {\scriptstyle S\lambda}\searrow \quad \swarrow{\scriptstyle \mu} & \\ & SF & \end{array}$$

commutes with the respective limit natural transformations $\lambda\colon \lim F \to F$ and $\mu\colon \lim SF \to SF$. In order to see that σ^S is natural we have to take another object $G\colon Y \to \mathfrak{A}$ of $\mathfrak{A}^{\mathfrak{D}}$ and a morphism $(\varkappa, f)\colon F \to G$ in $\mathfrak{A}^{\mathfrak{D}}$ and show the commutativity of the diagram

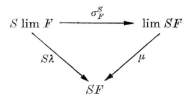

$$\begin{array}{ccc} S\,\mathrm{LIM}\,F & \xrightarrow{\sigma_F^S} & \mathrm{LIM}\,SF \\ \big\downarrow{\scriptstyle S\,\mathrm{LIM}\,(\varkappa, f)} & & \big\downarrow{\scriptstyle \mathrm{LIM}\,(S\varkappa, f)} \\ S\,\mathrm{LIM}\,G & \xrightarrow{\sigma_G^S} & \mathrm{LIM}\,SG \end{array}$$

This follows from the definition, the uniqueness in the universal property of the limit via the following diagram

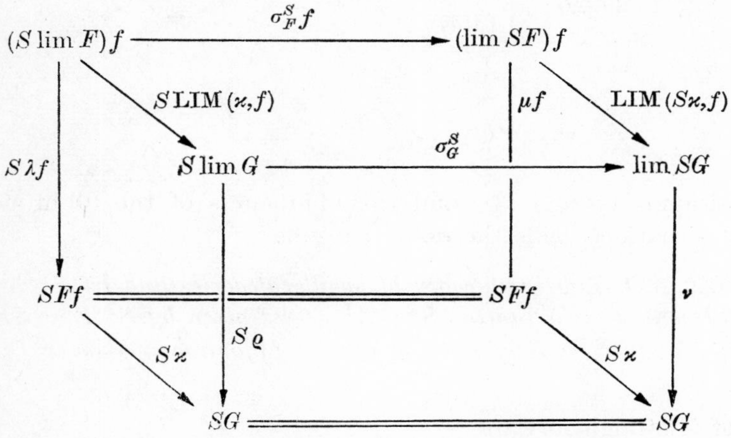

Definition 1.3. The functor S is called $\mathfrak{D}$-*continuous*, if σ^S is a natural isomorphism.

Remark. If a functor S is limit preserving, then it is $\mathfrak{D}$-continuous for any $\mathfrak{D}$. There are, however, many functors, which are not limit preserving but are still $\mathfrak{D}$-continuous relative to suitable categories $\mathfrak{D}$. If, e. g., $\mathfrak{A}$ is the category of compact spaces and $\mathfrak{B}$ the opposite category of the category of graded commutative rings, then the integral Čech cohomology is $\mathfrak{D}$-continuous for the category of directed sets; but since it is not product preserving, it is not limit preserving. If $\mathfrak{D}$ is the category of all finite categories, then a $\mathfrak{D}$-continuous functor is exactly a functor preserving finite limits.

We observe that if the category $\mathfrak{D}$ contains a one element small category $\mathbf{1}$, then there is a faithful functor $K: \mathfrak{A} \to \mathfrak{A}^{\mathfrak{D}}$, which associates with an onject A of $\mathfrak{A}$ the constant functor $K(A): \mathbf{1} \to \mathfrak{A}$ with value A and with a morphism $\varphi: A \to A'$ the obvious constant natural transformation

$$K(\varphi): K(A) \to K(A').$$

From now on we will assume throughout that $\mathbf{1} \in \mathfrak{D}$.

In the following we will make repeated use of the concept of a *comma category*. We remind the reader of its definition:

Definition 1.4. Let $S\colon \mathfrak{X} \to \mathfrak{A}$ and $T\colon \mathfrak{Y} \to \mathfrak{A}$ be functors. The comma category (S, T) has as its objects triples $(X, f, Y) \in (\text{ob } \mathfrak{X}) \times \mathfrak{A} \times (\text{ob } \mathfrak{Y})$ (where ob $\mathfrak{X}$ denotes the class of objects of $\mathfrak{X}$) with $f\colon SX \to TY$, and as its morphisms $(X, f, Y) \to (X', f', Y')$ such pairs $(\varphi, \psi) \in \mathfrak{X} \times \mathfrak{Y}$ of morphisms $\varphi\colon X \to X'$, $\psi\colon Y \to Y'$ for which

$$\begin{array}{ccc} SX & \xrightarrow{\ f\ } & TY \\ \big\downarrow{\scriptstyle S\varphi} & & \big\downarrow{\scriptstyle T\psi} \\ SX' & \xrightarrow{\ f'\ } & TY' \end{array}$$

commutes. Note that there are functors $M\colon (S, T) \to \mathfrak{A}^2$ into the category of morphisms of $\mathfrak{A}$ given by $M(X, f, Y) = f$ and $M(\varphi, \psi) = (S\varphi, T\psi)$ and $\Pr_{\mathfrak{X}}\colon (S, T) \to \mathfrak{X}$, $\Pr_{\mathfrak{Y}}\colon (S, T) \to \mathfrak{Y}$ (the natural projections), such that the diagram

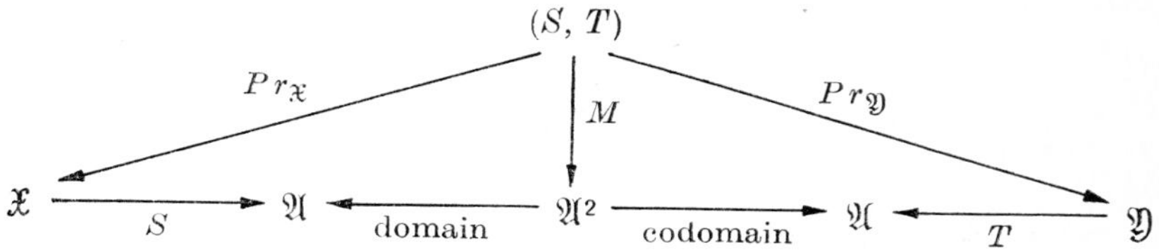

commutes (and (S, T) is in fact characterized as being a limit of the lower part of the diagram). Note that $\Pr_{\mathfrak{X}}$ and $\Pr_{\mathfrak{Y}}$ are faithful functors.

Let now $J\colon \mathfrak{A}_0 \to \mathfrak{A}$ be a functor; of most interest to us will be the case that J is the inclusion functor of a subcategory. Let $\mathfrak{D}$ be a category of small categories containing **1**. We now have two (inclusion) functors $K\colon \mathfrak{A} \to \mathfrak{A}^{\mathfrak{D}}$ and $J^{\mathfrak{D}}\colon \mathfrak{A}_0^{\mathfrak{D}} \to \mathfrak{A}^{\mathfrak{D}}$ so that we can form the comma category $(K, J^{\mathfrak{D}}) = (\mathfrak{A}, \mathfrak{A}_0^{\mathfrak{D}})$. It happens in applications, as we will see presently, that we are given a functor $C\colon \mathfrak{A} \to (K, J^{\mathfrak{D}})$ with the property that $\Pr_{\mathfrak{A}} \circ C\colon \mathfrak{A} \to \mathfrak{A}$ is a natural equivalence. The functor $\Pr_{\mathfrak{A}_0^{\mathfrak{D}}} \circ C\colon \mathfrak{A} \to \mathfrak{A}_0^{\mathfrak{D}}$ will be abbreviated as $L\colon \mathfrak{A} \to \mathfrak{A}_0^{\mathfrak{D}}$.

We immediately illustrate this by an example which will interest us in the applications.

Example 1.5. Let $\mathfrak{D}$ be the category of directed sets (which, of course, contains the one element category **1**). Let *Comp* denote the category of compact groups (or any full, $\mathfrak{D}$-complete subcategory of it, such as the category of abelian compact groups, connected compact groups, connected abelian compact groups etc.). Let *Lie* denote the full subcategory of *Comp* spanned by all Lie groups in *Comp*, and let J: *Lie* $\to$ *Comp* be the inclusion functor.

We now define C: *Comp* $\to$ (*Comp*, *Lie*$^{\mathfrak{D}}$). Let $G \in$ *Comp*. Then

$$C(G) = (G, \pi_G, L(G)),$$

where $L(G)$ is a functor from a directed set dom $L(G)$ into *Lie* and where π_G is a natural transformation from the constant functor G to $J^{\mathfrak{D}} L(G)$ defined as follows: Let dom $L(G)$ be the partially ordered set of all normal subgroups N of G such that G/N is a Lie group, and write $M \to N$ iff $M \subset N$; define

$$L(G)(N) = G/N,$$

and $L(N \to M)$ to be the natural projection $G/M \to G/N$. Then

$$L(G)\colon \operatorname{dom} L(G) \to \mathit{Lie}$$

is clearly a functor. We let

$$\pi_{G,N}\colon G \to J^{\mathfrak{D}} L(G)(N) = G/N$$

be the quotient map. Then considering G as the constant functor, clearly π_G is a natural transformation $G \to J^{\mathfrak{D}} L(G)$ of functors from dom $L(G)$ to *Comp*.

Now let $\varphi\colon G \to G'$ be a morphism of compact groups. We must define $C(\varphi) = (\varphi, (\varkappa, f))$, where $(\varkappa, f)\colon L(G) \to L(G')$ is a morphism in *Lie*$^{\mathfrak{D}}$ in such a way that

$$\begin{array}{ccc} G & \xrightarrow{\pi_G} & J^{\mathfrak{D}} L(G) \\ \downarrow & & \downarrow{\scriptstyle J(\varkappa,f)} \\ G' & \xrightarrow{\pi_{G'}} & J^{\mathfrak{D}} L(G') \end{array}$$

commutes.

First the definition of f: dom $L(G') \to$ dom $L(G)$: Let N' be a closed normal subgroup of G' such that G'/N' is a Lie group. Define $f(N') = \varphi^{-1}(N')$. Then $f(N')$ is a closed normal subgroup of G and $G/\varphi^{-1}(N')$ is isomorphic to the image of G in G/N' which is a closed subgroup of a Lie group and is therefore a Lie group. Clearly f preserves inclusion and is consequently a functor

$$\operatorname{dom} L(G') \to \operatorname{dom} L(G).$$

Secondly the definition of $\varkappa$: $L(G) f \to L(G')$: Let

$$N' \in \operatorname{dom} L(G') = \operatorname{dom} L(G) f.$$

Then we define

$$\varkappa_{N'}: L(G)\,f(N') \to L(G')\,(N')$$

by the diagram

$$\begin{array}{ccc} G & \xrightarrow{\quad \varphi \quad} & G' \\ \downarrow & & \downarrow \\ G/\varphi^{-1}(N') & \xrightarrow{\varkappa_{N'}} & G'/N' . \end{array}$$

The naturality of the $\varkappa$ so defined is readily checked.

Now let $N' \in \operatorname{dom} L(G')$; then we obtain a commutative diagram

$$\begin{array}{ccc} G & \xrightarrow{\pi_{G, f(N')}} & J^{\mathfrak{D}}\, L(G)\,(f(N')) = G/\varphi^{-1}(N') \\ \downarrow{\scriptstyle \varphi} & & \downarrow{\scriptstyle \varkappa_{N'}} \\ G' & \xrightarrow{\pi_{G', N'}} & J^{\mathfrak{D}}\, L(G')\,(N') = G'/N' \end{array}$$

which shows that $C(\varphi): C(G) \to C(G')$ is a morphism. That C is in fact a functor is an exercise of giving the diagram

$$\begin{array}{ccccc} G & \xrightarrow{\quad \varphi \quad} & G' & \xrightarrow{\quad \psi \quad} & G'' \\ \downarrow{\scriptstyle \pi_{G, ff'(N'')}} & & \downarrow{\scriptstyle \pi_{G', f'(N'')}} & & \downarrow{\scriptstyle \pi_{G'', N''}} \\ G/\varphi^{-1}\,\psi^{-1}(N'') & \xrightarrow{\varkappa_{\psi(N'')}} & G'/\psi^{-1}(N'') & \xrightarrow{\varkappa'_{N''}} & G''/N'' \end{array}$$

the appropriate interpretation.

We return now to the general theory and assume that we are given functors $C: \mathfrak{A} \to (\mathfrak{A}, \mathfrak{A}_0^{\mathfrak{D}})$, $L = \operatorname{Pr}_{\mathfrak{A}_0^{\mathfrak{D}}} C: \mathfrak{A} \to \mathfrak{A}_0^{\mathfrak{D}}$. We assume that $\mathfrak{A}$ is $\mathfrak{D}$-complete, so that the functor $\operatorname{LIM}_{\mathfrak{A}}: \mathfrak{A}^{\mathfrak{D}} \to \mathfrak{A}$ exists.

Then for each $A \in \mathfrak{A}$ we have $C(A) = (A, \bar{\eta}_A, L(A))$, and $\bar{\eta}: I_{\mathfrak{A}} \to J^{\mathfrak{D}} L$ is a natural transformation of functors. We immediately deduce, by the universal property of the limit, a natural transformation $\eta: I_{\mathfrak{A}} \to \operatorname{LIM}_{\mathfrak{A}} J^{\mathfrak{D}} L$ such that the following diagram commutes for each $A \in \mathfrak{A}$:

$$\begin{array}{ccc} A & \xrightarrow{\quad \eta_A \quad} & (\operatorname{LIM}_{\mathfrak{A}} J^{\mathfrak{D}} L)\,(A) = \lim_{\mathfrak{A}} JL(A) \\ & {\scriptstyle \bar{\eta}_A} \searrow \qquad \swarrow {\scriptstyle \lambda_A} & \\ & JL(A) & \end{array}$$

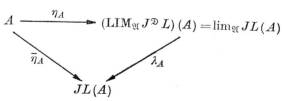

If $S\colon \mathfrak{A} \to \mathfrak{B}$ is an arbitrary functor, then in a similar way there is a unique natural transformation $\eta^S\colon S \to \mathrm{LIM}_{\mathfrak{B}}\, S^{\mathfrak{D}}\, J^{\mathfrak{D}}\, L$ such that the following diagram commutes for each $A \in \mathfrak{A}$.

$$\begin{array}{ccc} SA & \xrightarrow{\ \eta^S_A\ } & (\mathrm{LIM}_{\mathfrak{B}}\, S^{\mathfrak{D}}\, J^{\mathfrak{D}}\, L)\,(A) = \lim_{\mathfrak{B}} SJL(A) \\ & {\scriptstyle S\bar{\eta}_A} \searrow \quad \swarrow {\scriptstyle \mu_A} & \\ & SJL(A). & \end{array}$$

From the commutative diagram (see the discussion after Lemma 1.2)

$$\begin{array}{ccccc} SA & \xrightarrow{\ S\eta_A\ } & S \lim JL(A) & \xrightarrow{\ \sigma^S_{JL(A)}\ } & \lim SJL(A) \\ \downarrow{\scriptstyle S\bar{\eta}_A} & & \downarrow{\scriptstyle S\lambda_A} & & \downarrow{\scriptstyle \mu_A} \\ SJL(A) & \xrightarrow{\ =\ } & SJL(A) & \xrightarrow{\ =\ } & SJL(A) \end{array}$$

and the uniqueness in the limit property we can then conclude that

$$\eta^S = (\sigma^S\, JL)\,(S\eta).$$

Lemma 1.6. *Let $\alpha, \alpha'\colon S \to T$ be two natural transformations of functors from $\mathfrak{A}$ to $\mathfrak{B}$ and suppose that*

(i) $\alpha J = \alpha' J$,

(ii) η^T *is monic.*

Then $\alpha = \alpha'$.

Proof. We have a commutative diagram

$$\begin{array}{ccc} S & \xrightarrow{\ \varphi\ } & T \\ \downarrow{\scriptstyle \eta^S} & & \downarrow{\scriptstyle \eta^T} \\ \mathrm{LIM}(SJ)^{\mathfrak{D}}\, L & \xrightarrow{\ \mathrm{LIM}\,(\varphi J)^{\mathfrak{D}}\, L\ } & \mathrm{LIM}(TJ)^{\mathfrak{D}}\, L \end{array}$$

for $\varphi = \alpha, \alpha'$. The assertion follows.

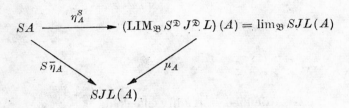

Definition 1.7. Let $\mathfrak{D}$ be a category of small categories with $\mathbf{1}$ and $\mathfrak{A}$ a $\mathfrak{D}$-complete category. A functor $J\colon \mathfrak{A}_0 \to \mathfrak{A}$ is said to have $\mathfrak{D}$-*dense range* if there is a functor $C\colon \mathfrak{A} \to (K, J^{\mathfrak{D}}) = (\mathfrak{A}, A_0^{\mathfrak{D}})$, with $L = \mathrm{Pr}_{\mathfrak{A}_0^{\mathfrak{D}}} \circ C$, such that

the natural transformation $\eta\colon I_{\mathfrak{A}} \to \mathrm{LIM}_{\mathfrak{A}}\, J^{\mathfrak{D}}\, L$ is an isomorphism. If $\mathfrak{A}_0$ is a subcategory and J an inclusion functor with $\mathfrak{D}$-dense range, we say that $\mathfrak{A}_0$ is $\mathfrak{D}$-*dense* in $\mathfrak{A}$.

Example 1.8. The inclusion functor J: *Lie* $\to$ *Comp* of Example 1.5 has $\mathfrak{D}$-dense range, where $\mathfrak{D}$ is the category of directed sets. Indeed this is just the well-known fact that each compact group is the projective limit of its Lie group quotients.

Suppose that $T_0\colon \mathfrak{A}_0 \to \mathfrak{B}$ is an arbitrary functor. We can define a new functor $T\colon \mathfrak{A} \to \mathfrak{B}$ by $T = \mathrm{LIM}_{\mathfrak{B}}\, T_0^{\mathfrak{D}}\, L$. There is a unique natural transformation $\eta_0^{T_0}\colon T_0 \to TJ$ such that the following diagram commutes for all $A_0 \in \mathfrak{A}_0$:

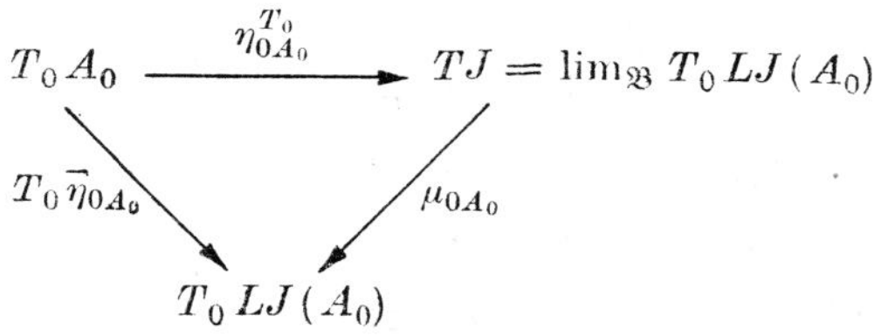

provided that there is a natural transformation $\bar{\eta}_0\colon A_0 \to LJ(A_0)$ of functors from dom $LJ(A_0)$ to $\mathfrak{A}_0$ for all $A_0 \in \mathfrak{A}_0$. This is clearly the case if J is faithful and full, i. e. if $\mathfrak{A}_0$ is a full subcategory of $\mathfrak{A}$.

Definition 1.9. We say that under the present conditions a functor $S_0\colon \mathfrak{A}_0 \to \mathfrak{B}$ is *extendable over* $J\colon \mathfrak{A}_0 \to \mathfrak{B}$ if the following two conditions are satisfied:

(i) There is a natural transformation $\bar{\eta}_0\colon I_{\mathfrak{A}_0} \to LJ$ such that $J\bar{\eta}_0 = \bar{\eta}J$.

(ii) The natural morphism $\eta_0^{S_0}\colon S_0 \to SJ$, $S = \mathrm{LIM}\, S_0^{\mathfrak{D}}\, L$ is an isomorphism.

The functor S is called the $\mathfrak{D}$-*extension* of S_0.

Remark. Condition (i) is automatic if J is faithful and full. Conditions (i) and (ii) are satisfied if the following condition holds

(iii) For each $A_0 \in \mathfrak{A}_0$ the category dom LJA_0 has an initial element i and $(\bar{\eta}_{0A_0})_i\colon A_0 \to LJA_0\, i$ is an isomorphism.

In Example 1.5, condition (iii) is satisfied.

If S_0 is extendable over J, then there are two natural transformations $\mathrm{LIM}\, S_0^{\mathfrak{D}}\, L = S \to \mathrm{LIM}\,(SJ)^{\mathfrak{D}}\, L$, namely, the transformation η^S and the natural isomorphism $\mathrm{LIM}\,(\eta_0^{S_0})^{\mathfrak{B}}\, L$. Because of Lemma 1.6, it is of great importance to know under which conditions these two natural transformations agree.

In order to investigate this question, we pick an object $A \in \mathfrak{A}$ and an object $x \in \operatorname{dom} LA$. We denote the limit natural transformations

$$\lim SJLA \to SJLA$$

with μ_A and $\lim S_0LA \to S_0LA$ with μ_A^0. Then η^S is characterized by the fact that the diagram

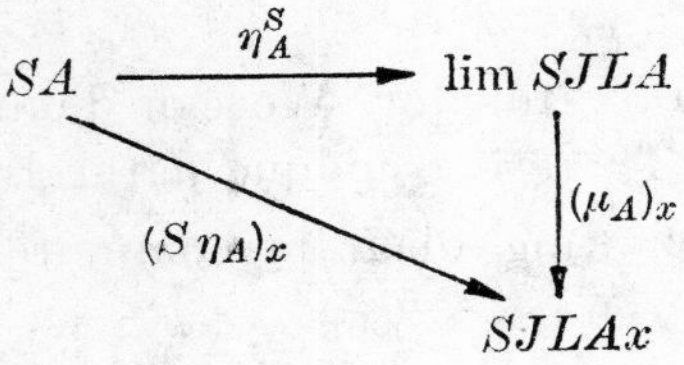

commutes for all $x \in \operatorname{dom} LA$. By the definition of the $\mathfrak{D}$-extension S, we have

$$(S\bar{\eta}_A)_x = (\lim S_0L\,(\bar{\eta}_A)_x)\colon \lim S_0LA \to \lim S_0L(JLAx)\,.$$

In order to characterize this morphism, we recall that the morphism $(\bar{\eta}_A)_x$: $A \to JLA\,(x)$ induces a morphism

$$L\,((\bar{\eta}_A)_x)\colon\; LA \to L\,(JLA\,(x))$$

in $\mathfrak{A}_0$ which is of the form $(\varkappa_x, f_x)$ with a functor f_x: $\operatorname{dom} L\,(JLA\,(x)) \to \operatorname{dom} LA$ of small categories and a natural transformation $\varkappa_x$: $(LA)\,f \to L\,(JLA\,(x))$. This notation enables us to recall that $(\lim S_0L\bar{\eta}_A)_x$ is characterized by the commuting of the diagram

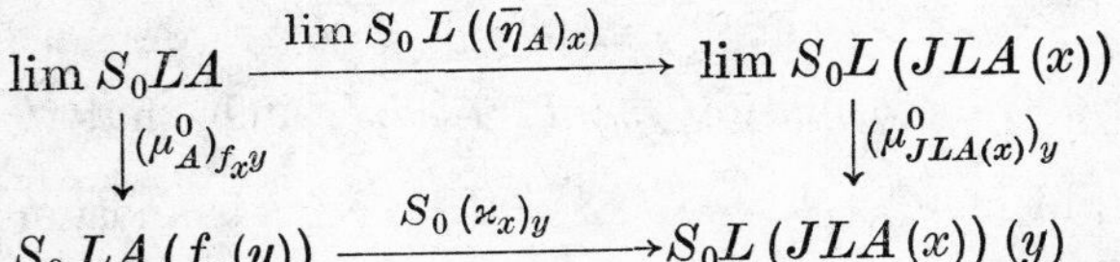

for all $y \in \operatorname{dom} L\,(JL(A)\,(x))$. This finishes the tracking of the definition of η_A^S. Now we turn to $\alpha_A = \operatorname{LIM}(\eta_0^{S_0})_{LA}$. This morphism is characterized by the fact that the following diagram commutes for all $x \in \operatorname{dom} LA$:

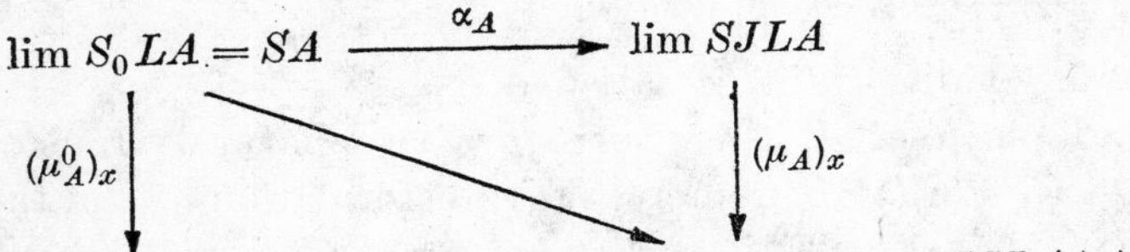

Now we recall the definition of $\eta_0^{S_0}$ and observe that $(\eta_0^{S_0})_{LA(x)}$ is the unique morphism making the following diagram commute for all $y \in \operatorname{dom} L\,(JLA\,(x))$:

$$\begin{array}{ccc} S_0\,LA\,(x) & \xrightarrow{(\eta_0^{S_0})_{LA(x)}} & \lim S_0(LJLA\,(x)) \\ & {\scriptstyle (S_0\bar{\eta}_{0LA(x)})_y} \searrow & \downarrow {\scriptstyle (\mu^0_{JLA(x)})_y} \\ & & S_0\,L(JLA\,(x))\,(y) \end{array}$$

This completes the full description of α_A. Now we compare the two: In order that $\eta_A^S = \alpha_A$ it is sufficient that

$$\lim S_0L(\eta_A)_x = \eta_0^{S_0}{}_{LA(x)}\,(\mu_A^0)_x$$

for all

$$x \in \operatorname{dom} LA\,.$$

In order that this latter condition be satisfied, it is sufficient that the following diagram commute

$$\begin{array}{ccc} \lim S_0LA & \xrightarrow{(\mu_A^0)\,x} & S_0LA\,(x) \\ \downarrow {\scriptstyle (\mu_A^0)_{f_xy}} & & \downarrow {\scriptstyle (S_0\,\bar{\eta}_{0LA(x)})_y} \\ S_0LA\,(f_x\,y) & \xrightarrow{S_0\,(\varkappa_x)_y} & S_0L\,(JLA\,(x))\,(y) \end{array}$$

for all $x \in \operatorname{dom} LA$, $y \in \operatorname{dom} L\,(JLA\,(x))\,(y)$, and this last condition is certainly satisfied, if we have the following statements

(a) for all $x \in \operatorname{dom} LA$, $y \in \operatorname{dom} L\,(JLA\,(x))$ there is a morphism $a_{xy}\colon x \to f_x\,y$ in dom LA,

(b) $(\varkappa_x)_y\,((LA)\,(a_{xy})) = (\bar{\eta}_{0LA(x)})_y$ for all $x \in \operatorname{dom} LA$.

This leads to the following definition:

Definition 1.10. Let $J\colon \mathfrak{A}_0 \to \mathfrak{A}$ be a functor into a $\mathfrak{D}$-complete category. A functor $C\colon \mathfrak{A} \to (K, J^{\mathfrak{D}}) = (\mathfrak{A}, \mathfrak{A}_0^{\mathfrak{D}})$ is called *compatible*, if the morphisms $L(\bar{\eta}_A)_x = (\varkappa_x, f_x)$ of $\mathfrak{A}_0^{\mathfrak{D}}$ for all $x \in \operatorname{dom} LA$ satisfy conditions (a) and (b) above for any $A \in \mathfrak{A}$.

We will first show that the functor C in our example is compatible.

Let G be an object in *Comp*; an element of dom LG is then a closed normal subgroup N of G such that G/N is in *Lie*. The morphism

$$(\bar{\eta}_G)_N\colon G \to (LG)\,(N) = G/N$$

is just the quotient morphism. Then an element of

$$\operatorname{dom} L(G/N) = \operatorname{dom} L(JLG(N))$$

is a closed normal subgroup of G/N modulo which G/N is a Lie group; since G/N is already a Lie group, any closed normal subgroup of G/N will do, and any such is of the form M/N with a closed normal subgroup M of G. The morphism

$$(\bar{\eta}_{LG(N)})_{M/N}\colon LG(N) = G/N \to L(JLG(N))(M/N) = (G/N)/(M/N)$$

is again the quotient map. The morphism

$$L(\bar{\eta}_G)_N = (\varkappa_N, f_N)\colon LG \to L(G/N)$$

is given by

$$f_N(M/N) = (\bar{\eta}_G)_N^{-1}(M/N) = M,$$

and we let $a_{N, M/N}\colon N \to f_N(M/N) = M$ be the unique morphism in $\operatorname{dom} LG$ representing the inclusion $N \subset M$. Now

$$(\varkappa_N)_{M/N}\colon (LG)(f_N\, M/N) = G/M \to L(G/N)(M/N) = (G/N)/(M/N)$$

is exactly the natural isomorphism given in the isomorphy theorem. The commutativity of the diagram

$$
\begin{array}{ccc}
 & G/N = LG(N) & \\
\scriptstyle(LG)(a_{N,\, M/N}) \swarrow & & \downarrow \scriptstyle(\bar{\eta}_{0LG(N)})_{M/N} \\
(LG)(f_N(M/N)) = G/M \xrightarrow[\scriptstyle(\varkappa_N)_{M/N}]{} & & (G/N)/(M/N) = L(JLG(N))(M/N)
\end{array}
$$

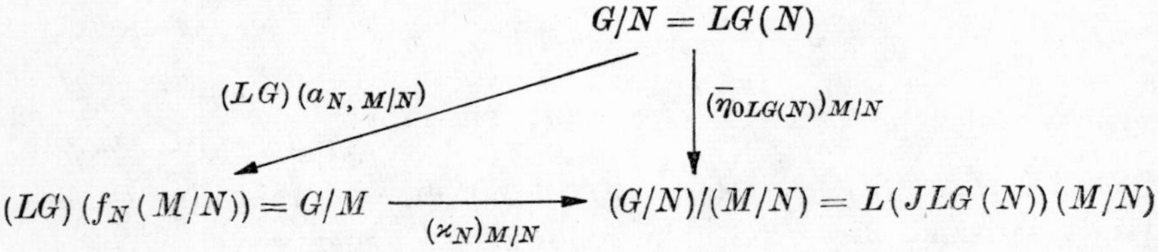

is the required condition (b).

This is our motivation for the following definition.

Definition 1.11. Let $\mathfrak{D}$ be a category of small categories containing $\mathbf{1}$. Let $\mathfrak{A}$ be a $\mathfrak{D}$-complete category and $J\colon \mathfrak{A}_0 \to \mathfrak{A}$ be a functor. We say that J has a *strictly $\mathfrak{D}$-dense range* (or, if J is the inclusion functor of a subcategory, that $\mathfrak{A}_0$ is *strictly $\mathfrak{D}$-dense* in $\mathfrak{A}$) if there is a compatible functor $C\colon \mathfrak{A} \to (K, J^{\mathfrak{D}})$ such that the associated natural transformation $\eta\colon I_{\mathfrak{A}} \to \operatorname{LIM}_{\mathfrak{A}} J^{\mathfrak{D}} L$ is an isomorphism.

We have just shown that indeed in Example 1.5, the category Lie is strictly dense in Comp.

One of the essential consequences of strict density is now the following proposition:

Proposition 1.12. *Let $\mathfrak{D}$ be a category of small categories containing* $\mathbf{1}$ *and let $J\colon \mathfrak{A}_0 \to \mathfrak{A}$ be a functor with strictly $\mathfrak{D}$-dense range. Let $\mathfrak{B}$ be a $\mathfrak{D}$-complete category and $S_0\colon \mathfrak{A}_0 \to \mathfrak{B}$ a functor which is extendable over J. Then we have the following conclusions*

(1) $\eta^S\colon S \to \mathrm{LIM}_{\mathfrak{B}}\, S^{\mathfrak{D}} J^{\mathfrak{D}} L$ *is an isomorphism.*

(2) *If $R\colon \mathfrak{A} \to \mathfrak{B}$ is any functor, then any natural transformation $\alpha_0\colon RJ \to S_0$ has a unique extension to a natural transformation $\alpha\colon R \to S$ (where extension means that $RJ \xrightarrow{\alpha J} SJ \xrightarrow{(\eta_0^{S_0})^{-1}} S_0$ agrees with α_0).*

Proof. We have just finished proving (1). From (1) and Lemma 1.6, the uniqueness of α follows. We have to establish the existence. For $A \in \mathfrak{A}$ we define α_A as the composition

$$\begin{array}{ccccc} RA & \xrightarrow{\eta^R_A} & \mathrm{LIM}_{\mathfrak{B}}\, R^{\mathfrak{D}} J^{\mathfrak{D}} LA & \xrightarrow{(\mathrm{LIM}_{\mathfrak{B}}\, \alpha_0^{\mathfrak{D}})_{LA}} & \mathrm{LIM}_{\mathfrak{B}}\, S_0^{\mathfrak{D}}\, LA = SA \\ & & \| & & \\ & & \lim\,(RJLA) & & \end{array}$$

Now let $A_0 \in \mathfrak{A}_0$. By the uniqueness in the universal property of the limit we have the equality of the two morphisms $\eta^R_{JA_0}$ and $\eta^{RJ}_{0A_0}$ from $R(JA_0) = (RJ)A_0$ to $\mathrm{LIM}\, R^{\mathfrak{D}} J^{\mathfrak{D}} LJA_0 = \mathrm{LIM}\,(RJ)^{\mathfrak{D}}\, LJA_0$. Consider the diagram

$$\begin{array}{ccc} RJA_0 & \xrightarrow{\quad \alpha_{0A} \quad} & S_0 A_0 \\ {\scriptstyle \eta^R_{JA} = \eta^{RJ}_{0A_0}} \downarrow & \searrow^{\alpha J_{A_0}} & \downarrow {\scriptstyle \eta^{S_0}_{0A_0}} \\ \mathrm{LIM}\,(RJ)^{\mathfrak{D}}\, LJA_0 & \xrightarrow[\mathrm{LIM}\,(\alpha_0^{\mathfrak{D}})_{LA}]{} & \mathrm{LIM}\, S_0^{\mathfrak{D}}\, LJA_0 = SJA_0 \end{array}$$

The outside rectangle commutes by the naturality of α_0 and the uniqueness in the universal property of the limit. The lower triangle commutes because of the definition of α. Hence the upper triangle commutes, and this is the assertion.

The universal property of the $\mathfrak{D}$-extension exhibited in Proposition 1.12 above is known in category theory as the defining property of *Kan extensions*. We formulate the definition in such a fashion that it suits our purposes.

Definition 1.13. Let $J\colon \mathfrak{A}_0 \to \mathfrak{A}$ and $P\colon \mathfrak{A}_0 \to \mathfrak{B}$ be functors. A functor $\mathfrak{A} \to \mathfrak{B}$ is called the *Kan extension of P along J*, and written $\mathrm{Kan}_J\, P$, if the following universal property holds:

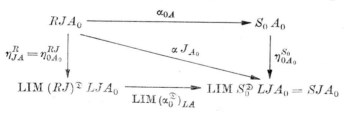

(Kan) There is a natural transformation ε_P: $(\mathrm{Kan}_J\, P)\, J \to P$ such that for any functor R: $\mathfrak{A} \to \mathfrak{B}$ and any natural transformation α_0: $RJ \to P$ there is a unique natural transformation α: $R \to \mathrm{Kan}_J\, P$ such that $\alpha_0 = \varepsilon_P(\alpha J)$.

If ε_P is an isomorphism, we shall say that $\mathrm{Kan}_J\, P$ is *true*. The situation is best apprehended at a glance by looking at the following illustration

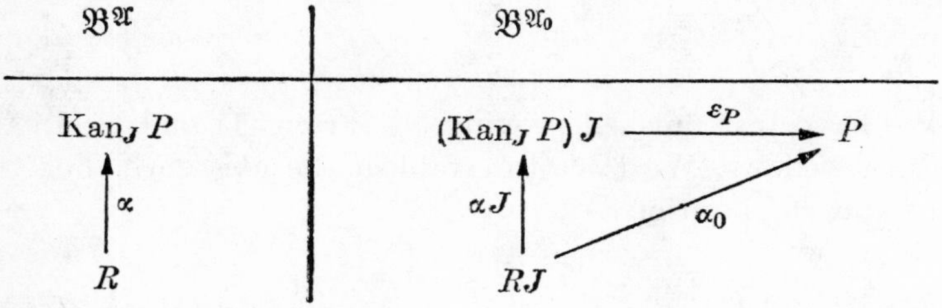

Note that the universal property sets up a natural equivalence of classes

$$\mathfrak{B}^{\mathfrak{A}_0}(RJ, P) \to \mathfrak{B}^{\mathfrak{A}}(R, \mathrm{Kan}_J\, P).$$

This puts into evidence that (apart from set theoretic apprehensions)

$$\mathrm{Kan}_J(-)\colon \mathfrak{B}^{\mathfrak{A}_0} \to \mathfrak{B}^{\mathfrak{A}}$$

is an adjoint of $(R \mapsto RJ)\colon \mathfrak{B}^{\mathfrak{A}} \to \mathfrak{B}^{\mathfrak{A}_0}$. For this reason, $\mathrm{Kan}_J\, P$ is also called the *right Kan extension of P along J*. The dual concept accordingly is then called the *left Kan extension*.

It is then also clear from the formalism of adjoint functors, that Kan extensions of a functor, if they exist, are unique up to natural isomorphism. It is therefore legitimate to speak of *the* Kan extension, if it exists.

With this convenient concept we can now summarize our discussion in the following

Theorem 1.14 (The $\mathfrak{D}$-Density Theorem). *Let $\mathfrak{D}$ be a category of small categories containing the one element category* 1. *Assume the following hypotheses:*

(1) *$\mathfrak{A}$ and $\mathfrak{B}$ are $\mathfrak{D}$-complete categories.*

(2) *J: $\mathfrak{A}_0 \to \mathfrak{A}$ is a functor with strictly $\mathfrak{D}$-dense range.*

(3) *$\mathfrak{S}_0$: $\mathfrak{A}_0 \to \mathfrak{B}$ is a functor which is extendable over J.*

Then $\mathfrak{S}_0$ has a true Kan extension, namely, $\mathrm{LIM}_{\mathfrak{B}}\, S_0^{\mathfrak{D}}\, L$ *where L: $\mathfrak{A} \to \mathfrak{A}_0^{\mathfrak{D}}$ is given by* (2). *Moreover, every $\mathfrak{D}$-continuous functor T: $\mathfrak{A} \to \mathfrak{B}$ is the Kan extension of TJ over J, and the extension is true.*

From this principal theorem and our discussion of the main motivating example we have the following which is the best functorial way that we know of to express the classical fact commonly expressed in the form: Compact groups are projective limits of Lie groups.

Theorem 1.15. (Density theorem for the category of compact groups.) *Let $\mathfrak{D}$ be the category of directed sets and isotone functions. Let Comp be any $\mathfrak{D}$-complete full subcategory of the category of compact groups (such as all compact groups, all compact abelian groups, all compact connected groups etc.), and let Lie be the corresponding full subcategory of Comp spanned by all Lie groups in Comp. Let $\mathfrak{B}$ be any $\mathfrak{D}$-complete category. Then any functor Q: Lie $\to \mathfrak{B}$ has a true Kan extension. In fact the Kan extension is given by* $\mathrm{LIM}_{\mathfrak{B}}\, Q^{\mathfrak{D}}\, L$, *where L: Comp $\to$ Lie is given in Example* 1.5.

Moreover, every $\mathfrak{D}$-continuous functor T: Comp $\to \mathfrak{B}$ is the Kan extension of TJ over J, and T is true over TJ.

It may be interesting to observe that the category of compact connected abelian groups is a $\mathfrak{D}$-complete full subcategory of the category of all compact groups, but that it is not a complete subcategory since the inclusion functor does not preserve kernels; it is, nevertheless a complete category, since it has products and kernels. The preceding theorem would not apply to compact connected groups if $\mathfrak{D}$ were replaced by, for example, the category of all small categories or even by the category of all finite categories.

Section 2

Multiplicative Hopf extensions

In our applications we shall need yet another theorem concerning Kan extensions of functors (in fact, to be more accurate, a slight variant of the concept of a Kan extension). In this context we are again concerned with multiplicative categories and multiplicative and exponential functors of categories of Hopf algebras.

Let $\mathfrak{B}$ be an arbitrary multiplicative category. In what follows, all multiplicative categories are to be pointed, and, of course, the point serves as an identity for multiplication. We associate with $\mathfrak{B}$ the category *Hopf* $\mathfrak{B}$ of all Hopf algebras $B \xrightarrow{c} B \otimes B \xrightarrow{m} B$ whose morphisms f: $(c, M) \to (c', m')$, of course, are morphisms $f\colon B \to B'$ with $c' f = (f \otimes f)\, c$ and $m'(f \otimes f) = f\, m$. The category $\mathit{Hopf}\,\mathfrak{B}$ is itself multiplicative; the product of two Hopf algebras $\Pi\{(c_i, m_i)\colon i = 1, 2\}$ is given by the diagram

$$\begin{array}{ccccc} B_1 \otimes B_2 & \xrightarrow{\;c_{12}\;} & B_1 \otimes B_2 \otimes B_1 \otimes B_2 & \xrightarrow{\;m_{12}\;} & B_1 \otimes B_2 \\ & \searrow^{c_1 \otimes c_2} & \big\downarrow \mu & \nearrow_{m_1 \otimes m_2} & \\ & & B_1 \otimes B_1 \otimes B_2 \otimes B_2 & & \end{array} \qquad (1)$$

where the natural isomorphism μ, the so-called middle four exchange, swaps the two middle factors. Also, $\mathit{Hopf}\,\mathfrak{B}$ has a zero algebra.

We obviously have a grounding functor $F\colon \mathit{Hopf}\,\mathfrak{B} \to \mathfrak{B}$ which sends a Hopf algebra $B \xrightarrow{\;c\;} B \otimes B \xrightarrow{\;m\;} B$ to B. Every multiplicative functor

$$S\colon (\mathfrak{A}, \otimes_{\mathfrak{A}}) \to (\mathfrak{B}, \otimes_{\mathfrak{B}})$$

(i. e. a functor $S\colon \mathfrak{A} \to \mathfrak{B}$ such that there is a natural isomorphism

$$\varepsilon_{A,A'}\colon S(A \otimes A') \to (SA) \otimes (SA'),$$

compatible with the coherence morphisms in $\mathfrak{A}$ and $\mathfrak{B}$) induces a multiplicative functor $\bar{S}\colon \mathit{Hopf}\,\mathfrak{A} \to \mathit{Hopf}\,\mathfrak{B}$ via the diagram

$$\begin{array}{ccccc} SA & \xrightarrow{\;Sc\;} & S(A \otimes A) & \xrightarrow{\;Sm\;} & SA \\ \| & & \big\downarrow \varepsilon_{A,A} & & \| \\ \bar{S}(c, m) = (SA & \longrightarrow & (SA) \otimes (SA) & \longrightarrow & SA) \end{array} \qquad (2)$$

and this functor is unique relative to the equation

$$SF_{\mathfrak{A}} = F_{\mathfrak{B}}\, \bar{S},$$

i. e. relative to the strict commuting of the functor diagram

$$\begin{array}{ccc} \mathit{Hopf}\,\mathfrak{A} & \xrightarrow{\;\bar{S}\;} & \mathit{Hopf}\,\mathfrak{B} \\ \big\downarrow F & & \big\downarrow F \\ \mathfrak{A} & \xrightarrow{\;S\;} & \mathfrak{B} \end{array} \qquad (3)$$

An arbitrary functor T: *Hopf* $\mathfrak{A} \to$ *Hopf* $\mathfrak{B}$ is said to be a *functor of Hopf algebras* if it is multiplicative and the diagram

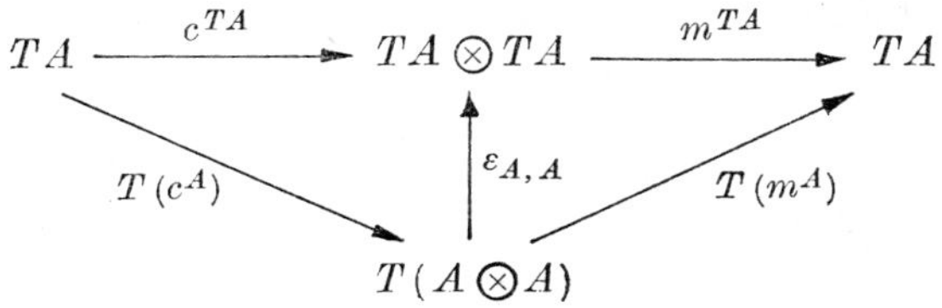
$TA \xrightarrow{c^{TA}} TA \otimes TA \xrightarrow{m^{TA}} TA$
$T(c^A)$
$\varepsilon_{A,A}$
$T(m^A)$
$T(A \otimes A)$

commutes for all $(c, m) \in$ Hopf $\mathfrak{A}$. Notice that with this definition, $\boldsymbol{S}$ above is a functor of Hopf algebras.

We say that a Hopf algebra is *commutative*, if both multiplication and comultiplication are morphisms of Hopf algebras. The full subcategory in *Hopf* $\mathfrak{A}$ of all commutative Hopf algebras will be called $Hopf_{ab}\,\mathfrak{A}$.

If $A \xrightarrow{c} A \otimes A \xrightarrow{m} A$ is a commutative Hopf algebra, then by induction, using coherence, we may define unique morphisms of Hopf algebras

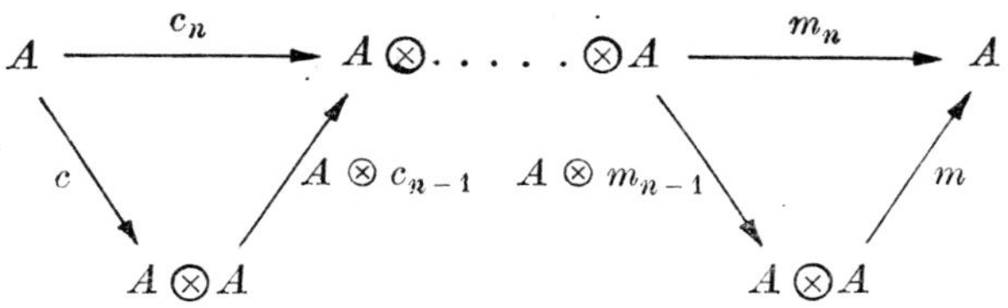
$A \xrightarrow{c_n} A \otimes \cdots \otimes A \xrightarrow{m_n} A$
c
$A \otimes c_{n-1}$
$A \otimes m_{n-1}$
m
$A \otimes A$
$A \otimes A$

(4)

We will abbreviate $A \otimes \cdots \otimes A$ (p factors) by A^p.

Suppose that f_{ji}: $A_i \to B_j$, $i = 1, \ldots, p$, $j = 1, \ldots, q$ is a matrix of morphisms of commutative Hopf algebras. Then we define a morphism

$$f\colon A \to B, \quad A = A_1 \otimes \cdots \otimes A_p, \quad B = B_1 \otimes \cdots \otimes B_q$$

via the diagram

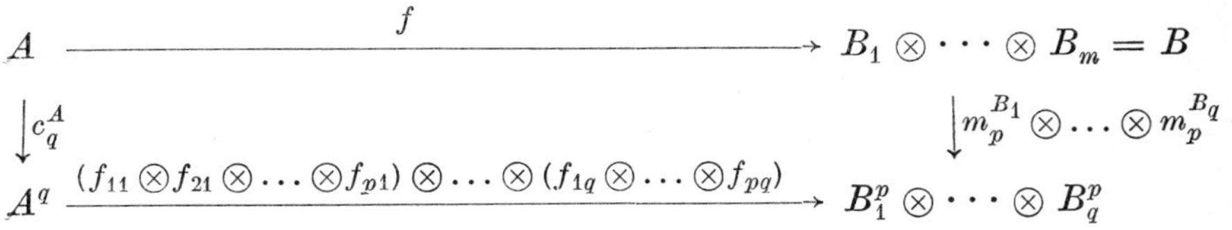
$A \xrightarrow{f} B_1 \otimes \cdots \otimes B_m = B$
c_q^A
$m_p^{B_1} \otimes \ldots \otimes m_p^{B_q}$
$A^q \xrightarrow{(f_{11} \otimes f_{21} \otimes \ldots \otimes f_{p1}) \otimes \ldots \otimes (f_{1q} \otimes \ldots \otimes f_{pq})} B_1^p \otimes \cdots \otimes B_q^p$

(5)

Suppose for the moment, that we have the more simple situation of matrices of morphisms f_{1i}: $A \to B_i$, g_{i1}: $B_i \to C$, $i = 1, \ldots, p$ of commutative Hopf algebras. Let f, resp. g be the morphisms defined by the matrices (f_{1i}) and

(g_{i1}) respectively. We then have the following commuting diagram

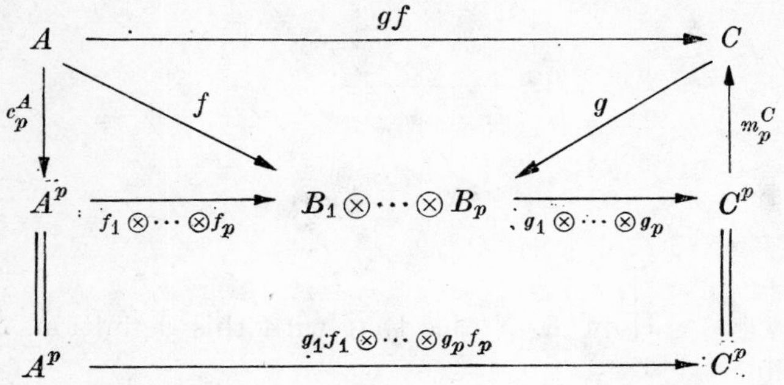

(6)

We introduce the notation $gf = \Sigma\{g_i f_i \colon i = 1, \ldots, p\}$. This notation is legitimate since one may see without undue difficulty that $\mathfrak{A}(A, B)$, for commutative Hopf algebras A, B in $\mathfrak{A}$, has a commutative semigroup structure defined by the operation

$$\begin{array}{ccc} A & \xrightarrow{f+g} & B \\ \downarrow{\scriptstyle c^A} & & \downarrow{\scriptstyle m^B} \\ A \otimes A & \xrightarrow{f \otimes g} & B \otimes B \end{array} \tag{7}$$

Finally assume that

$$f_{ji} \colon A_i \to B_j, \quad g_{kj} \colon B_j \to C_k, \quad i = 1, \ldots, p, \quad j = 1, \ldots, q, \quad k = 1, \ldots, r$$

are matrices of morphisms of commutative Hopf algebras. Define a new matrix

$$(h_{ki}) = (\Sigma\{g_{kj} f_{ji} \colon j = 1, \ldots, q\} \colon A_i \to C_k).$$

Then the morphism

$$h \colon A \to C,\ A = A_1 \otimes \cdots \otimes A_p,\ B = B_1 \otimes \cdots \otimes B_q,$$
$$C = C_1 \otimes \cdots \otimes C_r$$

associated with the matrix (h_{ki}) is exactly the composition

$$A \xrightarrow{f} B \xrightarrow{g} C.$$

Now let $\mathfrak{A}_1$ be a full subcategory of $\mathfrak{A} = \mathit{Hopf}_{ab}\,\mathfrak{B}$. Then $\mathfrak{M}(\mathfrak{A}_1)$ denotes the category, whose objects are p-tuples $(A_1, \ldots, A_p)$, $A_i \in \mathfrak{A}_1$ and whose morphisms $(A_1, \ldots, A_p) \to (B_1, \ldots, B_q)$ are matrices $(f_{ji} \colon A_i \to B_j)$ of morphisms of $\mathfrak{A}_1$ with composition given by matrix multiplication. We call $\mathfrak{M}(\mathfrak{A}_1)$ the *matrix category* over $\mathfrak{A}_1$ in $\mathfrak{A}$.

There is a functor $M\colon \mathfrak{M}(\mathfrak{A}_1) \to \mathfrak{A}$ given by

$$M(A_1, \ldots, A_p) = A_1 \otimes \cdots \otimes A_p,$$

$M((f_{ji})) = f$ (in the sense described above). There is an obvious injection $J_1\colon \mathfrak{A}_1 \to \mathfrak{M}(\mathfrak{A}_1)$ given by $J_1(A) = (A)$, $J_1(f) = (f)$, so that $MJ_1 = J$, the inclusion $\mathfrak{A}_1 \to \mathfrak{A}$.

Now let $\mathfrak{P}$ and $\mathfrak{Q}$ be two multiplicative categories. Let $\mathfrak{B}$ be a multiplicative subcategory of $\mathrm{Hopf}_{ab}\,\mathfrak{Q}$, $\mathfrak{A}$ a multiplicative subcategory of $\mathrm{Hopf}_{ab}\,\mathfrak{P}$, and let $\mathfrak{A}_1$ be a full subcategory of $\mathfrak{A}$. Let $S\colon \mathfrak{A} \to \mathfrak{B}$ be a functor of (commutative) Hopf algebras. We now produce a new functor

$$M(S)\colon \mathfrak{M}(\mathfrak{A}_1) \to \mathfrak{M}(\mathfrak{B})$$

by

$$M(S)\,(A_1, \ldots, A_p) = (SA_1, \ldots, SA_p)$$

and by

$$M(S)\,((f_{ji})) = (Sf_{ji}).$$

In order to check that this definition indeed yields a functor, one has to check that for any collection of morphisms $A \xrightarrow{f_i} B_i \xrightarrow{g_i} C$, $i = 1, \ldots, q$ in $\mathfrak{A}_1$ we have $S(\Sigma\, g_i f_i) = \Sigma\, Sg_i Sf_i$. We will indicate below under slightly more general circumstances, how this follows from diagram (6). Note that there is a commuting diagram of functors

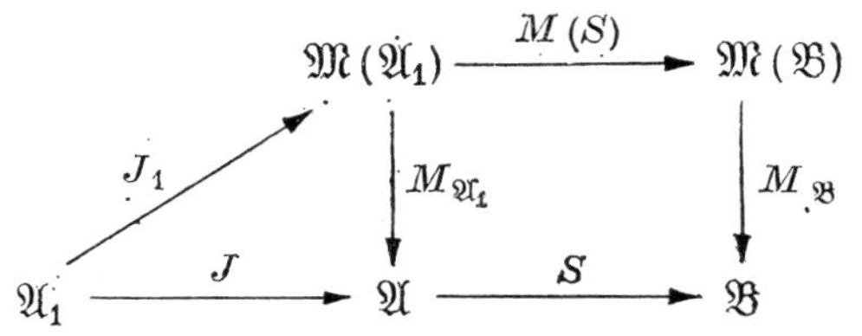

(8)

i.e. in particular, there is a natural isomorphism

$$SM_{\mathfrak{A}_1} \cong M_{\mathfrak{B}}\, M(S).$$

Now suppose that there is an analogous functor $T\colon \mathfrak{A} \to \mathfrak{B}$ and a natural transformation $\alpha_1\colon SJ_1 \to TJ_1$, where $J\colon \mathfrak{A}_1 \to \mathfrak{A}$ is the inclusion. Then we have a unique natural transformation $M(\alpha_1)\colon M(S) \to M(T)$ which is defined by

$$M(\alpha_1)_{(A_1, \ldots, A_p)} = \mathrm{diag}\,(\alpha_{1A_1}, \ldots, \alpha_{1A_p})\colon (SA_1, \ldots, SA_p) \to (TA_1, \ldots, TA_p),$$

such that $M_{\mathfrak{B}} M(\alpha_1) J_1 = \alpha_1$. (Here diag $(f_{11}, \ldots, f_{nn})$ denotes the matrix (f_{ij}) with $f_{ij} = 0$ if $i \neq j$.) To see that $M(\alpha_1)$ is indeed unique, let $\beta: M(S) \to M(T)$ be a natural transformation such that $M_{\mathfrak{B}} \beta J_1 = \alpha_1$, let $A = (A_1, \ldots, A_p) \subset \mathfrak{M}(\mathfrak{A}_1)$. Then $\beta^A = (\beta^A_{ji})$ is a $p \times p$ matrix of morphisms $\beta^A_{ji}: SA_i \to TA_j$. Consider the morphisms $e^i: (A_i) \to A$ given by the column matrix

$$e^i = (e^i_j), \quad e^i_j = 0 \quad \text{if} \quad i \neq 0, \quad e^i_j = 1_{A_i}.$$

Then we have a commuting diagram

$$\begin{array}{ccc}
(SA_i) = M(S)\,(A_i) & \xrightarrow{\begin{bmatrix} 0 \\ \vdots \\ 0 \\ 1_{SA_i} \\ 0 \\ \vdots \\ 0 \end{bmatrix}} & (SA_1, \ldots, SA_p) \\
\Big\downarrow {\scriptstyle \alpha_{1A_i} = \beta^{A_i}} & & \Big\downarrow {\scriptstyle \beta^A = (\beta^A_{ji})} \\
M(T)\,(A_i) & \xrightarrow{\begin{bmatrix} 0 \\ \vdots \\ 0 \\ 1_{TA_i} \\ 0 \\ \vdots \\ 0 \end{bmatrix}} & (TA_1, \ldots, TA_p)
\end{array}$$

Computing the products of these matrices in both directions we have

$$\begin{bmatrix} 0 \\ \vdots \\ 0 \\ \alpha_{1A_i} \\ 0 \\ \vdots \\ 0 \end{bmatrix} = \begin{bmatrix} \beta \\ \vdots \\ \cdot \\ \beta_{ii} \\ \cdot \\ \vdots \\ \beta_{pi} \end{bmatrix}$$

so $\beta_{ii} = \alpha_{iA_i}$ and $\beta_{ji} = 0$ if $j \neq i$.

Suppose that we are given a functor $S_1 : \mathfrak{A}_1 \to \mathfrak{B}$. Then there is an object preserving functor $\overline{M}(S_1) : \mathfrak{M}(\mathfrak{A}_1) \to \mathfrak{M}(\mathfrak{B})$ which is defined as follows: $\overline{M}(S_1)(A_1, \ldots, A_p) = (S_1 A_1, \ldots, S_1 A_p)$ and $\overline{M}(S_1)((f_{ji})) = (Sf_{ji})$. Now $\overline{M}(S_1)$ is a functor if and only if for any family

$$A \xrightarrow{f_i} B_i \xrightarrow{g_i} C, \quad i = 1, \ldots, p$$

of morphisms in $\mathfrak{A}_1$ we have

$$S_1(\Sigma g_i f_i) = \Sigma S_1 g_i S_1 f_i.$$

For this it suffices to know that the diagram

$$\begin{array}{ccc} S_1 A & \xrightarrow{S_1(h_1 + \cdots + h_p)} & S_1 C \\ \big\downarrow c_n^{S_1 A} & & \big\uparrow m_p^{S_1 C} \\ (S_1 A)^p & \xrightarrow{S_1 h_1 \otimes \cdots \otimes S_1 h_p} & (S_1 C)^p \end{array} \tag{9}$$

commutes for all families $h_i : A \to C$, $i = 1, \ldots, p$. By induction and coherence this follows from the commuting of all diagrams of the type of (9) with $p = 2$. For the case $p = 2$, observe that since S_1 is a morphism of Hopf algebras we have

$$S_1(A \xrightarrow{c^A} A \otimes A \xrightarrow{m^A} A) = S_1 A \xrightarrow{c^{S_1 A}} S_1 A \otimes S_1 A \xrightarrow{m^{S_1 A}} S_1 A$$

with a commutative diagram

$$\begin{array}{ccccc} S_1 A & \xrightarrow{c^{S_1 A}} & S_1 A \otimes S_1 A & \xrightarrow{m^{S_1 A}} & S_1 A \\ & {\scriptstyle S_1(c^A)} \searrow & \big\uparrow \varepsilon_{A,A} & \nearrow {\scriptstyle S_1(m^A)} & \\ & & S_1(A \otimes A) & & \end{array}$$

Now we can take the commuting diagram

$$\begin{array}{ccc} A & \xrightarrow{f+g} & C \\ \big\downarrow c^A & & \big\uparrow m^C \\ A \otimes A & \xrightarrow{f \otimes g} & C \otimes C \end{array}$$

and apply the functor S_1, obtaining the commuting diagram

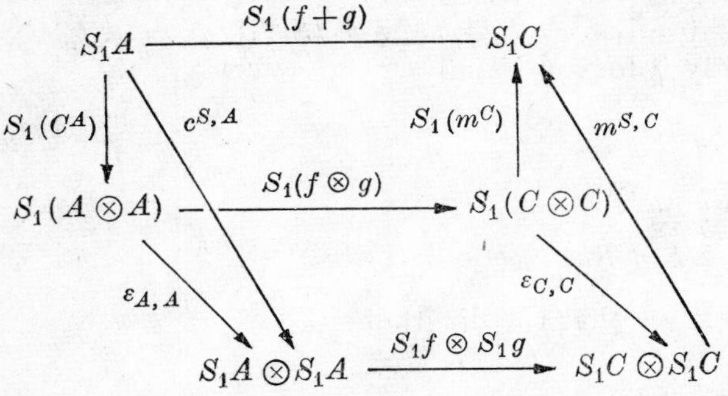

which establishes the commutativity of (9) for $p = 2$.

Definition 2.1. Let $\mathfrak{A}$ be a subcategory of $\mathit{Hopf}_{ab}\,\mathfrak{P}$ where $\mathfrak{P}$ is some multiplicative category. Let $\mathfrak{A}_1$ be a full subcategory of $\mathfrak{A}$. We say that $\mathfrak{A}$ is *freely generated* by $\mathfrak{A}_1$ if the functor $M_{\mathfrak{A}}\colon \mathfrak{M}(\mathfrak{A}_1) \to \mathfrak{A}$ is an equivalence.

We illustrate this concept by a series of examples.

Let $\mathfrak{P}$ be the category of finitely generated abelian groups with multiplication $\oplus$, the biproduct in the category. Let $\mathfrak{A} \subseteq \mathit{Hopf}_{ab}\mathfrak{P}$ be the full subcategory of all commutative Hopf algebras

$$A \xrightarrow{d^A} A \oplus A \xrightarrow{d_A} A$$

where d^A is the diagonal map and d_A the codiagonal map. One might remark that this is exactly the subcategory of commutative Hopf algebras with identity and coidentity. This category is obviously isomorphic to $\mathfrak{P}$ itself. Now let $\mathfrak{A}_1$ be the full subcategory of those Hopf algebras in $\mathfrak{A}$ given by the cyclic groups Z, $\mathsf{Z}/p^n\mathsf{Z}$, where p is a prime, $n = 1, 2, \ldots$ Then on the basis of elementary facts about finitely generated abelian groups, the functor M is faithful, full and representative, i. e. $\mathfrak{A}$ is freely generated by $\mathfrak{A}_1$.

The example may be repeated with finitely generated R-modules over a principal ideal domain R in place of abelian groups. The case of a field is particularly simple: the full subcategory spanned by the single object $R \to R \oplus R \to R$ generates freely in this case.

We now formulate the main theorem of the current discussion:

Theorem 2.2. *Let $\mathfrak{P}$ and $\mathfrak{Q}$ be pointed multiplicative categories and $\mathfrak{A} \subset \mathit{Hopf}_{ab}\,\mathfrak{P}$, $\mathfrak{B} \subset \mathit{Hopf}_{ab}\,\mathfrak{Q}$ multiplicative subcategories of commutative Hopf algebras in , respectively, $\mathfrak{Q}$. Let $J\colon \mathfrak{A}_1 \to \mathfrak{A}$ be the inclusion functor of a full subcategory such* *that* $\mathfrak{A}$ *is freely generated by* $\mathfrak{A}_1$. *Let* $S_1: \mathfrak{A}_1 \to \mathfrak{B}$ *be an arbitrary functor of Hopf algebras. Then there is a unique functor of Hopf algebras* $S: \mathfrak{A} \to \mathfrak{B}$ *such that*

(i) *there is a natural isomorphism* $\delta: SJ \to S_1$,

(ii) *if* $T: \mathfrak{A} \to \mathfrak{B}$ *is a multiplicative functor of Hopf algebras, and if* $\alpha_1: TJ \to S_1$ *is a natural transformation, then there is a unique natural transformation* $\alpha: T \to S$ *such that* $\delta\alpha J = \alpha_1$.

In particular, multiplicative Hopf algebra functors $\mathfrak{A} \to \mathfrak{B}$ *are uniquely determined up to natural isomorphism by their action on the subcategory* $\mathfrak{A}_1$.

Remark. The theorem applies particularly to the case that $\mathfrak{P}$ is a semi-additive category (a pointed category with finite biproducts), and that $\mathfrak{A}$ is the category of all Hopf algebras of the form $A \to A \oplus A \to A$, $A \in \mathfrak{P}$ with the diagonal and codiagonal map. If $\mathfrak{P}$ has a full subcategory $\mathfrak{P}_1$ such that every object in $\mathfrak{P}$ is isomorphic in an essentially unique fashion to a finite biproduct of objects from $\mathfrak{P}_1$, then $\mathfrak{A}$ is freely generated by the full subcategory $\mathfrak{A}_1$ of all $A \to A \oplus A \to A$ with $A \in \mathfrak{P}_1$.

Proof. We define the functor $S: \mathfrak{A} \to \mathfrak{B}$ to be $M_{\mathfrak{B}} \bar{M}(S_1) M'_{\mathfrak{A}}$, where $M'_{\mathfrak{A}}: \mathfrak{A} \to \mathfrak{M}(\mathfrak{A}_1)$ is a functor with $M_{\mathfrak{A}} M'_{\mathfrak{A}} \cong I_{\mathfrak{A}}$ and $M'_{\mathfrak{A}} M_{\mathfrak{A}} \cong I_{\mathfrak{M}(\mathfrak{A}_1)}$, which exists by hypothesis. We then have $SJ = M_{\mathfrak{B}} \bar{M}(S_1) M'_{\mathfrak{A}} J$; but $M'_{\mathfrak{A}} J \cong J_1$ since $J = M_{\mathfrak{A}} J_1$. Thus $\delta: SJ \xrightarrow{\cong} S_1$ exists.

We may, if we wish, for the moment assume that $\mathfrak{A}$ is a skeleton. Then every object of $\mathfrak{A}$ is of the form $A_1 \otimes \cdots \otimes A_p$ in a unique way with objects $A_i \in \mathfrak{A}_1$. Then we may assume that

$$M'_{\mathfrak{A}}(A_1 \otimes \cdots \otimes A_p) = (A_1, \ldots, A_p),$$

whence $S(A_1 \otimes \cdots \otimes A_p) = S_1 A_1 \otimes \cdots \otimes S_1 A_p$. From this we may deduce that S is a multiplicative functor of Hopf algebras. Returning to the general situation, we take an arbitrary multiplicative functor of Hopf algebras $T: \mathfrak{A} \to \mathfrak{B}$ and assume that there is a natural transformation $\alpha_1: TJ \to S_1$. Then there is a natural transformation $\alpha'_1 = \delta^{-1}\alpha_1: TJ \to SJ$, where $\delta: SJ \to S_1$ is the natural isomorphism which we established before. Then there is a unique natural transformation $\alpha' = M(\alpha'_1): M(T) \to M(S)$ of functors

$$\mathfrak{M}(\mathfrak{A}_1) \to \mathfrak{M}(\mathfrak{B})$$

as we have seen earlier, such that $M_{\mathfrak{B}}\, \alpha' J_1 = \alpha'_1$. We observe that there is a natural isomorphism $\gamma: M(S) \to M(S_1)$ given by

$$\gamma_{(A_1,\ldots,A_p)}: (SA_1, \ldots, SA_p) = (SJA_1, \ldots, SJA_p) \to (S_1 A_1, \ldots, S_1 A_p)$$

being diag $(\delta_{A_1}, \ldots, \delta_{A_p})$. Then $M_{\mathfrak{B}}\, \gamma\, \alpha' J_1 = \alpha_1$. Now we define the natural transformation $\alpha: T \to S$ as the composition

$$\begin{array}{ccccccc}
TJ = TM_{\mathfrak{A}} J_1 = M_{\mathfrak{B}} M(T) J_1 & \xrightarrow{M_{\mathfrak{B}} \alpha' J_1} & M_{\mathfrak{B}} M(S) J_1 & \xrightarrow{M_{\mathfrak{B}} \gamma J_1} & M_{\mathfrak{B}} \bar{M}(S_1) J_1 = \\
\downarrow \cong \qquad \downarrow \cong & & \downarrow \cong & & \downarrow \cong \\
TJ = TM_{\mathfrak{A}} M'_{\mathfrak{A}} J = M_{\mathfrak{B}} M(T) M'_{\mathfrak{A}} J & \xrightarrow{M_{\mathfrak{B}} \alpha' M'_{\mathfrak{A}} J} & M_{\mathfrak{B}} M(S) M'_{\mathfrak{A}} J & \xrightarrow{M_{\mathfrak{B}} \gamma M'_{\mathfrak{A}} J} & M_{\mathfrak{B}} \bar{M}(S_1) J = S
\end{array}$$

The uniqueness of α follows from the uniqueness of $\alpha' = M(\alpha'_1)$. That is, if $\beta\colon T \to S$ is another such map, we have $M(\beta J) = M(\alpha'_1) = M(\alpha J)$, and so $\beta M_{\mathfrak{A}} M'_{\mathfrak{A}} = \alpha M_{\mathfrak{A}} M'_{\mathfrak{A}}$. Since $M_{\mathfrak{A}} M'_{\mathfrak{A}}$ is an equivalence on $\mathfrak{A}$, $\beta = \alpha$.

Notice that the theorem sets up a natural bijection from the class

$$(\mathfrak{B}^{\mathfrak{A}_1})_{\text{Hopf}}(TJ, S_1)$$

of natural transformations of functors of commutative Hopf algebras $\mathfrak{A}_1 \to \mathfrak{B}$ and the class $(\mathfrak{B}^{\mathfrak{A}})_{\exp \text{Hopf}}(T, S)$ of natural transformations of multiplicative functors of Hopf algebras $\mathfrak{A} \to \mathfrak{B}$. In this sense $S_1 \to S$ is similar to a (right) adjoint of the operation $T \to TJ$ and is, therefore, a Kan extension of sorts. In fact, this example could provide a motivation for a suitable generalization of the concept of a Kan extension which would cover both Theorem 1.14 and Theorem 2.2 above. We leave this to the reader.

We draw some corollaries which will be needed for the applications.

Corollary 2.3. *Let $\mathfrak{A}$ be the category of compact abelian Lie groups, $\mathfrak{A}_1$ the full subcategory spanned by the objects R/Z, $\mathsf{Z}(p^n)$, p a prime, $n = 1, 2, \ldots$ Let $\mathfrak{Q}$ be any multiplicative category and $\mathfrak{B}$ a multiplicative subcategory of commutative Hopf algebras in $\mathfrak{Q}$. Then any functor of Hopf algebras $\mathfrak{A}_1 \to \mathfrak{B}$, where the category $\mathfrak{A}$ is considered as the category of commutative Hopf algebras*

$$A \xrightarrow{\text{diag}} A \oplus A \xrightarrow{\text{codiag}} A$$

in itself, has a unique exponential Hopf algebra functor as Kan extension.

In particular, every exponential Hopf algebra functor $\mathfrak{A} \to \mathfrak{B}$ is determined uniquely up to natural isomorphism by its operation on $\mathfrak{A}_1$.

Corollary 2.4. *Let $\mathfrak{A}$ be the category of compact connected abelian Lie groups and let $\mathfrak{A}_1$ be the category consisting of the single object R/Z and all of its $\mathfrak{A}$-endomorphisms. (Note that all these endomorphisms are of the form*

$$r + \mathsf{Z} \to z\,r + \mathsf{Z}$$

for some integer z.) Let $\mathfrak{Q}$ and $\mathfrak{B}$ be as in the preceding corollary, and consider $\mathfrak{A}$ as a category of commutative Hopf algebras in itself as explained there. Then any

Hopf algebra functor $\mathfrak{A}_1 \to \mathfrak{B}$ *has a unique exponential Hopf algebra functor as Kan extension in the sense of Theorem* 2.2.

In particular, every exponential Hopf algebra functor $\mathfrak{A} \to \mathfrak{B}$ *is determined up to natural isomorphism by its operation on* $\mathfrak{A}_1$.

Remark. Note that in this situation, prescribing a Hopf algebra functor $\mathfrak{A}_1 \to \mathfrak{B}$ means nothing but picking a particular Hopf algebra in $\mathfrak{B}$ together with a cyclic ring of endomorphisms of this Hopf algebra.

In the last corollary of this section we bring together parts of each of the two results about Kan extensions in this chapter.

Corollary 2.5. *Let* $\mathfrak{A}$ *be the category of compact abelian groups (resp., the category of compact connected abelian groups), and* $\mathfrak{A}_1$ *the full subcategory spanned by the groups* R/Z, $\mathsf{Z}(p^n)$, p *a prime,* $n = 1, \ldots$ *[respectively, the subcategory consisting of the single object* R/Z *and its endomorphisms* $g \to n\,g$, $n \in \mathsf{Z}$*]. Let* $\mathfrak{B}$ *be a multiplicative subcategory of commutative Hopf algebras over a suitable multiplicative category and* S, $T\colon \mathfrak{A} \to \mathfrak{B}$ *two functors such that the following conditions are satisfied:*

(i) $\mathfrak{B}$ *is* $\mathfrak{D}$*-complete with the category* $\mathfrak{D}$ *of directed sets.*

(ii) S, T *are* $\mathfrak{D}$*-continuous.*

(iii) S, T *are exponential Hopf algebra functors.*

(iv) *The restrictions of* S *and* T *to* $\mathfrak{A}_1$ *are isomorphic.*

Then S *and* T *are isomorphic.*

Remark. Condition (iii) is equivalent to the following

(iii′) S and T are exponential functors and for each $A \in \mathfrak{A}$ the diagram

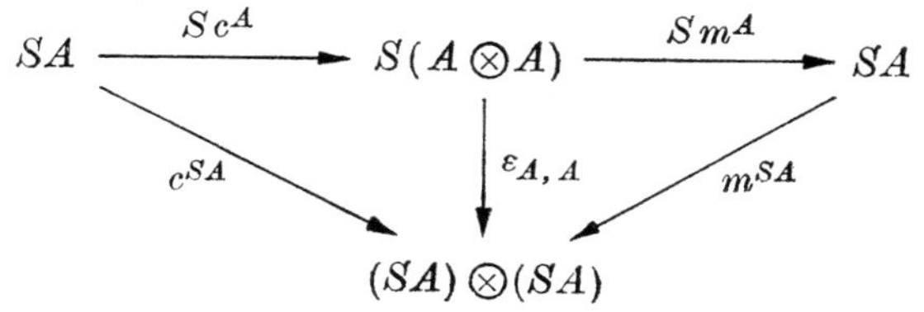

with the diagonal c^A and codiagonal (group operation) m^A on A commutes and each similar diagram with T in place of S commutes.

We will have occasion to use another Kan extension of a well-known exponential functor form the category of finite abelian groups to the category of graded commutative Hopf algebras, namely, the exterior algebra functor.

Proposition 2.6. *Let Comp denote the category of compact abelian groups and $\mathfrak{B}$ the category of graded commutative rings. Let Lie be the full category of all Lie groups in Comp. Then the functor E: Lie $\to \mathfrak{B}$ given by $EG = \bigwedge (G/G_0)$, where G_0 denotes the identity component of G, has a Kan extension to a functor Comp $\to \mathfrak{B}$ which we denote with $\bigwedge$.*

Note that $\bigwedge^i G = 0$ for $i > 0$ if G is connected.

The existence of $\bigwedge$ follows from Theorem 2.2. It is, however, of interest, actually to identify in more concrete terms the functor $\bigwedge$ for arbitrary compact abelian groups; this we shall do in the following.

Lemma 2.7. *The functor $\bigwedge$ from the category of topological abelian groups to the category of graded commutative topological rings with trivial zero component $\mathbb{Z}$ has the following properties:*

(1) *$\bigwedge$ is left adjoint to the functor L which associates with a graded commutative topological ring $A^0 \oplus A^1 \oplus \cdots$ the topological abelian group A^1.*

(2) *$\bigwedge$ is an exponential functor and extends the familiar exterior algebra over discrete groups.*

(3) *There is a multilinear alternating continuous function $G \times \cdots \times G \to \bigwedge^n G$, denoted $(g_1, \ldots, g_n) \to g_1 \wedge \cdots \wedge g_n$, such that for any multilinear alternating continuous function $f\colon G \times \cdots \times G \to H$, there is a unique morphism $f'\colon \bigwedge^n G \to H$ of topological abelian groups such that*

$$f(g_1, \ldots, g_n) = g_1 \wedge \cdots \wedge g_n.$$

(4) *If G is compact abelian, then so is $\bigwedge^n G$ for all $n > 0$. If $\pi\colon G \to G/G_0$ is the natural projection modulo the connected identity component, then*

$$\bigwedge \pi\colon \bigwedge G \to \bigwedge G/G_0$$

is an isomorphism. In particular, if G is connected, then $\bigwedge G = \mathbb{Z}$.

(5) *If G is totally disconnected compact abelian and $G = \varprojlim G/N$, where N runs through the directed family of subgroups with finite index, then*

$$\bigwedge^n G \cong \varprojlim \bigwedge^n G/N.$$

Proof. We mainly give indications for the proof of these facts and leave it to the reader to fill in the details. Let $\mathfrak{Q}$ be the category of topological abelian groups and $\mathfrak{A}$ the category of graded commutative topological algebras whose component of degree 0 is $\mathbb{Z}$. The category of topological abelian groups allows a tensor product for multilinear continuous morphisms $G_1 \times \cdots \times G_n \to G$ in the following sense: Let $\mathfrak{M}$ be a category whose objects are multilinear continuous maps $m\colon G_1 \times \cdots \times G_n \to G$ of topological abelian groups and whose morphisms are families of morphisms $\alpha_1 : G_i \to G_i'$, $\alpha : G \to G'$ giving rise to commutative diagrams

$$\begin{array}{ccc} G_1 \times \cdots \times G_n & \xrightarrow{\alpha_1 \times \cdots \times \alpha_n} & G_1' \times \cdots \times G_n' \\ \downarrow m & & \downarrow m' \\ G & \xrightarrow{\alpha} & G' \end{array}$$

Let D and R be the functors associating with m its domain and its range, respectively. Then one observes, using the coadjoint existence theorem, that D has a left adjoint functor T. One denotes $RT(G_1, \ldots, G_n)$ with $G_1 \otimes \cdots \otimes G_n$ and writes

$$T(G_1, \ldots, G_n)\,(g_1, \ldots, g_n) = g_1 \otimes \cdots \otimes g_n.$$

Then, by the universal properties of the left adjoint, for any continuous multilinear map f from the category $\mathfrak{M}$, there is a unique morphism

$$f' : G_1 \otimes \cdots \otimes G_n \to G \quad \text{such that} \quad f(g_1 \otimes \cdots \otimes g_n) = f(g_1, \ldots, g_n).$$

If the category $\mathfrak{M}$ consists of all alternating multilinear continuous morphisms $G \times \cdots \times G \to H$, then one writes $\wedge$ instead of $\otimes$. Later we shall see that this functor $\wedge$ is indeed the Kan extension of Proposition 2.6.

Relative to the tensor product of two abelian topological groups, $\mathfrak{G}$ is multiplicative, and relative to the tensor product of graded objects, $\mathfrak{A}$ is multiplicative. The functor L is a logarithmic functor, i. e.

$$L(A \otimes A') \cong LA \oplus LA'.$$

This functor has a left adjoint $\wedge'$ by Freyd's left adjoint existence theorem. The functor $\wedge'$ is exponential ([20], 4.4). Clearly it extends the usual exterior algebra functor for discrete groups. Thus (2) is established with $\wedge'$ in place of $\wedge$. We shall show that $\wedge' = \wedge$, where $\wedge$ is the functor given below extending the functors $\wedge^n$ defined above. In order to prove (3), we construct the functor $\wedge$ as follows: $\wedge^0$ is the constant functor with value $\mathbf{Z}$. For $n > 0$ we let

$$(g_1, \ldots, g_n) \to g_1 \wedge \cdots \wedge g_n : G \times \cdots \times G \to \wedge^n G = T(G, \ldots, G)$$

for the functor T described above, where $\mathfrak{M}$ is taken to be the category of all continuous alternating n-linear maps. Then the multilinear map

$$(g_1, \ldots, g_n) \to g_1 \wedge' \cdots \wedge' g_n : G \times \cdots \times G \to \wedge'^n G$$

factors uniquely through $\wedge^n G$ by the universal property of $\wedge^n$. Thus there is a unique natural transformation $\wedge \to \wedge'$. On the other hand, just as in the discrete case, $\bigwedge A$ is a commutative graded algebra with $\bigwedge^1 A = A$. Hence by the universal property of $\bigwedge'$, there is also a unique natural transformation $\bigwedge' \to \bigwedge$. And from uniqueness it then follows by standard arguments, that the two natural transformations are in fact inverses of each other.

In order to deal with (4), we observe that $\bigwedge^n G$ is obviously a quotient group of $\otimes^n G$. But it is easy to see that $G \otimes H$ is compact for compact G and H; in fact $G \otimes H$ is the character group of the discrete group Hom $(G, \widehat{H})$ (Hom in the category of topological groups) (see [21, 36]). Consequently $G_1 \otimes \cdots \otimes G_n$ is compact for compact G_i and thus so is $\bigwedge^n G$ if G is compact. Further, for compact G and H, we have the isomorphisms

$$G \otimes H = \text{Hom}\,(G, \widehat{H})\hat{} \cong \text{Hom}\,(G/G_0, \widehat{H})\hat{} \cong (G/G_0) \otimes H,$$

and by symmetry, then

$$G/G_0 \otimes H \cong G/G_0 \otimes H/H_0.$$

By induction this implies $\otimes^n G \cong \otimes^n (G/G_0)$, and thus $\bigwedge^n G \cong \bigwedge^n (G/G_0)$.

Finally we observe, that (5) is tantamount to saying that the functor $\bigwedge$ defined above and the Kan extension of the exterior algebra functor on finite abelian groups are naturally isomorphic. Let us, for the moment, denote the Kan extension defined in Proposition 2.6 by $\bigwedge^*$. Then $\bigwedge^{*n} G = \lim \bigwedge^n G/N$ for a totally disconnected compact abelian group G, where the limit is taken over all subgroups N of G with finite index, and where $n = 1, 2, \ldots$ From the universal property of a Kan extension there is a unique natural transformation $\bigwedge \to \bigwedge^*$ which reduces to the identity on finite abelian groups. Every alternating multilinear continuous function $G \times \cdots \times G \to F$ into a finite abelian group factors through a continuous morphism $\bigwedge^{*n} G \to F$, as is not hard to see. From this observation one can then derive the same universal property for a totally disconnected F in place of a finite one (compare [21]). This entails the existence of a natural transformation $\bigwedge^* \to \bigwedge$ which is seen to be the inverse of the one given above.

Suppose that R is a discrete ring, and that for a compact abelian group G we denote with Hom (G, R) the R-module of all *continuous* group morphisms $G \to R$, where R is given the discrete topology. For every such morphism f, the kernel, $\ker f$, is then an open subgroup of G with finite index. Let $\mathfrak{N}$ be the directed set of all subgroups N of G with finite index and let $F: \mathfrak{N} \to \mathit{Lie}$ be the functor given by $F(N) = G/N$. Note that $\varprojlim F \cong G/G_0$. ($\varprojlim$ denotes inverse limit, $\varinjlim$ direct limit over a directed set.) Now Hom $(F\text{-}, R)$ is a functor from $\mathfrak{N}$ into the opposite category of the category of R-modules. Each quotient map $G \to G/N$ induces an injection $\mathrm{Hom}\,(G/N, R) \to \mathrm{Hom}\,(G, R)$ and by the definition of Hom it is clear that $\mathrm{Hom}\,(G, R)$ is the union of the images of all of these maps. Thus $\mathrm{Hom}\,(G, R) \cong \varinjlim \mathrm{Hom}\,(F, R)$. In particular, we then have

$$\mathrm{Hom}\,(\wedge G, R) \cong \mathrm{Hom}\,(\wedge (G/G_0), R) \cong \mathrm{Hom}\,(\wedge \varprojlim G/N, R),$$

where the limit is taken over all subgroups n of G with finite index. By the preceding proposition we have

$$\wedge \varprojlim G/N \cong \varprojlim \wedge G/N$$

and after the preceding remarks we have

$$\mathrm{Hom}\,(\varprojlim \wedge G/N, R) \cong \varinjlim \mathrm{Hom}\,(\wedge G/N, R).$$

This colimit does not change if we let N range through the directed set of all closed subgroups N of G such that G/N is a Lie group since

$$\mathrm{Hom}\,(K, R) \cong \mathrm{Hom}\,(K/K_0, R).$$

But this shows then that the functor $\mathrm{Hom}\,(\wedge -, R)$ is the Kan extension of its restriction to *Lie*, since it was by this colimit that the Kan extension was defined. We formulate this in the following

Lemma 2.8. *Let R be an arbitrary discrete topological commutative ring with identity. The functor* $\mathrm{Hom}\,(\wedge -, R)$ *from the category of compact abelian groups into the opposite of the category of graded commutative R-algebras is the Kan extension of its own restriction to the subcategory of abelian Lie groups.*

If R is a field, then for finite G we have

$$\mathrm{Hom}\,(\wedge G, R) \cong \wedge_R \mathrm{Hom}\,(G, R)$$

by I—1.14. Hence the colimit over the R-vector spaces $\mathrm{Hom}\,(\wedge G/N, R)$ is then naturally isomorphic to $\lim \wedge_R \mathrm{Hom}\,(G/N, R)$. But $\wedge$, as a left adjoint commutes with colimits, so that the last colimit is actually isomorphic to

$$\wedge_R \varinjlim \mathrm{Hom}\,(G/N, R) \cong \wedge_R \mathrm{Hom}\,(G, R).$$

Thus we have

Lemma 2.9. *If R is a discrete topological field, then there is, for each compact abelian group G, a natural isomorphism*

$$\wedge_R \mathrm{Hom}\,(G, R) \to \mathrm{Hom}\,(\wedge G, R),$$

and both of these functors are the Kan extensions of their respective restrictions to Lie.

Recall that in II-Theorem V we developed other natural isomorphisms for the finite case which also lift without difficulty to the compact case, and indeed we have

Lemma 2.10. *For a discrete topological field R with prime field K, there is, for each compact abelian group G, a natural isomorphism*

$$R \otimes \operatorname{Hom}(G, K) \to \operatorname{Hom}(G, R),$$

and therefore a natural isomorphism

$$R \otimes \wedge \operatorname{Hom}(G, K) \to \wedge_R \operatorname{Hom}(G, R).$$

The natural isomorphism $\operatorname{Hom}(G, K) \cong \operatorname{Tor}(\hat{G}, K)$, *where $\hat{G}$ is the character group of G, yields another isomorphism*

$$R \otimes \wedge \operatorname{Hom}(G, K) \cong R \otimes \wedge \operatorname{Tor}(\hat{G}, K).$$

Chapter V
The cohomological structure of compact abelian groups

For compact abelian groups (as for arbitrary compact groups) there are at least two different cohomology theories. One of them arises from the topological, one from the algebraic structure of the group. The first one is the Čech cohomology H of the compact space underlying the group, the second is the functor h introduced in Chapter III (and the appendix, Chapter VI) which, when restricted to the discrete (hence finite) objects in the category yields the algebraic cohomology over a given module with the group operating trivially.

Section 1
The cohomologies of connected compact abelian groups

In this section we determine completely and functorially the cohomology Hopf algebras of a compact connected abelian group over a given ring R.

Throughout the section, $Comp_0$ denotes the $\mathfrak{D}$-complete category of compact connected abelian groups, where $\mathfrak{D}$ is the category of directed sets. The category $Comp_0$ has finite biproducts and may be identified with the category of commutative Hopf algebras $A \to A \times A \to A$ over itself. We let $\mathfrak{P}$ be the multiplicative category of graded abelian groups with the tensor product of graded abelian groups as multiplication. Let $\mathfrak{A}$ denote the category of all graded commutative Hopf algebras over Z.

Lemma 1.1. *The functor H: $Comp_0 \to (\mathfrak{A})^*$ of Čech cohomology of the underlying space is a $\mathfrak{D}$-continuous exponential functor.*

Proof. Firstly it is well known that Čech cohomology of compact spaces transforms limits over directed systems into colimits of graded groups. This means that H is $\mathfrak{D}$-continuous. Let A, B be two objects in $Comp_0$. By the Künneth theorem for Čech cohomology we have a natural exact sequence

$$0 \to HA \otimes HB \to H(A \times B) \to \text{Tor}\,(HA, HB) \to 0$$

in which the map into the Tor term is of degree $+1$. If $A \cong \mathsf{R}/\mathsf{Z}$ then $H^i A \cong \mathsf{Z}$ for $i = 0, 1$, and $= 0$ for $i > 1$; so HA is torsion free, so from our exact sequence we obtain the natural isomorphism

$$\wedge\, \mathsf{Z} \otimes HB \xrightarrow{\cong} H(\mathsf{R}/\mathsf{Z} \times B)$$

since the torsion term vanishes. By induction it follows now that $H(\mathsf{R}/\mathsf{Z})^n$ is torsion free for all $n = 1, 2, \ldots$; since the category Lie_0 of all compact connected Lie groups is dense in $Comp_0$ and direct limits of torsion free abelian groups are torsion free, then HA is torsion free for any A in $Comp_0$. The Künneth theorem then shows that H is an exponential functor.

Lemma 1.2. *The functor h: $Comp_0 \to (\mathfrak{A})^*$ of Čech cohomology of the Milnor classifying space, say, of a group in $Comp_0$ is a $\mathfrak{D}$-continuous exponential functor.*

Proof. That h is $\mathfrak{D}$-continuous was shown in III—1.11. By III—1.13 we have the Künneth sequence

$$0 \to hA \otimes hB \to h(A \times B) \to \text{Tor}\,(hA, hB) \to 0.$$

As in the proof of Lemma 1.1 it suffices now to show that $h(\mathsf{R}/\mathsf{Z})$ is torsion free. It is known that the cohomology of a classifying space of a circle group is a polynomial ring in one variable of degree 2 [see e. g. [28], pp. 54, 231]. Thus, as a graded abelian group, $h(\mathsf{R}/\mathsf{Z})$ is torsion free. This finishes the proof.

Lemma 1.3. *The functor $\bar{H}$: $Comp_0 \to (\mathfrak{A})^*$ given by $\bar{H}A = \wedge\,\text{Hom}\,(A, \mathsf{R}/\mathsf{Z})$, the graded commutative group underlying the exterior algebra over the character group of A, whereby all generating elements in the character group have degree 1, is a $\mathfrak{D}$-continuous exponential functor.*

Proof. The functor $\text{Hom}\,(-, \mathsf{R}/\mathsf{Z})$ is an additive contravariant functor from $Comp_0$ into the category Ab of discrete abelian groups which transforms limits into colimits. The functor $\wedge$ from Ab into the category of graded algebras is the left adjoint of the functor $(A^n) \to A^1$ which associates with a graded algebra the group of elements of degree 1. Hence it preserves colimits, in particular directed limits. Thus $\bar{H}$ is $\mathfrak{D}$-continuous. Since $\wedge$ is exponential, then $\bar{H}$ is exponential.

Lemma 1.4. *The functor* $\bar{h}$: *Comp*$_0 \to (\mathfrak{A})^*$ *given by* $\bar{h}A = P\,\mathrm{Hom}\,(A,\,\mathsf{R}/\mathsf{Z})$, *the graded commutative group underlying the symmetric algebra generated by the character group of* A, *where all elements of the generating group have degree* 2, *is a* $\mathfrak{D}$-*continuous exponential functor.*

Proof. Same as for Lemma 1.3.

By IV, Section 2 all of the functors H, h, $\bar{H}$, and $\bar{h}$ induce functors of commutative Hopf algebras, which we shall denote with the same letter. Corollary IV—2.4 then says that all of these functors are uniquely determined by their action on the object R/Z and its endomorphisms. Now $H(\mathsf{R}/\mathsf{Z}) = \mathsf{Z} + \mathsf{Z}$ as mentioned before, and multiplication under the cup product is zero except for multiplication with a scalar multiple of the identity. This makes $H(\mathsf{R}/\mathsf{Z})$ isomorphic to $\wedge\,\mathsf{Z} \cong \bar{H}(\mathsf{R}/\mathsf{Z})$. The endomorphism $g \to ng$, $n \in \mathsf{Z}$, $n \neq 0$ of R/Z is an n-fold covering map and induces multiplication with n on $H^1(\mathsf{R}/\mathsf{Z})$. This means that H and $\bar{H}$ operate in the same fashion on R/Z and its endomorphisms and therefore, by IV—2.4 are naturally isomorphic. Thus, we have

Lemma 1.5. $H \cong \bar{H}$ *on Comp*$_0$.

Next we turn to h and $\bar{h}$. We observed before that as a ring,

$$h(\mathsf{R}/\mathsf{Z}) \cong P\mathsf{Z} \cong \bar{h}(\mathsf{R}/\mathsf{Z}).$$

We have to show that the functors h and $\bar{h}$ agree on the endomorphisms of R/Z. It suffices to compare the effect of the endomorphism $g \to n\,g$, $n \neq 0$ on the ring generators of $P\mathsf{Z}$ and $h(\mathsf{R}/\mathsf{Z})$ in the first case, it is clearly multiplication with n. In the second, we look at the induced map $\varphi\colon B^2 \to B^2$ of the second Milnor classifying space $B^2_{\mathsf{R}/\mathsf{Z}}$, which is just the 2-sphere. Since the endomorphism $g \to n\,g$ of R/Z is an n-fold covering map of the circle, φ operates on the two sphere B^2 as does the map $z \to z^n$ on the Riemann sphere. This means that $h^2\,\varphi$ is just multiplication by n. Thus again h and $\bar{h}$ operate in the same fashion on R/Z and its endomorphisms. We now have

Lemma 1.6. $h = \bar{h}$ *on Comp*$_0$.

Now suppose that R is an arbitrary commutative ring with identity. Then by the universal coefficient theorem we have exact sequences

$$0 \to (HA) \otimes R \to H(A,\,R) \to \mathrm{Tor}(HA,\,R) \to 0$$

and a completely analogous one with h in place of H. Since HA and hA were torsion free as abelian graded groups, the torsion term vanishes again and we have the natural isomorphisms of the first two terms. Thus, we have

Lemma 1.7. *On* $Comp_0$ $R \otimes HA \cong H(A, R)$ *and* $R \otimes hA \cong h(A, R)$ *naturally in A and R.*

On the other hand, $R \otimes \wedge \hat{A} \cong \wedge_R (R \otimes \hat{A})$ and $R \otimes P\hat{A} = P_R(R \otimes \hat{A})$, whence

Lemma 1.8. *On* $Comp_0$ $H(A, R) \cong \wedge_R (R \otimes A)$ *and*

$$h(A, R) \cong P_R(R \otimes \hat{A})$$

naturally in A and R as Hopf algebras.

We finally observe that H, when restricted to the dense subcategory of Lie groups, agrees (up to natural isomorphism) with singular cohomology because of the axiomatic characterization of cohomology theories on compact manifolds. By IV—1.15, H is in fact the Kan extension of any space cohomology on Lie_0 satisfying the Eilenberg-Steenrod axioms. In a similar vein, if h is constructed via the Milnor resolution (or via other such resolutions as the Dold-Lashof [12] resolution in Chapter VI or the Milgram resolution [43]) then for computing the restriction to Lie groups of the functor h, we may have used singular cohomology in place of Čech cohomology, and h is indeed the Kan extension of the restriction of h to the subcategory Lie_0.

In any event, we have proved the following main theorem concerning the cohomology theories for compact connected abelian Lie groups:

Theorem 1.9. *If G is a compact connected abelian group and R a commutative ring with identity, then the Čech cohomology* $H(G, R)$ *is a Hopf algebra over R in a natural way and the Čech cohomology* $h(G, R)$ *of a classifying space* $B(G)$ *is a Hopf algebra over R in a natural way and there are natural isomorphisms of commutative graded Hopf algebras*

$$H(G, R) \cong R \otimes \wedge \hat{G} \cong \wedge_R (R \otimes \hat{G}),$$
$$h(G, R) \cong R \otimes P\hat{G} \cong P_R (R \otimes \hat{G}),$$

where $\hat{G}$ *denotes the character group of G, being placed in degree 1 in the case of H and in degree 2 in case of h.*

Moreover, $H(-, R)$ *is the right Kan extension of the functor* $H_{\text{sing}}(-, R)$ *of singular cohomology of compact connected abelian Lie groups, and* $h(-, R)$ *is the right Kan extension of* $H_{\text{sing}}(B(-), R)$ *of singular cohomology of a Milnor or Dold-Lashof or Milgram classifying space of a compact connected abelian Lie group.*

One might observe, that this result, among other things, says that the topology of a compact connected abelian group G completely determines its structure, since the topology determines $H^1G \cong \hat{G}$, and the character group of G determines G. Similarly, any classifying space of a compact connected abelian group (at least if it is of the type of classifying space we have discussed) determines the group.

It should be pointed out that singular and Čech cohomology do not agree on $Comp_0$, though they do agree on Lie_0.

Indeed let us discuss, if somewhat informally, the singular cohomology of a compact connected abelian group.

Firstly we observe, that the functorial construction of a universal space $E(G)$ and a classifying space $B(G)$ of a topological group G is not restricted to compact groups; in particular it may be carried out for discrete groups; however, the topologies introduced on the $E^n(G)$ would have to be taken to be the so-called coordinate function topology, which is, in general, coarser than the quotient topology used by us; note that in the compact case both topologies agree. (For the Milnor construction with the coordinate function topology see e. g. [28]; also compare the appendix in Chapter VI.) When we talk about $E(G)$ and $B(G)$ in the following discussion we will understand that these spaces are endowed with their coordinate function topology.

One fact which we did not discuss in talking about the Milnor resolution is that $E(G)$ is contractible.

For a compact group G, $E(G) \to B(G)$ is a Serre fibration ([23], p. 317), and thus has an exact homotopy sequence ([26], p. 152) with $\pi_n(E(G)) = 0$ for all n. For A discrete, $E(A) \to B(A)$ is the simply connected covering map of the locally connected space $B(A)$ and thus we actually have a bundle where again the exact homotopy sequence exists. Thus, we obtain the isomorphism $\pi_{n+1}(B(A)) \cong \pi_n(A)$. In particular then, if A is a discrete group, we have

$$\pi_n(B(A)) = \begin{cases} A & \text{for} \quad n = 1, \\ 0 & \text{for} \quad n \neq 1. \end{cases}$$

Thus $B(A)$ is then an Eilenberg-MacLane space $K(A, 1)$.

The method indicated in III—2 will still show that the algebraic cohomology of A (over Z with trivial action) is naturally isomorphic to the space cohomology of $B(A)$ in the sense of $hA = \varprojlim H(B^n(A))$, where for the space cohomology we take singular cohomology. The telescope device of III—1 is still capable of showing then that hA is the singular cohomology of $B(A)$. It is known that the singular cohomology of an arcwise connected $K(A, n)$-space depends only on A. (See Eilenberg-MacLane [14].) Thus the singular cohomology of any $K(A, 1)$ is isomorphic to hA, the algebraic cohomology of A in case that A is discrete.

Now let G be a compact connected abelian group and let G_a be the arc component of the identity. Clearly $\pi_n(G) = \pi_n(G_a)$ for $n > 0$. The homotopy groups of a compact abelian group are known to be

$$\pi_n(G) = \begin{cases} \operatorname{Ext}(\hat{G}, \mathsf{Z}) & \text{for } n = 0, \\ \operatorname{Hom}(\hat{G}, \mathsf{Z}) & \text{for } n = 1, \\ 0 & \text{for } n > 1 \end{cases}$$

(see e. g. [19]). This means, that G_a is a $K(\operatorname{Hom}(\hat{G}, \mathsf{Z}), 1)$. Hence, after the preceding remarks, the singular cohomology of G_a is naturally isomorphic to $h(\operatorname{Hom}(\hat{G}, \mathsf{Z}))$. The arc component G_a is the image of the map

$$\operatorname{Hom}(\mathsf{R}, G) \to \operatorname{Hom}(\mathsf{Z}, G) \cong G$$

(see e. g. [19]) and is, therefore, divisible. Thus, as an algebraic subgroup of G, it splits; in other words, algebraically, we have a (non-canonical) group isomorphism $G \cong G_a \oplus \operatorname{Ext}(\hat{G}, \mathsf{Z})$. This gives a non-natural isomorphism $H_{\text{sing}}\, G \cong (H_{\text{sing}}\, G_a)^{\operatorname{Ext}(\hat{G}, \mathsf{Z})}$ for the functor H_{sing} of singular cohomology. Thus

$$H_{\text{sing}}\, G \cong (h \operatorname{Hom}(\hat{G}, \mathsf{Z}))^{\operatorname{Ext}(\hat{G}, \mathsf{Z})}.$$

We turn for a moment to the question of the algebraic cohomology hA of a discrete torsion free abelian group A over Z. Suppose, for the moment that A is finitely generated free. Then the natural map

$$A \to \operatorname{Hom}(\operatorname{Hom}(A, \mathsf{Z}), \mathsf{Z})$$

is an isomorphism, and with the abbreviation $G = \operatorname{Hom}(A, \mathsf{Z})\hat{}$ (which is, in this case, a finite dimensional torus), we obtain

$$hA \cong H_{\text{sing}}\, G \cong HG \cong \wedge G \cong \wedge \operatorname{Hom}(A, \mathsf{Z}),$$

and this isomorphism is natural in A and is an isomorphism of Hopf algebras. This isomorphism need not prevail for arbitrary torsion free groups — even for rank 1 groups: Take $A = Q$ the additive group of rationals. Then

$$\operatorname{Hom}(A, \mathsf{Z}) = 0,$$

but hA always contains an isomorphic copy of $\operatorname{Ext}(A, \mathsf{Z})$ which in this case is not zero. From this point of view, the conjecture, that

$$H_{\text{sing}}\, G_a \cong h(\operatorname{Hom}(\hat{G}, \mathsf{Z}))$$

might actually be naturally isomorphic to $\wedge \operatorname{Hom}(\operatorname{Hom}(\hat{G}, \mathsf{Z}), \mathsf{Z})$ (which is the case for the respective restrictions to the subcategory of abelian connected Lie groups) would seem not to be warranted. However, the counterexample above has no counterpart to decide *this* conjecture in the negative, since no group of the form $\mathrm{Hom}(G, \mathsf{Z})$ ever has elements of infinite height, which was exactly the property of Q that made the counterexample work.

Let us collect the gist of our discussion in the following theorem (noting that connectivity of G was not used).

Theorem 1.10. *Let G be a compact abelian group. Then the integral singular cohomology $H_{\mathrm{sing}}\, G$ is isomorphic (nonnaturally) to $H_{\mathrm{alg}}(\mathrm{Hom}(\hat{G}, \mathsf{Z}))^{\mathrm{Ext}(\hat{G}, \mathsf{Z})}$ where H_{alg} denotes the algebraic cohomology. If G_a is the arc component of the identity in G, then there is a natural isomorphism*

$$H_{\mathrm{sing}}\, G_a \cong H_{\mathrm{alg}}(\mathrm{Hom}(\hat{G}, \mathsf{Z}))\,.$$

This result maintains if the integral coefficient ring is replaced by an arbitrary coefficient ring or module.

We also have the following corollary complementing our extensive discussion of the cohomology of finitely generated abelian groups.

Corollary 1.11. *Let G be a finitely generated abelian group. Then*

$$G = G_1 \times G_2$$

with a finite group G_1 and a finitely generated free group G_2. Then for any principal ideal domain R, we have

$$\begin{aligned} H_{\mathrm{alg}}(G, R) &\cong H_{\mathrm{alg}}(G_1, R) \otimes \wedge \mathrm{Hom}(G_2, \mathsf{Z}) \\ &\cong H_{\mathrm{alg}}(G_1, R) \otimes_R \wedge_R \mathrm{Hom}(G_2, R)\,. \end{aligned}$$

Remark. $\wedge \mathrm{Hom}(G_2, \mathsf{Z})$ is the exterior algebra over Z in n generators, where $n = \mathrm{rank}\, G_2$.

Proof. By our preceding discussion we know that

$$H_{\mathrm{alg}}(G_2, \mathsf{Z}) \cong \wedge \mathrm{Hom}(G_2, \mathsf{Z})$$

naturally. As a graded abelian group this is torsion free, hence by the universal coefficient theorem, we have

$$\begin{aligned} H_{\mathrm{alg}}(G_2, R) &\cong R \otimes \wedge \mathrm{Hom}(G_2, \mathsf{Z}) \cong \wedge_R (R \otimes \mathrm{Hom}(G_2, \mathsf{Z})) \\ &\cong \wedge_R \mathrm{Hom}(G_2, R)\,. \end{aligned}$$

Thus Corollary II—1.9 applies and yields the result.

Section 2

The space cohomology of arbitrary compact abelian groups

For a compact space X and a discrete commutative ring R with identity, we let $C(X, R)$ denote the ring of all continuous, i. e. locally constant functions. This means that we have $C(X, R) = H^0(X, R)$ relative to Čech cohomology. If $\sim$ denotes the connectivity relation on X, then it is clear that the quotient map $X \to X/\sim$ induces an isomorphism $C(X/\sim, R) \to C(X, R)$. The natural isomorphism

$$C(X, R) \otimes C(Y, R) \to C(X \times Y, R)$$

may be easily proved directly or may be considered as a special case of the Künneth formula. Thus $C(-, R)$ is a multiplicative functor from the category of compact spaces into the category of R-algebras. If G is a compact abelian group, the $C(G, R)$ is a Hopf algebra, because

$$C(G \oplus G', R) \cong C(G, R) \otimes C(G', R).$$

Whence $C(-R)$ is an exponential functor from the category of all compact abelian groups into the category of R-algebras. In view of the equality

$$C(X, R) = H^0(X, R),$$

it is clear that $C(-, R)$ is $\mathfrak{D}$-continuous for the category $\mathfrak{D}$ of directed sets as a functor from the category of compact abelian groups into the opposite category of the category of R-algebras.

Theorem VI (The Structure Theorem for Topological Cohomology). *For compact abelian groups G and commutative rings R with identity there are natural isomorphisms of Hopf algebras*

$$H(G, R) \cong R \otimes C(G, \mathsf{Z}) \otimes \wedge \hat{G}_0 \cong C(G, R) \otimes \wedge \hat{G}_0$$
$$\cong \wedge_{C(G, R)} (C(G, R) \otimes \hat{G}_0).$$

Note that $\hat{G}_0 \cong \hat{G}/\hat{G}_t$ where $\hat{G}_t$ is the torsion subgroup of $\hat{G}$, where G_0 is the component of the identity of G.

Proof. (a) Assume $R = \mathsf{Z}$. We compare the two $\mathfrak{D}$-continuous exponential functors $H, \bar{H}$ from the category of compact abelian groups into the category of graded commutative rings given by integral Čech cohomology and by

$$\bar{H}G = C(G, \mathsf{Z}) \otimes \wedge \hat{G}_0.$$

By IV—2.5, we have to check that these functors are naturally isomorphic on the full subcategory Lie_1 spanned by the objects R/Z and $\mathsf{Z}(p^n)$, p a prime,

$n = 1, 2, \ldots$ Their natural isomorphy on the category of connected compact groups was already established in Section 1. Their natural isomorphy on the category of finite groups is clear after the initial remarks. Since all morphisms $\mathsf{R}/\mathsf{Z} \to \mathsf{Z}(p^n)$ are zero, we have to investigate morphisms $f\colon \mathsf{Z}(p^n) \to \mathsf{R}/\mathsf{Z}$. In positive dimensions, both $H^i(f)$ and $\overline{H}^i(f)$ are zero. There remains the case of degree 0: we have to show that the diagram

$$\begin{array}{ccccc}
H^0(\mathsf{R}/\mathsf{Z}) & \xrightarrow{\cong} & C(\mathsf{R}/\mathsf{Z}, \mathsf{Z}) & \xleftarrow{i} & \mathsf{Z} \\
\downarrow{\scriptstyle H^0(f)} & & \downarrow{\scriptstyle C(f, \mathsf{Z})} & & \| \\
H^0(\mathsf{Z}(p^n)) & \xrightarrow{\cong} & C(\mathsf{Z}(p^n), \mathsf{Z}) & \xleftarrow{j} & \mathsf{Z}
\end{array}$$

commutes where i and j are the injections of Z onto the constant functions. The left rectangle commutes by definition of the zero-th Čech cohomology. The injection i is surjective, since R/Z is connected. The function $\varphi\colon \mathsf{R}/\mathsf{Z} \to \mathsf{Z}$ with constant value z has as image $C(f, \mathsf{Z})(\varphi)$, the constant function φf with constant value z. Hence the right rectangle commutes. Thus the functors H and $\overline{H}$ restrict to isomorphic functors on the category Lie_1. By IV—2.5, then they are naturally isomorphic on the category of all compact abelian groups.

(b) Let R be arbitrary. By (a), as a graded abelian group,

$$HG \cong C(G, \mathsf{Z}) \otimes \wedge \hat{G}_0$$

is torsion free for all G. Hence, by the universal coefficient theorem, we have $\mathrm{H}(G, R) \cong R \otimes H(G, \mathsf{Z})$. Thus

$$H(G, R) \cong R \otimes C(G, \mathsf{Z}) \otimes \wedge \hat{G}_0.$$

The remaining isomorphisms are routine.

Corollary 2.2. *For any compact abelian group there is a natural isomorphism*

$$H^1(G, \mathsf{Z}) \cong C(G, \mathsf{Z}) \otimes \hat{G}_0 \cong C(G, \hat{G}_0).$$

It should be noted that the space cohomology can retrieve the complete structure of a connected abelian group from the geometry alone, and it can retrieve the complete structure of a totally disconnected abelian group, if the Hopf algebra structure of the cohomology ring is given (which is just $C(G, \mathsf{Z})$ in this case). However, if G is a compact abelian group in which G_0 does not split (and there are such compact groups (see e. g., [19], p. 155), then the space cohomology, even if the Hopf algebra structure is furnished, cannot distinguish between G and $G_0 \oplus G/G_0$, which may well be different abelian groups on the same space.

Section 3

The canonical embedding of $\hat{G}$ in hG

In Section 2 we saw that the character group $\hat{G}$ can be partially recovered from the topological cohomology. In the current section we will show that it can be completely recovered if the cohomology of a universal space is known. In fact we prove the following theorem:

Theorem 3.1. (a) *Let G be a compact abelian group. There is a natural isomorphism $\hat{G} \to h^2(G, \mathsf{Z})$. There is a natural injective morphism of graded algebras $P\hat{G} \to h(G, \mathsf{Z})$; in particular, $h(G, \mathsf{Z})$ is a graded $P\hat{G}$-algebra in a natural fashion.*

(b) *Let R be a commutative ring with identity. Then there is a natural injection $n_{G,R}$: $R \otimes \hat{G} \to h^2(G, R)$. It is an isomorphism if $\hat{G}$ has no p-torsion for any prime dividing the characteristic of R (in particular when R has characteristic zero).*

(c) *If $a \in R$ is such that $a\,x = 0$ if and only if $x = 0$, and if d_a is the Bockstein differential on $h(G, R/aR)$ as in III—1.18, then $d_a^2\, n_{G,R/aR} = 0$. In particular, d_a vanishes on the image of $P_R(R/aR \otimes \hat{G}) \to h(G, R/aR)$ and*

$$(h(G, R/aR), d_a)$$

is a differential $P_R(R/aR \otimes \hat{G})$-module.

Proof. (a) Suppose that G is a compact abelian Lie group. Then

$$G = G_0 \oplus K$$

with a finite group K. By Theorem 1.9, we have $hG_0 = P\hat{G}_0$, which is a polynomial ring in n variables if the dimension of the torus G_0 is n and is torsion free. Thus III—1.14 is applicable to yield $hG \cong P\hat{G}_0 \otimes hK$. But

$$hK = H_{\text{alg}}(K, \mathsf{Z})$$

is the algebraic cohomology of K over Z. Then $h^1K = 0$ since K is finite (II—4.1). It follows that $h^1G = 0$. Since h is $\mathfrak{D}$-continuous for the category $\mathfrak{D}$ of directed sets, we deduce that $h^1G = 0$ for all compact abelian groups G. Since $h^0G = \mathsf{Z}$ for all G,

$$\operatorname{Tor}(hG_1, hG_2) = \Sigma\{\operatorname{Tor}(h^i G_1, h^{3-i} G_2)\colon i = 0, 1, 2, 3\} = 0.$$

Thus from III—1.13 we derive

$$h^2(G_1 \oplus G_2) \cong h^2G_1 \oplus h^2G_2$$

for all compact abelian groups G_1, G_2. Thus h^2 is a $\mathfrak{D}$-continuous additive functor from the category *Comp* of all compact abelian groups into the category *Ab* of discrete abelian groups, as is the functor $G \to \hat{G}$. By IV—2.5, in order to show that these functors are naturally isomorphic, it suffices to show that their restrictions to the full category $Lie_1 \subset Comp$ spanned by the objects R/Z, $\mathsf{Z}(p^n)$, p prime, $n = 1, 2, \ldots$ are naturally isomorphic.

We discuss an equivalent proposition: Let I: $Ab \to Ab$ be the identity functor and F: $Ab \to Ab$ the functor given by $FG = h^2\hat{G}$. We have to show that these functors are naturally isomorphic when restricted to the full subcategory Ab_1 spanned by the objects Z and $\mathsf{Z}(p^n)$. By III—2.4 (saying that for finite groups K we have $FK = H^2_{\text{alg}}(\hat{K}, \mathsf{Z}))$ and by II—4.1 (saying that $H^2_{\text{alg}}(\hat{K}, \mathsf{Z}) \cong K$) the restrictions of I and F to the full category of finite groups are naturally isomorphic. Moreover, the isomorphism of their restrictions to the full subcategory of all torsion free groups is guaranteed by Section 1. Since all morphisms $\mathsf{Z}(p^n) \to \mathsf{Z}$ are zero, we have to be concerned with the morphisms f: $\mathsf{Z} \to \mathsf{Z}(p^n)$ only. Let ν_G: $G \to FG$ be an isomorphism which is natural on the full subcategory of finite (hence on the full subcategory of torsion) groups and on the full subcategory of torsion free groups.

We now construct a natural isomorphism μ: $I \to F$ such that $\nu_G = \mu_G$ for torsion free groups. To this end, define $\mu_G = \nu_G$ for torsion free groups and for $G = \mathsf{Z}(n)$ by the diagram

$$\begin{array}{ccccccccc} 0 & \longrightarrow & \mathsf{Z} & \xrightarrow{\;n\;} & \mathsf{Z} & \xrightarrow{\;\pi_n\;} & \mathsf{Z}(n) = Z/n\mathsf{Z} & \longrightarrow & 0 \\ & & \downarrow{\scriptstyle \nu_{\mathsf{Z}}} & & \downarrow{\scriptstyle \nu_{\mathsf{Z}}} & & \downarrow{\scriptstyle \mu_G} & & \\ 0 & \longrightarrow & F\mathsf{Z} & \xrightarrow{\;Fn = n\;} & F\mathsf{Z} & \xrightarrow{\;F\pi_n\;} & F\mathsf{Z}(n) & \longrightarrow & 0 \end{array}$$

Since the left hand rectangle commutes, μ_G is uniquely determined and is an isomorphism. By the general extension theorem it suffices to show that $\mu \mid Ab_1$ is natural; then there is a unique extension to an isomorphism μ: $I \to F$.

Case (i). Let f: $\mathsf{Z}(m) \to \mathsf{Z}(n)$ be an arbitrary morphism. Then there exists a morphism f': $\mathsf{Z} \to \mathsf{Z}$ such that $f\,\pi_m = \pi_n f'$, and consequently

$$(Ff)\,(F\pi_m) = (F\pi_n)\,(Ff').$$

Now (with the abbreviation $\mu_k = \mu_{\mathsf{Z}(k)}$) we have

$$\begin{aligned} \mu_n f\,\pi_m &= \mu_n\,\pi_n f' = (F\pi_n)\,\nu_{\mathsf{Z}} f' = (F\pi_n)\,(Ff')\,\nu_{\mathsf{Z}} \\ &= (Ff)\,(F\pi_m)\,\nu_{\mathsf{Z}} = (Ff)\,\mu_m\,\pi_m. \end{aligned}$$

Since π_m is epic, we have $\mu_n f = (Ff)\,\mu_m$.

Case (ii). Let $f: \mathsf{Z} \to \mathsf{Z}(n)$ be arbitrary. Then there is an $f': \mathsf{Z}(m) \to \mathsf{Z}(n)$ such that $f = f' \pi_m$. By case (i) above we have $(Ff')\mu_m = \mu_n f'$. Hence the diagram commutes, showing $\mu_n f = (Ff)\nu_{\mathsf{Z}}$.

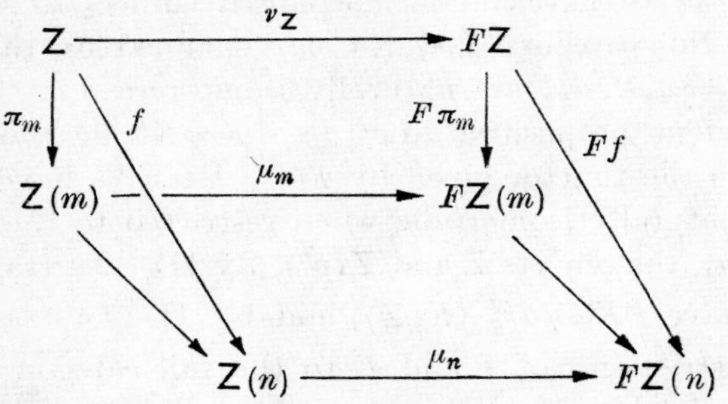

Now μ gives a natural isomorphism $I\,|\,Ab_1 \to F\,|\,Ab_1$. As said before, by IV—2.4 there is a unique extension to a natural isomorphism $\mu: I \to F$.

(b) From the Universal Coefficient Theorem, we have the exact sequence

$$0 \to R \otimes h^2(G, \mathsf{Z}) \to h^2(G, R) \to \mathrm{Tor}\,(R, h^3(G, \mathsf{Z})) \to 0,$$

which with (a) will give the assertion provided the Tor-term vanishes. For this it suffices to see the claim for any Lie group quotient $\bar{G}$ of G, and for $\bar{G}$ we have

$$h\bar{G} \cong P(\bar{G}_0)\hat{} \otimes H_{\mathrm{alg}}(K, \mathsf{Z})$$

with $\hat{K} \cong \bar{G}/\bar{G}_0$; but since the polynomial ring contributes only to the even degrees and $H^1_{\mathrm{alg}}(K, \mathsf{Z}) = 0$, we have

$$h^3\bar{G} \cong H^3_{\mathrm{alg}}(K, \mathsf{Z}) = \wedge^2 \hat{K}$$

(see II—4.1). It follows that

$$h^3 G \cong \wedge^2 (G/G_0)\hat{} = \wedge^2 \hat{G}_t,$$

where $\hat{G}_t$ is the torsion subgroup of $\hat{G}$. Thus a sufficient condition for the Tor-term to vanish is that $\hat{G}$ has no p-torsion where p is any prime dividing the characteristic of R.

(c) We apply Lemma III—1.20 with $F(R/I) = R/I \otimes \hat{G}$ for any ideal I of R, and with the natural transformation

$$n_{G,R/I}: R/I \otimes \hat{G} \to h^2(G, R/I)$$

defined in (b). Since tensoring is a right exact functor, the hypotheses of Lemma III—1.20 are satisfied, and the Lemma proves the assertion

$$d_a^2\, n_{G,\,R/aR} = 0.$$

The remainder is straightforward.

By comparison with a remark at the end of Section 2 in which we comment on the extent to which the *space* cohomology determines the full structure of the group, we may now point out that the functor h carries more information than does the functor H. Indeed if the topological structure of a classifying space $B(G)$ alone is given, it is sufficient to derive the full structure of G from this information, since $h^2G = H^2B(G)$ (with integral coefficients) is naturally isomorphic to the character group $\hat{G}$, which of course determines G completely and functorially.

Section 4

Cohomology theories for compact groups over fields as coefficient domains

Throughout this section, R denotes a commutative field with prime field K. In the present section, *Comp* stands for the category of all compact abelian groups and *Lie* for the full subcategory of all Lie groups in this category. The category $\mathfrak{D}$ will always denote the category of directed sets with order preserving maps.

Again $\mathfrak{A}$ denotes the category of all graded commutative Hopf algebras, but this time over R (not over $\mathbf{Z}$).

In view of the results of Section 2, which clarified completely the space cohomology of a compact abelian group (for any ring of coefficients), we may now concentrate on the functor h, except that, as mentioned before, R is a field.

Lemma 4.1. *The functor* h: *Comp* $\to (\mathfrak{A})^*$ *of Čech cohomology of (say) the Milnor classifying space with coefficients in* R *is a* $\mathfrak{D}$*-continuous exponential functor.*

Proof. The functor h transforms projective limits into direct limits (see III—1.11). Moreover, all R modules are free, hence flat, so the Künneth Theorem III—1.13 finishes the assertion.

Lemma 4.2. *The functor* $\bar{h}$: *Comp* $\to (\mathfrak{A})^*$ *given by*

$$\bar{h}G = P_R(R \otimes \hat{G}) \otimes \wedge \operatorname{Tor}(G, K) \cong R \otimes P\hat{G} \otimes \wedge \operatorname{Tor}(\hat{G}, K),$$

where K is the prime field of R and the Tor is taken over Z, is a $\mathfrak{D}$-continuous exponential functor.

Proof. Straightforward.

By IV, Section 2 the functors h and $\bar{h}$ induce functors of commutative Hopf algebras. Corollary IV—2.5 says that these functors are uniquely determined by their action on the full subcategory Lie_1 of Lie spanned by the objects R/Z, $\mathsf{Z}(p^n)$, p a prime, $n = 1, 2, \ldots$

In order to show that $\bar{h} \cong h$ we have to show that $\bar{h}\,|\,Lie_1 \cong h\,|\,Lie_1$. The R algebra

$$\bar{h}G \cong R \otimes P\hat{G} \otimes \wedge \operatorname{Tor}(\hat{G}, K)$$

is generated by the degree 1 and 2 components

$$R \otimes 1 \otimes \operatorname{Tor}(\hat{G}, K) \oplus R \otimes \hat{G} \otimes 1 .$$

We use the natural isomorphism

$$\psi_{G,R}\colon R \otimes 1 \otimes \operatorname{Tor}(\hat{G}, K) \to h^1(G, R)$$

for finite groups G which was discussed in II—3, Theorem V and the isomorphism

$$n_{G,R}\colon R \otimes \hat{G} \otimes 1 \to h^2(G, R)$$

for connected compact groups which was given in Theorem 3.1 of this Chapter to define a natural isomorphism

$$\nu'_G\colon \bar{h}^1(G, R) \oplus \bar{h}^2(G, R) \to h^1(G, R) \oplus h^2(G, R)$$

of R-algebras for $G \in Lie_1$. Since $\bar{h}G$ is a free graded commutative ring generated by its degree 1 and 2 components, ν'_G extends to a unique morphism $\nu_G\colon \bar{h}G \to hG$ of graded commutative R-algebras. By II—3, Theorem V, and Theorem 1.9 of this Chapter we know that ν_G is an isomorphism on Lie_1. Finally, by the density argument IV—2.5, we can now conclude that $\bar{h} \cong h$.

In Chapter IV, Lemma 2.9 and 2.10, we saw that the naturally isomorphic cofunctors

$$\begin{aligned}\operatorname{Hom}(\wedge -, R) &\cong \wedge_R \operatorname{Hom}(-, R) \cong R \otimes \wedge \operatorname{Hom}(-, K)\\ &\cong R \otimes \wedge \operatorname{Tor}(-, K)\end{aligned}$$

were the Kan extensions of their restrictions to Lie. Since the tensor product preserves direct limits, the cofunctors

$$G \mapsto (R \otimes P\hat{G}) \otimes_R \operatorname{Hom}(\wedge G, R) \cong P\hat{G} \otimes \operatorname{Hom}(\wedge G, R),$$
$$G \mapsto P_R(R \otimes \hat{G}) \otimes_R \wedge_R \operatorname{Hom}(G, R),$$
$$G \mapsto R \otimes P\hat{G} \otimes \wedge \operatorname{Hom}(G, K),$$
$$G \mapsto R \otimes P\hat{G} \otimes \wedge \operatorname{Tor}(\hat{G}, K) = \bar{h}G$$

are all the Kan extensions of their restriction to Lie, all of which are naturally isomorphic. Hence all of these functors are naturally isomorphic. We therefore have the following theorem:

Theorem VII (Cohomology with Field Coefficients). *Let G be an arbitrary compact group and R an arbitrary field with prime field K. Then there are natural isomorphisms of Hopf algebras*

$$h(G, R) \cong R \otimes P\hat{G} \otimes \wedge \operatorname{Tor}(\hat{G}, K) \cong P_R(R \otimes \hat{G}) \otimes_R \wedge_R \operatorname{Hom}(G, R)$$
$$\cong P\hat{G} \otimes \operatorname{Hom}(\wedge G, R).$$

Thus, we have a complete structure theory for h when the coefficient ring is a field. For rings, even for $R = \mathsf{Z}$ this is not the case, since then $h(G, R)$ ceases to be a Hopf algebra in any natural way. However, considerable information is still available as we shall see in the next section. But first we draw a few corollaries from the main result of this section.

Corollary 4.3. *Let G be a compact abelian group and* $\operatorname{Hom}(\mathsf{R}, G) = LG$ *its Lie algebra (see* [19], 4.59). *Let $(LG)^*$ be the dual of LG. Then*

$$h(G, \mathsf{R}) \cong P_R(LG)^* \cong h(G_0, \mathsf{R}).$$

Moreover, $h(G, \mathsf{R})$ is a Hopf algebra and these isomorphisms are isomorphisms of Hopf algebras.

Proof. One has $(LG)^* \cong \mathsf{R} \otimes \hat{G}$ ([19], 4.59). The assertions then follow from Theorem 4.3.

For the remainder of the section we take $R = K = GF(p)$. Then we have for any abelian group A a natural isomorphism $\operatorname{Tor}(R, A) = R \otimes A_t$ where A_t denotes the torsion subgroup of A. Thus we have in this case a natural isomorphism of graded commutative Hopf algebras over $GF(p)$ of the form

$$\psi_{G,p}\colon P_R(R \otimes \hat{G}) \otimes_R \wedge_R (R \otimes (G/G_0)\hat{}\,) \to h(G, R)$$

since the domain of $\psi_{G,p}$ is naturally isomorphic to $R \otimes P\hat{G} \otimes \wedge \operatorname{Tor}(K, \hat{G})$. Note that there is an injection $\hat{\pi}_G\colon (G/G_0)\hat{} \to \hat{G}$ where $\pi_G\colon G \to G/G_0$ is the quotient map.

What in this context goes beyond the previous discussion is that

$$P_R(R \otimes \hat{G}) \otimes_R \wedge_R (R \otimes (G/G_0)\hat{}\,)$$

is the bigraded Hopf algebra underlying the differential graded algebra $E_2(R \otimes \pi_G)$ with

$$R \otimes \hat{\pi}_G\colon R \otimes (G/G_0)\hat{} \to R \otimes \hat{G}$$

as in I—2. On the other hand, $h(G, R)$, because $R = GF(p)$, is a differential graded algebra relative to the Bockstein differential, and the question arises naturally whether or not $\psi_{G,p}$ is in fact a morphism of differential graded algebras. That this is indeed the case is asserted in the following

Theorem 4.4. *For any compact abelian group G, there is a natural isomorphism of differential graded algebras*

$$\psi_{G,p}: E_2(GF(p) \otimes \hat{\pi}_G) \to (h(G, GF(p)), d_p).$$

If $Sh(G, \mathsf{Z})$ denotes the ideal of $h(G, \mathsf{Z})$ of all $x \in h(G, \mathsf{Z})$ with $px = 0$, then there is an exact sequence

$$0 \to p\, h^+(G, \mathsf{Z}) \cap Sh(G, \mathsf{Z}) \to Sh^+(G, \mathsf{Z}) \to \operatorname{im}(GF(p) \otimes d) \to 0,$$

where d is the differential of $E_2(R \otimes \hat{\pi}_G)$.

Proof. By II—3.19 the isomorphism $\psi_{G,p} = \psi_G$ is an isomorphism of differential graded algebras if G is finite. By III—1.19 and the functorial continuity argument using the exactness of the direct limit functor we derive that ψ_G is an isomorphism of differential graded algebras if G is totally disconnected. By the naturality of ψ we have a commutative diagram

$$\begin{array}{ccc}
P_R(R \otimes (G/G_0)\hat{}\,) \otimes_R \wedge_R (R \otimes (G/G_0)\hat{}\,) & \xrightarrow{\psi_{G/G_0}} & h(G/G_0, R) \\
\big\downarrow \alpha = P_R(R \otimes \hat{\pi}) \otimes_R \wedge_R (R \otimes 1) & & \big\downarrow h(\pi, R) \\
P_R(R \otimes \hat{G}) \otimes_R \wedge_R (R \otimes (G/G_0)\hat{}\,) & \xrightarrow{\psi_G} & h(G, R)
\end{array}$$

All maps with the possible exception of ψ_G are morphisms of differential graded algebras; the horizontal maps are isomorphisms, the vertical maps are injections. From Theorem 3.1 (c) we know that the Bockstein d_G of $h(G, R)$ vanishes on $\psi_G(P_R(R \otimes \hat{G}) \otimes 1)$ and that the differential d vanishes on $P_R(R \otimes \hat{G}) \otimes 1$.

The commuting of the diagram further shows that on the image of α we have $\psi_G d = d_p \psi_G$. But then, since as an algebra, $E_2(R \otimes \pi)$ is generated by $P_R(R \otimes \hat{G}) \otimes 1$ and im α, it follows that $\psi_G d = d_p \psi_G$, i. e. that ψ is a morphism of differential graded algebras.

The exact sequence in the second part of the Theorem comes from the first part together with I—4.15.

Section 5

The structure of h for arbitrary compact abelian groups and integral coefficients

We first make the following observation:

Proposition 5.1. *Let R be any commutative ring with identity and of characteristic zero (i. e. with torsion free additive group). Let G be a compact connected and G' an arbitrary compact abelian group. Then there is a natural morphism of graded commutative R-algebras*

$$(R \otimes P\hat{G}) \otimes_R h(G', R) \to h(G \times G', R).$$

Proof. By Theorem 1.9 there is a natural isomorphism $R \otimes P\hat{G} \to h(G, R)$. Thus $h(G, R)$ is torsion free by our hypothesis about R. The assertion then follows from III—1.15 which clearly maintains for rings with torsion free additive group in place of a principal ideal domain.

Remark. The preceding proposition holds true for a field R as coefficient ring by the results of Section 4.

Proposition 5.2. *Let G be a compact abelian Lie group. Then there are integers $m, z_1, \ldots, z_n$ such that*

$$\hat{G} \cong \mathsf{Z}^m \oplus \cdots \oplus \mathsf{Z}(z_n),$$

with $z_1 | \cdots | z_n$, and if $\varphi = \mathrm{Hom}\,(f, R)$ is defined as in Definition I—3.13 *and* II—3.1, *with G/G_0 in place of G, then*

$$h(G, R) \cong h(G_0, R) \otimes h\,(G/G_0, R) \cong P\hat{G}_0 \otimes E_3(\varphi)$$

for any principal ideal domain R. The natural morphism of commutative graded R-algebras

$$\tau_{G,R}: P_R\,(R \otimes \hat{G}) \to h(G, R)$$

of 3.1 is a coretraction.

Proof. By Corollary II—1.8 and the comment thereafter, and Proposition 5.1 above,

$$h\,(G, R) \cong h\,(G_0, R) \otimes_R h\,(G/G_0, R) \cong P\hat{G}_0 \otimes h\,(G/G_0, R).$$

By III—2.4 we have $h(G/G_0, R) \cong H\,(G/G_0, R)$ where H denotes the standard algebraic cohomology. Then by Theorem III of Chapter II, Section 2, we obtain $H(G/G_0, R) \cong E_3(\varphi)$. Hence the first assertion. From Chapter II we know that $\tau_{G/G_0, R}$ is a coretraction (Theorem 2.13);

$$\tau_{G_0, R}: P_R(R \otimes \hat{G}_0) \to h(G, R)$$

is an isomorphism by 1.9. Thus $\tau_{G_0, R} \otimes \tau_{G/G_0, R}$ is a coretraction, whence $\tau_{G, R}$ is a coretraction.

With Proposition 5.2 and the results of Section 3 of Chapter II, the ring $h(G, R)$ is, in principle, known for a principal ideal domain R. Because of the unnaturality of the above splitting, this result does not extend directly to all compact abelian groups via the functorial continuity argument. However, a little more can be said even in the general case.

Proposition 5.3. *For any compact abelian group G, the natural morphism*

$$h(\pi, \mathsf{Z}): h(G/G_0, \mathsf{Z}) \to h(G, \mathsf{Z})$$

of graded rings with the quotient map $\pi: G \to G/G_0$ is an injection. The assertion remains valid for a principal ideal domain in place of Z.

Proof. The assertion is true on the dense subcategory of Lie groups by Proposition 5.2. Since the direct limit functor is exact, the assertion follows.

Proposition 5.4. *For any compact abelian group G, the injection $i: G_0 \to G$ induces a surjection*

$$h(i, \mathsf{Z}): h(G, \mathsf{Z}) \to h(G_0, \mathsf{Z}).$$

The assertion remains true for a principal ideal domain in place of Z.

Proof. Same as for Proposition 5.3.

Remark. Note that there is in fact an isomorphism of exact sequences

$$\begin{array}{ccccccccc}
0 & \longrightarrow & h^2(G/G_0, \mathsf{Z}) & \xrightarrow{h^2(\pi, \mathsf{Z})} & h^2(G, \mathsf{Z}) & \xrightarrow{h^2(i, \mathsf{Z})} & h^2(G_0, \mathsf{Z}) & \longrightarrow & 0 \\
 & & \downarrow & & \downarrow & & \downarrow & & \\
0 & \longrightarrow & G_0^{\perp} = \hat{G}_t & \longrightarrow & \hat{G} & \longrightarrow & \hat{G}_0 & \longrightarrow & 0
\end{array}$$

where $G_0^{\perp}$ is the annihilator of G_0 in $\hat{G}$, $\hat{G}_t$ the torsion group of $\hat{G}$. Recall $G_0^{\perp} \cong (G/G_0)\hat{}$. Since there are compact abelian groups of as low a dimension as 1 in which the component of the identity does not split (i. e. $\hat{G}_t$ does not split in $\hat{G}$) (see [19], p. 154), there are already one dimensional compact abelian groups G such that

$$h(G, \mathsf{Z}) \not\cong h(G_0, \mathsf{Z}) \otimes h(G/G_0, \mathsf{Z}),$$

as one observes by considering the homogeneous component of degree 2.

Now we can formulate a counterpart of Theorem 3.1 and thereby obtain information about a subgroup of $h(G, \mathsf{Z})$ which generates $h(G, \mathsf{Z})$ as a $P\hat{G}$-module. First we observe that there is a natural transformation

$$b^i: h^i(G, \mathsf{R}/\mathsf{Z}) \to h^{i+1}(G, \mathsf{Z}), \quad i = 0, 1, \ldots,$$

namely, the connecting morphism in the long exact sequence derived from the coefficient sequence $0 \to \mathsf{Z} \to \mathsf{R} \to \mathsf{R}/\mathsf{Z} \to 0$. Further, we observe, that for totally disconnected G, this morphism b^i is an isomorphism when $i > 0$, for then $h^+(G, \mathsf{R}) = 0$ by Corollary 4.3.

Theorem VIII (Principal Theorem for $h(G, \mathsf{Z})$). *Let G be a compact abelian group and R an abelian group (resp. a commutative ring with identity).*

(a) *There is a natural coretraction*

$$\varrho_{G,R}: \mathrm{Hom}\,(\wedge G, R) \to h\,(G, R),$$

which factors through $h(\pi, R): h(G/G_0, R) \to h(G, R)$ (Proposition 5.3) and is a morphism of algebras if R is a ring.

(b) *There is a natural coretraction*

$$\tau_{G,R}: P_R\,(R \otimes \hat{G}) \to h(G, R)$$

making $h\,(G, R)$ into an augmented $P_R(R \otimes \hat{G})$-algebra, if R is a ring. If b is the connecting morphism

$$b^i: h^i(G, \mathsf{R}/\mathsf{Z}) \to h^{i+1}\,(G, \mathsf{Z}),$$

then

$$M(G) = \mathsf{Z} \oplus \mathrm{im}\, b\varrho_{G,\mathsf{R}/\mathsf{Z}}$$

is a subgroup generating $h(G, \mathsf{Z})$ as a $P\hat{G}$-module and $M(G)$ is a minimal subgroup generating $h(G, \mathsf{Z})$ as a ring. Moreover, $b\varrho_{G,\mathsf{R}/\mathsf{Z}}$ is injective, so

$$M(G) \cong (\wedge G)\hat{} \cong \wedge\,(G/G_0)\hat{}$$

with a shift in dimension. Also, $h(G, \mathsf{Z})$ is a torsion free $P\hat{G}$-module.

Remark. *Recall that $h^2(G, \mathsf{Z}) \cong \hat{G}$, and that the exterior functor $\wedge$ as defined for finite abelian groups has a Kan extension to a functor (also denoted by $\wedge$) defined on all compact abelian groups* (IV—2.6).

Proof. (a) By II—2.13 and III—2.4, $\varrho_{G,R}$ exists for finite G, hence, by a functorial limit argument (Chapter IV) for totally disconnected compact G. The diagram

$$\begin{array}{ccc}
\mathrm{Hom}\,(\wedge\,(G/G_0), R) & \xrightarrow{\varrho_{G,R}} & h(G/G_0, R) \\
\downarrow{\scriptstyle \mathrm{Hom}\,(\wedge \pi, R)} & & \downarrow{\scriptstyle h\,(\pi, R)} \\
\mathrm{Hom}\,(\wedge G, R) & \cdots\cdots\cdots\to & h(G, R)
\end{array}$$

has a unique fill-in at the bottom because Hom $(\wedge \pi, R)$ is an isomorphism by Lemma IV—2.7 (4). This is the desired morphism.

(b) By Theorem IV of II—3, the assertion is true for finite G, and then via a functorial continuity argument, for totally disconnected groups. For Lie groups the assertion follows from Proposition 5.2, and then for arbitrary groups by passage to the limit.

Remark. Recall that in II—4.2, it was proved that for a finite group G, $(\wedge^n G)^\wedge$ is naturally isomorphic to $\wedge^n \hat{G}$, and by Lemma IV—2.6 (5), also for totally disconnected groups. However, if G is connected, $\wedge^n G = 0$, $n > 1$, whereas $\wedge^n \hat{G} \neq 0$.

In certain special cases, additional information can be provided.

Lemma 5.5. *Let G be a compact abelian group. Then there are cardinals a, $a(p)$, p prime, and a compact subgroup $K \subset G$ such that $G \cong \hat{\mathsf{Q}}^a \times \prod \mathsf{Z}_p^{a(p)} \times K$, where Z_p denotes the additive group of p-adic integers, and where the character group of K is reduced (i. e. contains no nontrivial divisible subgroup). As usual, Q denotes the discrete additive group of rationals.*

Proof. The character group $\hat{G}$ of G may be written as a direct sum of the additive group of an a-dimensional rational vector space and some groups $\mathsf{Z}(p^\infty)^{(a(p))}$, one for each prime, and some reduced group $\hat{K}$ ([19], p. 148). The assertion then follows by duality.

Lemma 5.6. $h(\mathsf{Z}_p, \mathsf{Z}) \cong \mathsf{Z} \oplus \mathsf{Z}(p^\infty)$, *where the elements of $\mathsf{Z}(p^\infty)$ have degree 2 and the product of two elements of degree 2 is zero.*

Proof. We have that

$$h(\mathsf{Z}(p^n), \mathsf{Z}) = H(\mathsf{Z}(p^n), \mathsf{Z}) = P\mathsf{Z}(p^n)$$

is a natural isomorphism of functors. Moreover, $h(-, \mathsf{Z})$ transforms projective limits into direct limits (Proposition III—1.11). Thus,

$$h(\mathsf{Z}_p, \mathsf{Z}) = \varinjlim H(\mathsf{Z}(p^n), \mathsf{Z}).$$

Since $H(\mathsf{Z}(p^n), \mathsf{Z})$ is generated by Z and $H^2(\mathsf{Z}(p^n), \mathsf{Z})$, $h(\mathsf{Z}_p, \mathsf{Z})$ is generated by

$$h^2(\mathsf{Z}_p, \mathsf{Z}) = \varinjlim \{\mathsf{Z}(p^n) \to \mathsf{Z}(p^{n+1}) : n = 1, 2, \ldots\} = \mathsf{Z}(p^\infty).$$

Since $P^m \mathsf{Z}(p^\infty) = 0$ for $m \neq 0, 2$, and $P^2 \mathsf{Z}(p^\infty) = \mathsf{Z}(p^\infty)$, the result follows.

Proposition 5.7. *For a compact abelian group G, the following conditions are equivalent:*

(a) *G is a direct product of p-adic groups.*

(b) *G has a compact classifying space.*

Proof. (a) implies (b): Floyd has observed, that a p-adic group has a compact two dimensional classifying space. (See Williams [45].) This implies that a product of p-adic groups has a compact classifying space.

(b) implies (a): In the Čech cohomology ring of a compact space, every element of positive degree is nilpotent. By Theorem 3.1, the cohomology ring of any classifying space of G contains a subring isomorphic to $P\hat{G}$. Thus (b) implies that in $P\hat{G}$, every element of positive degree is nilpotent. By the exponentiality of the functor P, by Lemma 5.5, we have

$$P\hat{G} \cong P\mathbf{Q}^{(a)} \otimes P \oplus \{\mathbf{Z}(p^{\infty})^{a(p)}\} \otimes P\hat{K},$$

where $\hat{K}$ is reduced. Clearly, if $a \neq 0$, $P\mathbf{Q}^{(a)}$ contains elements of infinite order. Thus we are left only with the consideration of $\hat{K}$. If $\hat{K}$ contains an element of infinite order, then $\hat{K}/\hat{K}_t$ is non-zero and torsion free, where $\hat{K}_t$ denotes the torsion subgroups of $\hat{K}$. Moreover, $P(\hat{K}/\hat{K}_t)$ is a quotient ring of $P\hat{K}$, and every element of positive degree is nilpotent, a contradiction. Thus, $\hat{K}$ is torsion. But any reduced torsion group contains a cyclic direct factor (see e. g. Fuchs [18], p. 80). Hence, if $\hat{K} = K' \oplus C$ with cyclic C, then $PK \cong P\hat{K}' \otimes PC$ since P is an exponential functor. But PC is a polynomial ring in one variable over a finite cyclic ring and therefore contains elements which are not nilpotent. This is a contradiction.

Proposition 5.9. *Let G be a totally disconnected compact group. Then*

$$h^+(G, \mathbf{Z}) \cong \oplus \{h^+(G_p, \mathbf{Z}) : p \text{ prime}\},$$

where G_p is the p-primary factor of G (i. e. the annihilator of the direct sum of all q-Sylow subgroups of G with $q \neq p$).

Proof. The assertion holds for finite G with a natural isomorphism (see II—3.16), and the direct sum decomposition is the decomposition into the sum of the Sylow subgroups. This decomposition is compatible with direct limits and the decomposition of G into its p-primary factors is compatible with projective limits. Hence, using III—1.11, we may pass to limits and obtain the result.

The computation of $h(G, \mathbf{Z})$ for totally disconnected compact groups G is therefore reduced to the computation of $h(G, \mathbf{Z})$ for reduced pro-p-groups.

Chapter VI
Appendix

Another construction of the functor h

(by Eric C. Nummela)

The purpose of this appendix is to show that, via a construction slightly different from that of Chapter III above, the cohomology functor h defined in Chapter III for compact groups can be extended to the category of compact monoids. The results presented here are contained in the author's doctoral dissertation [1]), about which a research announcement entitled *Algebraic cohomology of compact monoids* (Semigroup Forum 1, 1970) has appeared.

In general, if S is a topological semigroup acting on a topological space X (with action denoted by the juxtaposition of a semigroup element and a space element), we define a preorder $\prec$ (reflexive and transitive) on X by setting $x \prec y$ if and only if $x \in y \cup Sy$. We then define an *orbit* in X to be a maximal $\prec$-directed set. Note that an element of X may belong to more than one orbit, and that orbits need not be of the form $x \cup Sx$ for some $x \in X$.

The following proposition is easily proved.

Proposition 1. *Let S be a topological monoid acting on a topological space X.*

(1) *If S is compact, then the graph of $\prec$ is closed in $X \times X$.*

(2) *If the graph of $\prec$ is closed and X is compact, then orbits in X are closed.*

(3) *If S and X are compact, then each orbit P in X contains a maximal (with respect to $\prec$) element z; i. e., $P = Sz$.*

In the event that the orbits in X are pairwise disjoint, we say that the action of S on X is *clean*, and we have a naturally defined orbit space X/S. If the

---

1) *Algebraic cohomology of compact monoids*, Dissertation, Tulane University, 1970.

orbits are not disjoint, it is not clear just how the set of orbits should be topologized, although two possibilities are suggested in the author's dissertation and a third is implicitly described by Borrego and DeVun in a manuscript entitled *Maximal semigroup orbits* (not yet published). Naturally, all three yield the canonical orbit space for a compact monoid acting cleanly on a compact space.

In this appendix we shall have occasion to consider only the following kind of action: a topological monoid S with identity e acts cleanly on a space X, and given an orbit P in X there exists $z \in P$ such that the function

$$s \mapsto s\,z : S \to P$$

is a homeomorphism. Such actions will be called *free.*

Our main concern is to give a method of constructing a spectrum of universal spaces for an arbitrary compact monoid S. Although it is clear that S acts freely on itself (by multiplication), the Milnor spectrum used in Chapter III fails for the following reason. Even though S acts freely on the spaces X and Y, the coordinate-wise action

$$(s, (x, y)) \mapsto (s\,x, s\,y) : S \times (X \times Y) \to X \times Y$$

is not necessarily clean, hence not necessarily free.

Example. Let $S = X = Y = [0, 1]$ with the usual topology and multiplication. Then orbits in $X \times Y$ under the coordinate-wise action are segments between $(0, 0)$ and (x, y), where $(x, y) \in X \times \{1\} \cup \{1\} \times Y$. Hence $(0, 0)$ is common to all orbits. (We suggest that the orbit space of this action should be homeomorphic to $X \times \{1\} \cup \{1\} \times Y$.)

The spectra constructed by Dold and Lashof [12] and Milgram [1]) [43] being designed for use in the context of H-spaces, are better suited for monoids. However, application of the "telescope" device [2]) (see Remark after III—1.9) requires that the connecting morphisms in the spectrum be cofibrations, and this can be obtained for the Milgram spectrum if and only if the inclusion $\{e\} \to S$ is a cofibration (i. e. $\{e\}$ is a neighborhood deformation retract of S, which implies that the identity e has a countable system of neighborhoods). Thus we will use the Dold-Lashof spectrum. For completeness, we give below our interpretation of this construction.

---

[1]) See Milgram's original paper, *The bar construction and abelian H-spaces*, Ill. J. Math. **7** (1967), 242—250.

[2]) This device is due to J. Milnor, *On axiomatic homology theory*, Pacif. J. Math. **12** (1962), 337—341.

Given a compact monoid S, we set $E^0 = S$. Given E^n $(n \geqq 0)$, we construct the mapping cylinders of the continuous functions

$$P_S\colon S \times E^n \to S \text{ (projection)}$$

and

$$\omega_n\colon S \times E^n \to E^n \text{ (action)}.$$

We then "glue" these mapping cylinders together along their common base to obtain E^{n+1}. We define ε^n to be the canonical embedding of E^n into the mapping cylinder of ω_n, and we define the action of S on E^{n+1} to be that induced by the canonical action of S on $S \times E^n$ (i. e., just multiplication on the first coordinate).

Equivalently, we set

$$E^{n+1} = (S \times E^n \times [0, 1])/R,$$

where $(a, b, t)\ R\ (a', b', t')$ if and only if

(i) $0 = t = t', \quad a = a'$;

(ii) $0 < t = t' < 1, \quad a = a', \quad b = b'$;

or

(iii) $t = t' = 1, \quad a\,b = a'\,b'$.

We denote by $|a, b, t|$ the R-equivalence class of (a, b, t). We define

$$\varepsilon^n\colon b \longrightarrow |e, b, 1|\colon E^n \to E^{n+1},$$

$$\omega_{n+1}\colon (s, |a, b, t|) \longrightarrow |s\,a, b, t|\colon S \times E^{n+1} \to E^{n+1}.$$

Then ω_{n+1} is a free action of S on E^{n+1}. We let $P_n\colon E^n \to B^n$ $(0 \leqq n \leqq \infty)$ denote the quotient mapping to the orbit space, where E^∞ is the colimit of the sequence

$$E^0 \xrightarrow{\varepsilon^0} E^1 \xrightarrow{\varepsilon^1} \cdots \xrightarrow{\varepsilon^{n-1}} E^n \xrightarrow{\varepsilon^n} \cdots.$$

Proposition 2. *Let S be a compact monoid, and let $E^n(S) = E^n$ be the n-th space in the Dold-Lashof spectrum.*

(1) *Each E^n is compact $(0 \leqq n < \infty)$.*

(2) *Each ε^n is a cofibration.*

(3) *$H^i(E^n) = 0$ for $0 < i < n$ (Čech cohomology).*

(4) *Each B^n is compact $(0 \leqq n < \infty)$.*

(5) *Each $\beta^n\colon B^n \to B^{n+1}$ (the continuous function induced by ε^n) is a cofibration.*

(6) *$H^i(\beta^n)\colon H^i(B^{n+1}) \to H^i(B^n)$ is an isomorphism for $1 < i < n$ and an epimorphism for $i = 1$.*

(7) $H^i(E^\infty) \cong \varprojlim_n H^i(E^n)$, *hence* E^∞ *is acyclic*;

$H^i(B^\infty) \cong \varprojlim_n H^i(B^n)$.

Proof (Indication). Parts (1) and (4) are obvious from the construction. To prove part (2), use a retraction of $[0, 1] \times [0, 1]$ onto

$$([0, 1] \times \{0\}) \cup (\{1\} \times [0, 1])$$

which collapses $\{0\} \times [0, 1]$ onto $\{(0, 0)\}$ to construct a retraction of $E^{n+1} \times [0, 1]$ onto $\varepsilon^n(E^n) \times [0, 1] \cup E^{n+1} \times \{1\}$. Then use Theorem 7.1 of Steenrod's *A convenient category of topological spaces*, Michigan Math. J. **14** (1967), 133—152. Part (5) follows by the analog of (2) on the orbit space level. Part (3) follows by applying the Mayer-Vietoris sequence to the triples

$$(E^{n+1}, \text{mapping cylinder of } P_S, \text{mapping cylinder of } \omega_n).$$

Part (6) follows by the analog of (3) on the orbit space level. Part (7) follows from the "telescope" argument and (2) and (5).

Thus E^∞ is a universal space for S, and B^∞ is a classifying space for S. Moreover, we can define

$$h(S, -) = H(B^\infty(S), -).$$

Since the Dold-Lashof spectrum and the Milnor spectrum yield equivalent classifying spaces for a compact group (see Remark after III—1.9), we have extended the functor h from compact groups to compact monoids. Moreover, using the monoid analog of the argument outlined in Section 2 of Chapter III, we can show that h extends the usual algebraic cohomology functor for finite monoids. (This last argument is presented in full detail in Chapter 3 of the author's dissertation.)

Remarks. (1) Borel's proof that the classifying space of a compact group is cohomologically unique (Remark after III—1.9) depends on the coordinatewise action of the group on a product of two spaces, and hence is not adaptable to monoids (recall the previous Example). The author does not know if the classifying space of a compact monoid is cohomologically unique.

(2) The Dold-Lashof spectrum can be constructed in any Cartesian closed subcategory of the category of topological spaces, thus in particular in the category $\mathfrak{K}$ of compactly generated Hausdorff spaces. (See Steenrod's Michigan Math. Journal article cited above.) Suppose there exists a cohomology theory H on $\mathfrak{K}$, with coefficients in an arbitrary but fixed abelian group, which

(a) agrees with Čech theory at least on compact spaces;

(b) is continuous (i. e., transforms projective limits of spaces into direct limits of cohomology groups);

(c) takes topological sums of spaces into direct products of cohomology groups; and

(d) satisfies the following theorem: if X and Y are $\mathfrak{K}$-spaces and if $H^i(Y) = 0$ for $0 < i < n$, then the projection $p_X \colon X \times Y \to X$ (product in $\mathfrak{K}$) induces isomorphisms

$$H^i(p_X) \colon H^i(X) \to H^i(X \times Y) \quad \text{for} \quad 0 < i < n.$$

(The author does not know if, in fact, such a cohomology theory exists.) Then the Dold-Lashof spectrum yields a universal space for any $\mathfrak{K}$-monoid, and the corresponding cohomology theory for such monoids extends the existing theory for discrete monoids.

Bibliography

[1] Atiyah, M. F., Characters and cohomology of finite groups, Inst. des Hautes Études Scient. Publ. Math. **9** (1961), 2–64.

[2] Atiyah, M. F., and T. C. Wall, Cohomology of groups, Proceedings on Algebr. Number Theory, J. W. Cassels and A. Fröhlich ed.; Thompson Book Co., Washington, D. C., 1967, 94–113.

[3] Borel, A., Sur la cohomologie des espaces fibrés principaux et des espaces homogènes de groupes de Lie compacts, Ann. Math. **54** (1953), 115–207.

[4] Borel, A. (ed.), Seminar on Transformation Groups, Annals of Math. Studies No. 46, Princeton University Press, Princeton 1960.

[5] Bourbaki, N., Éléments de Mathématique, Livre II, Chap. 3 (Algèbre multilinéaire), 2me éd. 1958; Livre II, Chap. 2 (Algèbre linéaire), 3me éd. 1962, Hermann, Paris.

[6] Bourbaki, N., Éléments de Mathématique, Livre II, Chap. 6 (Modules sur les anneaux principaux), Hermann, Paris 1952.

[7] Bourgin, N., Modern Algebraic Topology, MacMillan Co., New York 1963.

[8] Bredon, G. E., Sheaf Theory, McGraw-Hill, New York 1967.

[9] Cartan, H., La transgression dans un groupe de Lie et dans un espace fibré principal, Colloque de Topologie, Bruxelles 1950, 57–71.

[10] Cartan, H., Algèbres d'Eilenberg-MacLane et Homotopie, Séminaire Henri Cartan, 7e année 1954/55, Paris.

[11] Cartan, H., and S. Eilenberg, Homological Algebra, Princeton Univ. Press, Princeton 1956.

[12] Dold, A. E., and R. K. Lashof, Principal quasifibrations and fibre homotopy equivalences, Ill. J. Math. **3** (1959), 285–305.

[13] Eilenberg, S., and G. M. Kelly, Closed categories, Proc. Conf. on Categorical Algebra, La Jolla 1965, Springer-Verlag, New York 1966.

[14] Eilenberg, S., and S. MacLane, On the groups $H(\Pi, n)$ I, II, III, Ann. of Math. **58** (1953), 55–106; **60** (1954), 49–139; **60** (1954), 513–557.

[15] Eilenberg, S., and J. C. Moore, Homological algebra and fibrations, Colloque de Topologie, Bruxelles 1964, 81–90.

[16] Eilenberg, S., and J. C. Moore, Homological algebra and fibrations I, Comment. Math. Helv. **40** (1966), 199–236.

[17] Evens, L., The cohomology ring of a finite group, Trans. Amer. Math. Soc. **101** (1961), 224–239; Erratum: ibid **102** (1922), 545.

[18] Fuchs, L., Abelian Groups, Pergamon Press, New York 1960.

[19] Hofmann, K. H., Introduction to the Theory of Compact Groups I, Lecture Notes, Tulane University, 1966–1967.

[20] Hofmann, K. H., Categories with convergence, exponential functors, and the cohomology of compact abelian groups, Math. Z. **104** (1968), 106–144.

[21] Hofmann, K. H., Tensorprodukte lokal kompakter abelscher Gruppen, J. reine angew. Math. **216** (1964), 134–149.

[22] Hofmann, K. H., Der Schur'sche Multiplikator topologischer Gruppen, Math. Z. **79** (1962), 389–421.

[23] Hofmann, K. H., and P. S. Mostert, Elements of Compact Semigroups, Chas. E. Merrill Books, Columbus 1966.

[24] Hofmann, K. H., and P. S. Mostert, The cohomology of compact abelian groups, Bull. Amer. Math. Soc. **74** (1968), 975–978.

[25] Hofmann, K. H., and P. S. Mostert, About the cohomology ring of a finite abelian group, Bull. Amer. Math. Soc. **75** (1969), 391–395.

[26] Hu, S. T., Homotopy Theory, Academic Press, New York 1959.

[27] Huppert, B., Endliche Gruppen I, Springer-Verlag, Berlin-Heidelberg-New York 1967.

[28] Husemoller, D., Fibre Bundles, McGraw-Hill, New York 1966.

[29] Koszul, J. L., Sur un type d'algèbres differentielles en rapport avec la transgression, Colloque de Topologie, Bruxelles 1950, 73–81.

[30] Lang, S., Rapport sur la Cohomologie des Groupes, W. A. Benjamin, New York 1966.

[31] Leray, J., L'anneau spectral et l'anneau filtré d'homologie d'un espace localement compact et d'une application continue, J. Math. Pure Appl. **29** (1950), 1–139.

[32] MacLane, S., Homology, Springer-Verlag, Berlin-Göttingen-Heidelberg 1963.

[33] MacLane, S., Categorical algebra, Bull. Amer. Math. Soc. **71** (1965), 40–106.

[34] McCord, M. E., Classifying spaces and infinite symmetric products, Trans. Amer. Math. Soc. **146** (1969), 273–298.

[35] Mitchell, B., Theory of Categories, Academic Press, New York 1965.

[36] Moscowitz, M., Homological algebra in locally compact abelian groups, Trans. Amer. Math. Soc. **127** (1967), 361–404.

[37] Palermo, F. P., The cohomology ring of product complexes, Trans. Amer. Math. Soc. **86** (1957), 174–196.

[38] Petrie, T., The Eilenberg-Moore, Rothenberg-Steenrod spectral sequence for K-theory, Proc. Amer. Math. Soc. **19** (1968), 193–194.

[39] Priddy, S., Koszul resolutions, Trans. Amer. Math. Soc. **152** (1970), 39–60.

[40] Rothenberg, M., and N. E. Steenrod, The cohomology of classifying spaces of H-spaces. Bull. Amer. Math. Soc. **71** (1965), 872–875.

[41] Shafer, T., On the homology ring of an abelian group, Dissertation, University of Chicago 1965.

[42] Spanier, E. H., Algebraic Topology, McGraw-Hill, New York 1966.

[43] Steenrod, N., Milgram's classifying space of a topological group, Topology **7** (1968), 349–368.

[44] Steenrod, N., and D. B. A. Epstein, Cohomology Operations, Ann. of Math. Studies, No. 50, Princeton 1962.

[45] Williams, R. F., The construction of certain 0-dimensional transformation groups, Trans. Amer. Math. Soc. **129** (1967), 140–156.

List of notation

Index